"I'm offering you the choice of life or death. You can choose either blessings or curses."

—DEUTERONOMY CHAPTER 30, VERSE 19
(NEW INTERNATIONAL READER'S VERSION)

OUR CHOICE

A Plan to Solve the Climate Crisis

AL GORE

Few have ever seen this historic first image of the earth above the moon's horizon. Captured by the unmanned Lunar Orbiter 1 two years before the famous "Earthrise" photo taken in 1968 by Bill Anders on the Apollo 8 mission—this photo was transmitted in such low resolution it had little impact. Forgotten for 42 years, it was digitally upgraded in 2008. Similarly, the climate crisis has been visible—faintly—for many years, but for a long time had little impact on our thinking.

Published by Rodale Inc.

LIVE YOUR WHOLE LIFE™

33 East Minor Street
Emmaus, PA 18098
www.rodale.com

Rodale books may be purchased for business or promotional use or for special sales. For information, please write to: Special Markets Department, Rodale Inc., 733 Third Avenue, New York, NY 10017.

This book was produced by

124 West 13th Street
New York, NY 10011
www.melcher.com

Publisher: Charles Melcher
Associate Publisher: Bonnie Eldon
Editor in Chief: Duncan Bock
Production Director: Kurt Andrews
Project Editors: David E. Brown and Jessi Rymill
Editor: Megan Worman
Associate Editor: Lindsey Stanberry
Production Associate: Daniel del Valle

Illustrations by Don Foley
Design by mgmt. design

The typefaces used in this book are Akkurat designed at lineto in Zürich, Switzerland, and Brioni STD designed at Typotheque in The Hague, The Netherlands.

Printed in the USA

ISBN-13: 978-1-59486-734-7
ISBN-10: 1-59486-734-8
Library of Congress control number is available upon request.

09 10 11 12 / 10 9 8 7 6 5 4 3 2 1

To Karenna, Kristin, Sarah, and Albert

CONTENTS

THE CRISIS

      INTRODUCTION 10

1   WHAT GOES UP MUST COME DOWN 30

OUR SOURCES OF ENERGY

2   WHERE OUR ENERGY COMES 50
    FROM AND WHERE IT GOES

3   ELECTRICITY FROM THE SUN 62

4   HARVESTING THE WIND 76

5   SOAKING UP GEOTHERMAL ENERGY 92

6   GROWING FUEL 112

7   CARBON CAPTURE AND SEQUESTRATION 134

8   THE NUCLEAR OPTION 150

LIVING SYSTEMS

9   FORESTS 170

10   SOIL 196

11   POPULATION 224

HOW WE USE ENERGY

12   LESS IS MORE 242

13   THE SUPER GRID 272

THE OBSTACLES WE NEED TO OVERCOME

14   CHANGING THE WAY WE THINK 298

15   THE TRUE COST OF CARBON 318

16   POLITICAL OBSTACLES 348

GOING FAR QUICKLY

17   THE POWER OF INFORMATION 370

18   OUR CHOICE 392

      INDEX 406

      ACKNOWLEDGMENTS 411

      CREDITS 415

THE CRISIS

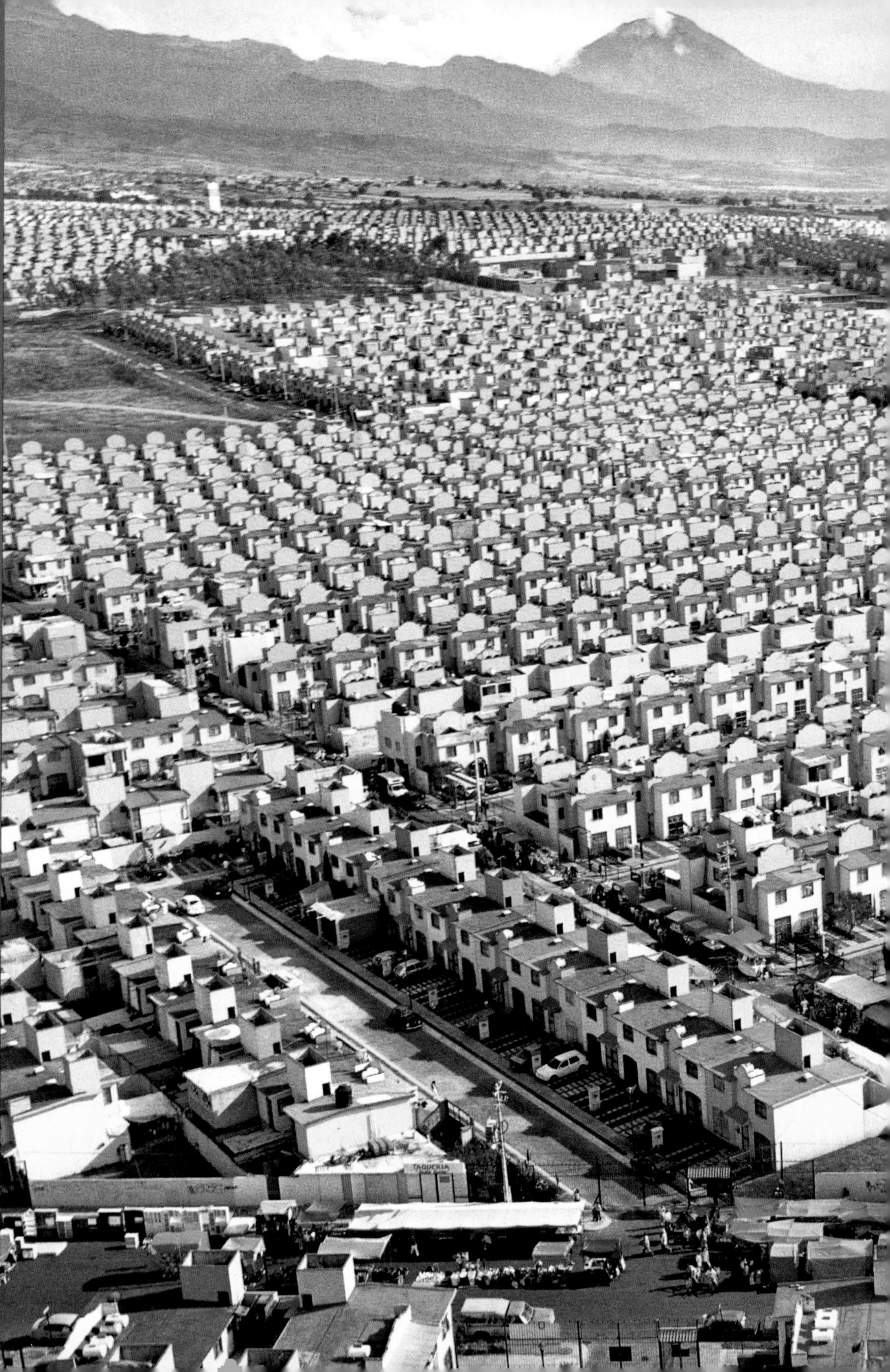

INTRODUCTION

IXTAPALUCA, A NEW DEVELOPMENT AT THE EDGE OF MEXICO CITY. THE RAPID GROWTH OF THE WORLD'S POPULATION IS ONE OF THE FACTORS THAT HAS CHANGED OUR RELATIONSHIP WITH THE EARTH'S ECOLOGICAL SYSTEM.

Almost 20 years ago, the late American novelist Kurt Vonnegut wrote, "Is there nothing about the United States of my youth, aside from youth itself, that I miss sorely now? There is one thing I miss so much that I can hardly stand it, which is freedom from the certain knowledge that human beings will very soon have made this moist, blue-green planet uninhabitable by human beings."

With his trademark blend of surrealism, dark humor, and cynicism, Vonnegut continued, "If flying-saucer creatures or angels or whatever were to come here in a hundred years, say, and find us gone like the dinosaurs, what might be a good message for humanity to leave for them, maybe carved in great big letters on a Grand Canyon wall?"

His suggestion for the message our civilization ought to leave was:

"WE PROBABLY COULD HAVE SAVED OURSELVES, BUT WERE TOO DAMNED LAZY TO TRY VERY HARD.... AND TOO DAMNED CHEAP."

The knowledge that serious damage has already been done to the global environment and to the healthy climate balance on which our civilization depends can become a cause of paralyzing despair. The danger is that this despair may render us incapable of reclaiming control of our destiny in time to avert the unimaginable catastrophe that would unfold on this planet if we don't start making dramatic changes quickly.

Yet the majority of experts on the climate crisis agree that we probably still do have time to avert the worst of the impacts and set the stage for a long but ultimately successful recovery of the climate balance and ecological integrity that are so crucial for the survival of our civilization.

In any case, despair serves no purpose when reality still offers hope. Despair is simply another form of denial, and it invites inaction. We don't have time for despair. The solutions are available to us! We need to make our choice to act now.

An old African proverb says, "If you want to go quickly, go alone; if you want to go far, go together."

We have to go far...quickly.

This book is about the solutions to the climate crisis. During the three and a half years since the publication and release of *An Inconvenient Truth*, I have organized and moderated more than 30 lengthy and intensive "Solutions Summits," where leading experts from around the world have come to discuss and share their knowledge of and experience in subjects relevant to the construction of a plan to solve this crisis. In addition to hosting these group meetings, I have engaged in a large number of one-on-one sessions with other leading experts around the world in an extended effort to find the most effective courses of action.

From neuroscience to economics, information technology to agriculture, many seemingly diverse subjects relevant to the effort to understand and

"If you want to go quickly, go alone; if you want to go far, go together."

AFRICAN PROVERB

INDIAN SCHOOLCHILDREN PARTICIPATE IN A MASS TREE PLANTING IN HYDERABAD, 2008. TREE PLANTING PROGRAMS HAVE HELPED TO OFFSET DEFORESTATION, WHICH IS A MAJOR CONTRIBUTOR TO GLOBAL WARMING.

map out a successful blueprint for action on a global scale have been generously and patiently explained by the foremost global leaders in their respective disciplines. *Our Choice* is the result of the groundbreaking insights offered by the participants in this multiyear dialogue. These experts made it possible to construct a fresh and unique approach I have not seen before.

That's why I have written this book, chosen the pictures, and commissioned the illustrations—to gather in one place all of the most effective solutions that are available now and that, together, will solve this crisis. It is meant to inspire readers to take action—not only on an individual basis but

We can solve the climate crisis. It will be hard, to be sure, but if we choose to solve it, I have no doubt whatsoever that we can and will succeed.

Moreover, we should feel a sense of joy that those of us alive today have a rare privilege that few generations in history have known: the chance to undertake an historic mission worthy of our best efforts. It should be seen as an honor to live in a time when the future of human civilization will be shaped forever by what we do now.

In rising to this challenge, we will find fresh evidence that the fate of our civilization depends on effective, cooperative, global measures to save the habitability of the earth and build the

We can solve the climate crisis. It will be hard, to be sure, but if we choose to solve it, I have no doubt whatsoever that we can and will succeed.

as participants in the political processes by which every country, and the world as a whole, makes the choice that now confronts us.

For me, this has been an exciting and illuminating journey, because it is now abundantly clear that we have at our fingertips all of the tools we need to solve three or four climate crises—and we only need to solve one. The only missing ingredient is collective will. But we are getting closer to a political tipping point, beyond which enough people in all of the key countries recognize the reality of this global emergency and accept the challenge of working together to rescue our civilization.

foundation for a more just, humane, and prosperous world.

Properly understood, the climate crisis is an unparalleled opportunity to address, at long last, many persistent causes of suffering and misery that have long been neglected and to transform the prospects for future generations to live healthier, more prosperous lives with a greater chance of success in each new generation's pursuit of happiness.

The good news about making a definitive choice to solve the climate crisis is that the scale of systemic transformation necessary will bring, as collateral benefits, highly effective solutions to many of these long-lasting problems. Extreme poverty, threatening diseases, widespread hunger, and malnutrition are among the scourges that have beset large parts of the human population throughout history. Indeed, our success in transforming the global economy to a low-carbon pattern will bring about needed solutions for problems that have been allowed to fester for too many centuries.

The key first step toward a solution is this: we must make a choice. By *we,* I mean our global civilization. And therein lies, as Shakespeare named it, "the rub"—because it seems absurd to imagine that we as a species are capable of making a conscious collective decision. And yet that is the task we are now confronting.

This book is focused on the collective decision that we now face: to make the rescue of civilization the central organizing principle of our politics, economics, and social action.

We have arrived at a moment unlike any other in all of human history. Our home is in grave danger. What is at risk of being destroyed is not the earth itself, of course, but the conditions that have made it hospitable for human beings.

While the choice we must make at the simplest level is very clear, the decision to embark on this new course will be difficult, precisely because the scale of the changes necessary is completely unprecedented—and the speed with which we will have to begin making them is also unparalleled in the history and experience of our civilization.

It is a mistake to believe that we can maintain the traditional silos of knowledge in all the different disciplines that hold essential components of the solution. This issue crosses disciplines, national boundaries, ideologies, and politics.

It is also naive to place the burden of solutions on individuals alone. In order to solve the climate

THE HOUSE WHERE MINA WEYIOUANNA GREW UP IN SHISHMAREF, ALASKA, COLLAPSED BECAUSE OF MELTING PERMAFROST—WHICH ALSO RELEASES METHANE AND CO_2 INTO THE ATMOSPHERE.

crisis, we must recognize the necessity of concerted global action. Each of us as an individual has a part to play, of course, and the actions we take in our own lives, households, and businesses are extremely important. They add up, and they reinforce the hope and sustained commitment necessary for success. But we must change more than our lightbulbs and windows. We have to change our laws and policies.

Individuals eager to become a part of the solution must become active as citizens in advocating and fighting for the new laws and treaties that will ultimately lead to the necessary global-scale solutions. This book, therefore, does not address the agenda for individual options as much as it focuses on providing a road map to the large solutions that require our shared commitment.

I have heard often from individuals who find it hard to believe that such a shared commitment will ever be possible even in the United States, much less on a global scale. However, even now, we are beginning to move powerfully in the right direction. Whether it is corporations responding to perceived market opportunities or the determination by a group of fifth graders to help "end global warming," the signs of the change we need are now beginning to appear all around us.

For example, global population growth is already slowing and beginning to stabilize, albeit at a number of people that a few decades from now will probably be five and a half times more than the population at the beginning of the 20th century. Another encouraging new development of the past three years has been the beginning of a major shift throughout the world in humankind's awareness of the climate crisis and its connections to the other major challenges we confront.

The emergence of a full-blown global economic crisis in the fall of 2008 has underscored the value of a coordinated global stimulus to create jobs and bring an end to the effects of the unusually deep, synchronized global recession. Simultaneously, the worsening military conflict in Afghanistan and continuing struggle to stabilize Iraq are both ongoing reminders of the security threats that will likely continue to emanate from the Persian Gulf region so long as the United States and the rest of the global economy are so dependent on oil—the largest reserves of which are controlled by sovereign states in the Middle East.

Although geologists and economists continue to debate the exact timing of "peak oil," the preponderance of evidence now suggests that we may be at or near the peak of global oil production—at least from supplies recoverable at a cost approximating what the world is used to paying. The International Energy Agency, in its first in-depth analysis of the 800 largest oil fields worldwide, reported last year that the majority of the largest fields are already past their peak production rate, and that the predicted declines in production are now accelerating at twice the rate predicted in 2007.

T. Boone Pickens, the highly successful U.S. oil and gas veteran, stated in August 2008 that, in his view, global oil production actually peaked in 2006. Whether he and those who agree with him are right or not, it is now only a matter of time before the world will be forced to adjust to the reality of a growing mismatch between the declining rate of new discoveries and the growing demand for oil from emerging economies like China and India.

So long as the United States spends close to half a trillion dollars per year for foreign oil, the U.S. current account deficit will be impossible to manage, and the value of the dollar will become increasingly vulnerable. And the longer the global economy is hostage to energy reserves located in what is arguably the most unstable region in the

HURRICANE KATRINA PUSHED THIS OFFSHORE OIL
RIG MORE THAN 60 MILES, GROUNDING IT OFF THE
ALABAMA COAST. THE NUMBER OF RECORD STORMS
HAS MADE THE CLIMATE CRISIS TANGIBLE.

DUNES AT THE EDGE OF CHINA'S GOBI DESERT
THREATEN FERTILE FARMLAND NEAR THE YELLOW
RIVER. ACCORDING TO THE U.N., GLOBAL WARMING
IS A PRIMARY CAUSE OF THE SPREAD OF DESERTS.

world, we will continue to experience periodic price hikes like the one that drove the price of oil to $145 a barrel in the summer of 2008.

For its part, China has been using its vast current account surpluses to buy up energy companies and controlling stakes in oil fields in many parts of the world. It has also embarked upon an ambitious plan to dominate the production of solar panels, which the Chinese intend to build for their own and the world's transition to low-carbon energy production. China will also soon become the largest source of wind power in the world, and is building an 800-kilovolt super grid connecting all parts of the country to the most sophisticated smart transmission and distribution network for electricity the world has ever seen.

The U.S. current account deficit, by contrast, is being driven by ridiculously high dependence on foreign oil toward what the Peterson Institute, founded by former Secretary of Commerce Pete Peterson, describes as an economic catastrophe in the making. In the first quarter of 2009, the U.S. current account deficit stood at $101.5 billion (lower than usual because of the recession)—driven in part by $46 billion in imports of foreign petroleum and petroleum products.

All three of these crises—the security crisis, the economic crisis, and the climate crisis—seem unsolvable in isolation. Yet when we look closely, we can see the common thread running through them, deeply ironic in its simplicity: our dangerous overreliance on carbon-based fuels is at the core of all three of these challenges.

If we grab hold of that thread and pull it hard, all of these complex problems begin to unravel, and we will find that we are holding the answer to all of them right in our hand: we need an historic commitment to put people to work building the infrastructure and technology base for a massive and speedy shift away from coal, oil, and gas to renewable forms of energy.

President Barack Obama's initial stimulus package made impressive steps toward building a renewable-energy infrastructure in the United States, but as of this writing, the oil, coal, and natural gas companies—in combination with fossil fuel–dependent electric utilities and ideologically driven climate deniers—have exercised their control over congressional decision-making by blocking climate and energy legislation in the U.S. Senate.

For now, the United States is still borrowing money from China to buy oil from the Persian Gulf to burn in ways that destroy the planet. Every bit of that's got to change.

The latest scientific studies that measure the impact of the climate crisis continue a pattern that has been obvious for the last 20 years at least. Each new in-depth assessment has found that earlier projections of the worst-case outcome have understated how serious this crisis is and how rapidly it is growing. The world authority on the climate crisis, the Intergovernmental Panel on Climate Change, after 20 years of detailed study and four unanimous reports, now says that the evidence is "unequivocal."

But the golden thread of reason that used to be stretched taut to mark the boundary between the known and the unknown is now routinely disrespected. We are living in a political culture driven partially mad by the transformation of the "public forum" that emerged in the wake of the printing press, which brought us newspapers, books, mass literacy, the "rule of reason," egalitarianism, and representative democracy.

What philosophers of the early Enlightenment described as "the Republic of Letters" has been subjugated by electronic images that carelessly blend news with entertainment, advocacy with

CARS AND TRUCKS AT RUSH HOUR ON
I-75 IN ATLANTA. TRANSPORTATION
ACCOUNTS FOR AT LEAST 10 PERCENT OF
GLOBAL WARMING POLLUTION.

The United States is still borrowing money from China to buy oil from the Persian Gulf to burn in ways that destroy the planet. Every bit of that's got to change.

advertisements, and the public interest with self-interest.

The late German philosopher Theodor Adorno first described this transformation in a very different context 58 years ago: "The conversion of all questions of truth into questions of power…has attacked the very heart of the distinction between true and false."

Here is one seemingly trivial but illustrative example of this phenomenon: last summer in the United States, some political opponents of President Obama claimed that he was actually not born in the United States, and thus was not eligible to continue serving as president. The governor of

resembles the willful refusal of many so-called climate skeptics to accept the truth about the climate crisis, as exhaustively documented by the Intergovernmental Panel on Climate Change in four comprehensive assessments over the past 20 years—the conclusions of which were unanimously endorsed by the national academies of science in all the world's leading nations, including the United States, China, the United Kingdom, India, Russia, Brazil, France, Italy, Canada, Germany, and Japan.

The late Senator Pat Moynihan once famously said, "Everyone is entitled to his own opinions, but not to his own facts." In order to arrive at a global consensus sufficient to support the bold solutions

"Everyone is entitled to his own opinions, but not to his own facts."

SENATOR PAT MOYNIHAN

Hawaii, where Obama was born, though not of his political party, personally examined and publicly verified the official certificate of his birth 48 years ago. Libraries in Hawaii provided copies of contemporaneous birth announcements in two Honolulu newspapers, confirming the facts as stated on his birth certificate. Nevertheless, millions of people continued to dispute these established facts; some even suggested the possibility of a dark, multigenerational, well-planned conspiracy, at one point offering a sloppy and patently bogus forgery of a Kenyan birth certificate.

This odd faux controversy is hardly worth mentioning—except for the fact that it so closely

now essential to our survival, we have to find ways to search for the best available evidence, test it in vigorous and open discussion, but then, when reasonable people agree in good faith that the facts thus established are far more likely to be true than alternative explanations, we should accept them and move on to address their implications.

The second reason for noting the "birther" episode is to draw a second parallel to the treatment of climate science. Disturbingly, some established news organizations—perhaps because they were confused about the distinction between reporting the news and providing fictional entertainment to boost their ratings by fanning the

IN NOVEMBER 2007, THE WORST FLOODING IN
50 YEARS HIT THE CITY OF VILLAHERMOSA IN
SOUTHERN MEXICO.

WOMEN COLLECT WATER IN GOUROUKOUN, CHAD, A VILLAGE THAT HAS BECOME HOME TO REFUGEES FROM THE CLIMATE-INFLUENCED CONFLICT IN THE DARFUR REGION OF SUDAN.

flames of a counterfeit controversy—pretended to a mass audience that the facts were still in dispute and warranted further fevered investigation.

It is in this context that some opponents of progressive change have grown weary of warnings of catastrophe as a basis of mobilizing support for policy changes. As a consequence, they regularly discount the scientific evidence. This is one of the reasons it has been so difficult to convince civic and business leaders who ought to know better that the climate crisis is not exaggerated.

The integrity of the deliberative process on which self-government depends is put at risk by the continuing willful promotion of false controversy over long- and well-established facts delineating the magnitude and gravity of the climate crisis. Sometimes, bearing false witness can be murderous, and self-deception can be suicidal. In the Book of Proverbs, King Solomon offered a warning about people committed to violence as a way of life: "They set an ambush for themselves."

In the 6th century B.C., Aesop told his fable about the young shepherd boy who watched helplessly as the sheep entrusted to his care were eaten because he had cried "Wolf!" falsely on too many previous occasions in order to mischievously amuse himself with the spectacle of nearby farmers running to the sound of his voice.

Two hundred years earlier, an unknown Chinese fabulist told of an emperor who repeatedly rang the alarm bell intended to signal an invasion of the city in order to amuse his favored concubine with the immediate fevered preparations for imminent battle by his soldiers. When the real invasion came, the alarm bell was met with languor, the city was overrun, and the emperor was killed.

In our generation, the decision by powerful ideologues and self-interested corporate advocates to convert "questions of truth into questions of power" has produced a similar lassitude in reaction to genuine, fact-based warnings of an onrushing tragedy with no parallel in all of history.

Some who reject the scientific consensus and minimize the crisis argue that our best course is to simply adapt to the changes being wrought and acknowledge that stopping it is beyond our ability. Others have viewed the task of adaptation as a dangerous diversion from the overriding challenge of preventing the destruction of the conditions on our planet that have nurtured the development and rise of humanity and are essential for the continuation of civilization as we know it.

But in reality, this is a false choice. We must undertake both challenges simultaneously, rescuing those in harm's way while saving the future of human civilization. Any other strategy would rightly meet condemnation.

Moreover, compassion and assistance for those already being hurt by the early impacts of the climate crisis are a practical necessity for building and strengthening the global consensus necessary for the larger task of solving the crisis and avoiding the worst impacts. The overriding reality overlooked by those who wish to focus only on adaptation is that unless we take bold steps to stop the destruction of the earth's environment, adaptation would prove to be totally impossible.

Were we not to take bold action, the worst impacts of the climate crisis would unfold over many generations, escalating in their destructive power decade by decade. But we cannot wait for the full fury of the crisis in order to mobilize a response, because by then it would already be too late to stop the process that we have set in motion.

By that point, the generation that finally realized that humans had been condemned to an endless degradation of their prospects for the entirety of their lives and the lives of their children and their children's children would be entirely justified in looking backward at us in our time as a criminal generation that they would curse endlessly as the architects of humanity's destruction.

In practical terms, the rescue of future generations must start right now. Even as we extend our hands to those suffering in the present generation, we must reject the argument that this rescue is any more than the beginning of what we must do.

One thin September soon
A floating continent disappears
In midnight sun

Vapors rise as
Fever settles on an acid sea
Neptune's bones dissolve

Snow glides from the mountain
Ice fathers floods for a season
A hard rain comes quickly

Then dirt is parched
Kindling is placed in the forest
For the lightning's celebration

Unknown creatures
Take their leave, unmourned
Horsemen ready their stirrups

Passion seeks heroes and friends
The bell of the city
On the hill is rung

The shepherd cries
The hour of choosing has arrived
Here are your tools

—Al Gore, Nashville, Tennessee, 2009

CHAPTER ONE

WHAT GOES UP MUST COME DOWN

THE NIEDERAUSSEM COAL-FIRED POWER PLANT IN GERMANY RANKS AS THE THIRD WORST EMITTER IN EUROPE IN TERMS OF CO_2 PER KILOWATT-HOUR PRODUCED.

Human civilization and the earth's ecological system are colliding, and the climate crisis is the most prominent, destructive, and threatening manifestation of this collision. It is often lumped together with other ecological crises, such as the destruction of ocean fisheries and coral reefs; the growing shortages of freshwater; the depletion of topsoil in many prime agricultural areas; the cutting and burning of ancient forests, including tropical and subtropical rain forests rich in species diversity; the extinction crisis; the introduction of long-lived toxic pollutants into the biosphere and the accumulation of toxic waste from chemical processing, mining, and other industrial activities; air pollution; and water pollution.

These manifestations of the violent impact human civilization has on the earth's ecosystem add up to a worldwide ecological crisis that affects and threatens the habitability of the earth. But the deterioration of our atmosphere is by far the most serious manifestation of this crisis. It is inherently global and affects every part of the earth; it is a contributing and causative factor in most of the other crises; and if it is not quickly addressed, it has the potential to end human civilization as we know it.

For all its complexity, however, its causes are breathtakingly simple and easy to understand.

All around the world, we humans are putting into the atmosphere extraordinary amounts of six different kinds of air pollution that trap heat and raise the temperature of the air, the oceans, and the surface of the earth.

These six pollutants, once emitted, travel up into the sky quickly. But all six of them eventually come back down to earth, some quickly, others very slowly. And as a result, the oft-cited aphorism "What goes up must come down" will work in our favor when we finally decide to solve the climate crisis.

Indeed, the simplicity of global warming

causation points toward a solution that is equally simple, even if difficult to execute: we must sharply reduce what goes up and sharply increase what comes down. That's what this book is about.

The biggest global warming cause by far—carbon dioxide—comes primarily from the burning of coal for heat and electricity, from the burning of oil-based products (gasoline, diesel, and jet fuel) in transportation, and from the burning of coal, oil, and natural gas in industrial activity. Carbon dioxide produced in the burning of these fossil fuels accounts for the single largest amount of the air pollution responsible for the climate crisis. That is why most discussions of how to solve the climate crisis tend to focus on producing energy in ways that do not at the same time produce dangerous emissions of CO_2.

At this point, however, the burning of coal, oil, and natural gas is not only the largest source of CO_2 but also far and away the most rapidly increasing source of global warming pollution.

After fossil fuels, the next largest source of human-caused CO_2 pollution—almost a quarter of the total—comes from land use changes—

EXCESS NATURAL GAS IS FLARED OFF AT A
GAS PLATFORM OFF THE COAST OF THAILAND.
FLARING PRODUCES CO_2, BUT MINIMIZES
THE RELEASE OF METHANE, AN EVEN MORE
POTENT GREENHOUSE GAS. IT IS WASTEFUL
NOT TO CAPTURE THE METHANE.

WHAT GOES UP: GREENHOUSE GASES

The pollutants that produce global warming come from many different activities, especially electricity generation, industry, agriculture, deforestation, and transportation. Carbon dioxide, the most prevalent greenhouse gas, enters the atmosphere from the processing and burning of coal (and other fossil fuels) for electricity and heat; burning forests and agricultural waste; land, air, and sea transportation; and frozen carbon just beginning to be released from the thawing of permafrost, to name just a few sources. The best scientists say we must reduce CO_2 to 350 parts per million in the atmosphere. Methane, which is less abundant but has a much stronger greenhouse effect, comes from sources such as livestock, rice cultivation, decaying waste in landfills, and "fugitive emissions" from coal, oil, and gas processing. Black-carbon pollution, now believed to be an extremely important contributor to global warming, is produced by burning forests and grasslands, cooking fires, and other man-made sources. Some industries and businesses emit very powerful greenhouse gases known as halocarbons, some of which are thousands of times more powerful molecule per molecule than CO_2. Industrial agriculture is also the largest source of nitrous oxide, carbon monoxide, and volatile organic compounds (VOCs).

MELTING PERMAFROST

COAL MINING

COAL PLANTS

INDUSTRIAL PROCESSES

INDUSTRIAL AGRICULTURE

OIL PRODUCTION

CROP BURNING

FOREST BURNING

FERTILIZATION

LAND TRANSPORTATION

LANDFILLS

predominantly deforestation, the burning of trees and vegetation. Since the majority of forest burning is in relatively poorer developing countries and the majority of industrial activity is in relatively wealthier developed countries, the negotiators of proposed global agreements to solve the climate crisis generally try to strike a balance between measures that sharply reduce the burning of fossil fuels on the one hand and sharply reduce deforestation on the other.

There's good news and bad news about CO_2. Here is the good news: if we stopped producing excess CO_2 tomorrow, about half of the man-made CO_2 would fall out of the atmosphere (to be absorbed by the ocean and by plants and trees) within 30 years.

Here is the bad news: the remainder would fall out much more slowly, and as much as 20 percent of what we put into the atmosphere this year will remain there 1,000 years from now. And we're putting 90 million tons of CO_2 into the atmosphere every single day!

The good news should encourage us to take action now, so that our children and grandchildren will have reason to thank us. Although some harmful consequences of the climate crisis are already under way, the most horrific consequences can still be avoided. The bad news should embolden us to a sense of urgency, because—to paraphrase the old Chinese proverb—a journey of a thousand years begins with a single step.

The second most powerful cause of the climate crisis is methane. Even though the volume of methane released is much smaller than the volume of CO_2, over a century-long period, methane is more than 20 times as potent as CO_2 in its ability to trap heat in the atmosphere—and over a 20-year period, it is about 75 times as potent.

Methane is different from CO_2 in one other key

SYNCRUDE TAR SANDS PROCESSING PLANT, ALBERTA, CANADA. OVER ITS LIFE CYCLE, FUEL MADE FROM TAR SANDS EMITS MUCH MORE CO_2 THAN EITHER COAL OR OIL. A TOYOTA PRIUS RUNNING ON GASOLINE MADE FROM TAR SANDS HAS THE CARBON FOOTPRINT OF A HUMMER.

FEEDLOT NEAR BAKERSFIELD, CALIFORNIA. ABOUT
HALF OF OUR DIET-RELATED GREENHOUSE GAS
EMISSIONS COME FROM THE PRODUCTION OF MEAT.

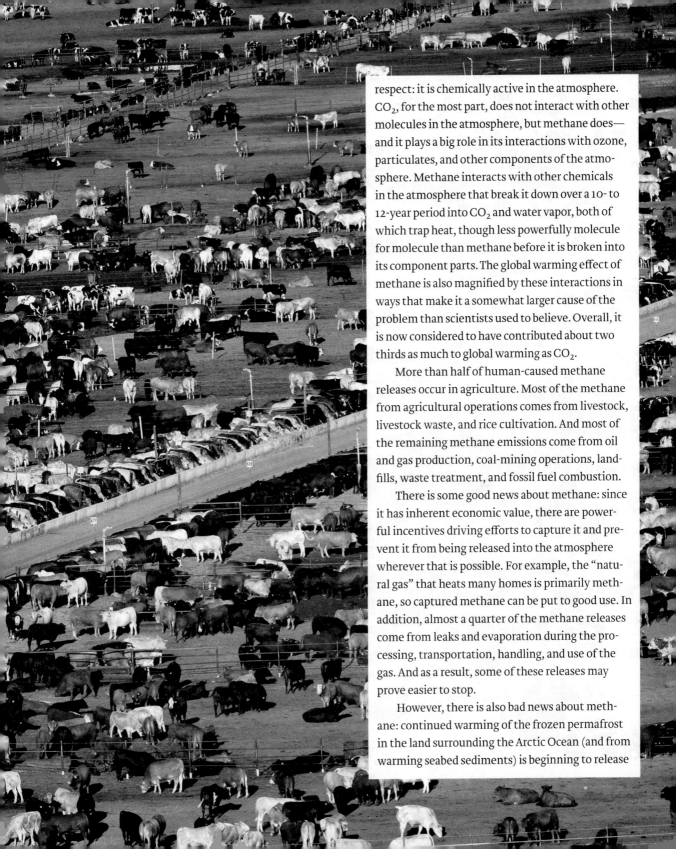

respect: it is chemically active in the atmosphere. CO_2, for the most part, does not interact with other molecules in the atmosphere, but methane does—and it plays a big role in its interactions with ozone, particulates, and other components of the atmosphere. Methane interacts with other chemicals in the atmosphere that break it down over a 10- to 12-year period into CO_2 and water vapor, both of which trap heat, though less powerfully molecule for molecule than methane before it is broken into its component parts. The global warming effect of methane is also magnified by these interactions in ways that make it a somewhat larger cause of the problem than scientists used to believe. Overall, it is now considered to have contributed about two thirds as much to global warming as CO_2.

More than half of human-caused methane releases occur in agriculture. Most of the methane from agricultural operations comes from livestock, livestock waste, and rice cultivation. And most of the remaining methane emissions come from oil and gas production, coal-mining operations, landfills, waste treatment, and fossil fuel combustion.

There is some good news about methane: since it has inherent economic value, there are powerful incentives driving efforts to capture it and prevent it from being released into the atmosphere wherever that is possible. For example, the "natural gas" that heats many homes is primarily methane, so captured methane can be put to good use. In addition, almost a quarter of the methane releases come from leaks and evaporation during the processing, transportation, handling, and use of the gas. And as a result, some of these releases may prove easier to stop.

However, there is also bad news about methane: continued warming of the frozen permafrost in the land surrounding the Arctic Ocean (and from warming seabed sediments) is beginning to release

BURNING SUGARCANE, BRAZIL. BURNING
AGRICULTURAL LAND AND VEGETATION IS A
MAJOR SOURCE OF BLACK-CARBON POLLUTION

large amounts of methane into the atmosphere as the frozen structures containing it melt, and as microbes digest the thawing carbon buried in the tundra. The only practical way to prevent these releases is to slow and then halt global warming itself—while there is still time.

The third largest source of the climate crisis is black carbon, also called soot. Black carbon is different from the other air pollutants that cause global warming. First, unlike the others, it is technically not a gas but is made up of tiny carbon particles like those you can see in dirty smoke, only smaller. That is one reason it only recently became a major focus for scientists, who discovered the surprisingly large role it was playing in warming the planet. Second,

unlike the other five causes of global warming, which absorb infrared heat radiated by the earth back toward space, black carbon absorbs heat from incoming sunlight. It is also the shortest lived of the six global warming culprits.

The largest source of black carbon is the burning of biomass, especially the burning of forests and grasslands, mostly to clear land for agriculture. This problem is disproportionately concentrated in three areas: Brazil, Indonesia, and Central Africa. Forest fires and seasonal burning of ground cover in Siberia and Eastern Europe also produce soot that is carried by the prevailing winds into the Arctic, where it settles on the snow and ice and has contributed greatly to the progressive disappearance

CLEANING THE AIR AFTER THE GREAT SMOG OF 1952

In December 1952, a lethal smog descended on London, immersing the city in a thick blanket of pollution for five dark days. Four thousand people died that week, and 8,000 more in the following months, from respiratory infections as well as asphyxiation.

The tragedy was the result of increased coal burning prompted by a stretch of especially cold weather. The city's million-plus coal-heated households added pollutants to the already thick industrial smog produced by local factories. Unusual weather conditions—including a temperature inversion—kept the hugely elevated levels of black soot and tar particles close to the ground, reducing visibility and bringing the city to a virtual standstill.

After this disaster, the government took action to improve the country's air quality. In 1956, the British Parliament introduced the Clean Air Act, outlawing the burning of coal in open-hearth fires and incentivizing the replacement of coal with cleaner sources of energy, such as electricity, gas, and oil. Soon thereafter, a determined environmental movement also emerged in the United States and beyond.

Daytime air quality in London's Trafalgar Square in 1952

of the Arctic's sea ice cover. Indeed, one estimate is that black carbon is responsible for an estimated 1°C (1.8°F) of the 2.5°C (4.5°F) of warming that has already occurred in the Arctic. Large amounts of black carbon are also produced by forest fires in North America, Australia, Southern Africa, and elsewhere. In addition to biomass burning, as much as 20 percent of the black carbon comes from the burning of wood, cow dung, and crop residues in South Asia for cooking and heating homes, and from China, where the burning of coal for home heating is also a major source.

Black carbon also poses a particular threat to India and China, partly because of the unusual seasonal weather pattern over the Indian subcontinent, which typically goes without much rain for six months of the year between monsoon seasons. The temperature inversion that forms over much of South Asia during that period traps the black carbon above the glaciers and snow, causing air pollution high in the Himalayas and on the Tibetan Plateau. In some of these areas, air pollution levels are now comparable to those of Los Angeles. So much black carbon settles on the ice and snow that the melting already triggered by atmospheric warming has accelerated. By some estimates, 75 percent of all Himalayan glaciers less than 15 square kilometers could disappear in as little as 10 years.

BLACK CARBON AND HIMALAYAN GLACIERS

As much as 20 percent of the black carbon in our atmosphere comes from the burning of wood, dung, and crop residues for household cooking and heating in India, and from lumps of coal burned by households in China. Between monsoons, brown clouds of pollution are trapped above the Himalayas. The black carbon falls on the glaciers, darkening their surface and causing the ice and snow to absorb sunlight instead of reflecting it, accelerating the rate of melting. Partly as a result, scientists expect many Himalayan glaciers will disappear by 2020.

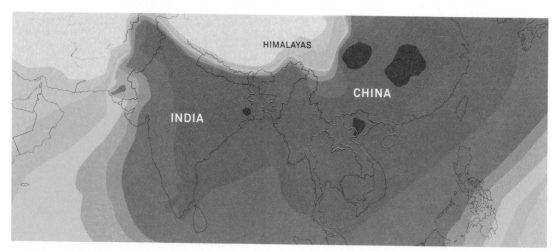

AMOUNT OF BLACK CARBON IN THE ATMOSPHERE

LESS   MORE

SOURCE: *New York Times*, "Third-World Soot Is Target in Climate Fight," April 15, 2009

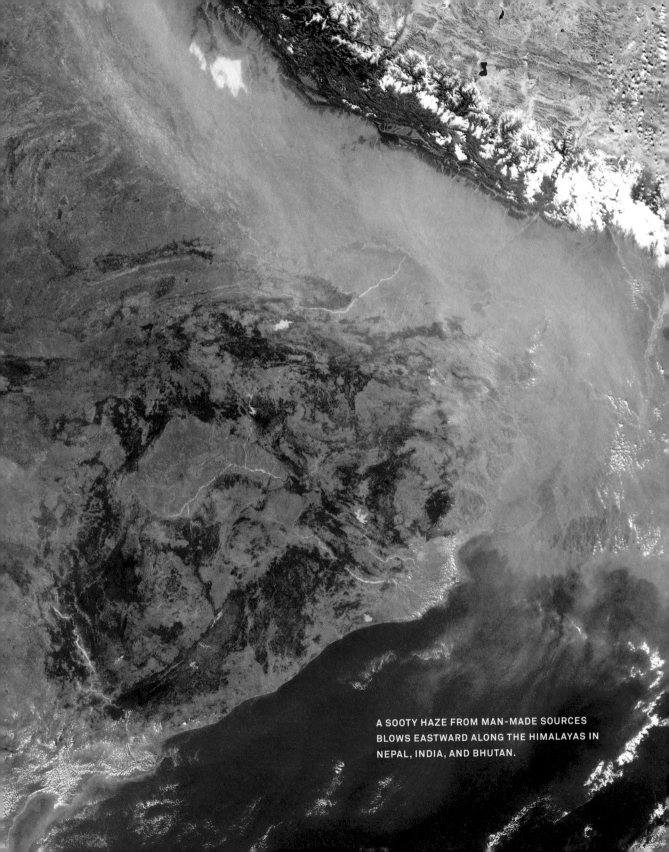

A SOOTY HAZE FROM MAN-MADE SOURCES BLOWS EASTWARD ALONG THE HIMALAYAS IN NEPAL, INDIA, AND BHUTAN.

Since half of the drinking water and agricultural water in India and much of China and Indochina comes from the seasonal melting of these same glaciers, the human consequences could soon become catastrophic. For example, 70 percent of the water flowing in the Ganges River comes from the melting of ice and snow in the Himalayas.

Black carbon is also produced by the burning of agricultural waste, such as residue from sugarcane (bagasse) and residue from corn (stover), and from burning firewood throughout the world.

More than a third of the black carbon in the atmosphere comes from the burning of fossil fuels, primarily from diesel trucks not equipped with devices to trap emissions as they exit the tailpipe. Though these devices have recently been introduced, they are not yet widely used.

It is noteworthy that so much of the black-carbon pollution comes from activities that simultaneously produce CO_2, including inefficient engines for small vehicles in Asia and wasteful coal-fired power plants. But this need not be the case. For example, coal burning in industrial countries produces CO_2 without producing much black carbon due to measures taken in the past several decades to make fuel combustion more efficient and to curb local air pollution.

Most of the global warming caused by black carbon comes from its absorption of incoming sunlight. It is a primary component of the large brown clouds that cover vast areas of Eurasia and drift eastward across the Pacific Ocean to North America and westward from Indonesia across the Indian Ocean to Madagascar. These clouds—like some other forms of air pollution—partially mask global warming by blocking some of the sunlight that would otherwise reach lower into the atmosphere. Black carbon typically does not linger in the atmosphere for long periods of time, because

it is washed out of the air by rain. (That may be yet another reason why it was traditionally not included in the list of greenhouse gas pollutants.) As a result, once we stop emitting black carbon, most of it will stop trapping heat in the atmosphere in a matter of weeks. Right now, however, we put such enormous quantities of black carbon into the air every day, the supply is continually replenished. And scientists have taken note that in areas of the world that experience long, dry seasons with no rainfall, black-carbon concentrations build up to extraordinarily high levels.

Moreover, scientists are increasingly concerned about black carbon because it also causes the earth to warm up in a second way: when it falls on ice and snow, it darkens the white reflective surface so much that sunlight that used to bounce off is absorbed instead, causing more rapid melting.

The overall reflectivity of the earth is an important factor in understanding the problem of global warming. The more sunlight bounces off the tops of clouds and the highly reflective parts of the earth's surface, the less solar radiation is absorbed as heat. The less heat absorbed, the less trapped by global warming pollution when it is re-radiated toward space as infrared radiation.

This has led some scientists to suggest painting millions of roofs white and other steps to increase the reflectivity of the earth's surface. These ideas are worthy of serious consideration. But, in the meantime, we are losing much of the earth's natural reflectivity (or *albedo*, as scientists refer to it) with the melting of ice and snow—particularly in the Arctic and the Himalayas.

The fourth most significant cause of global warming is a family of industrial chemicals called halocarbons—including the notorious chlorofluorocarbons (CFCs). Many are already being regulated and reduced under a 1987 treaty (the Montreal

ALBEDO: MEASURING THE SUN'S REFLECTION

Albedo is a measure of the reflectivity of different objects and surfaces on the earth: the lower the number, the more energy is absorbed, contributing to global warming. The most reflective surfaces are snow and ice, which send as much as 90 percent of the sun's energy back toward space. As sea ice melts, it exposes the far more energy-absorbent ocean.

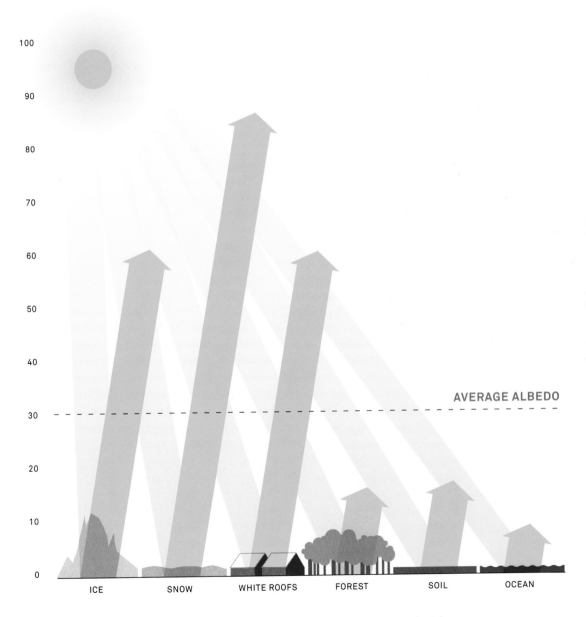

SOURCE: National Center for Atmospheric Research; Lawrence Berkeley National Laboratory; C. Donald Ahrens, *Meteorology Today*

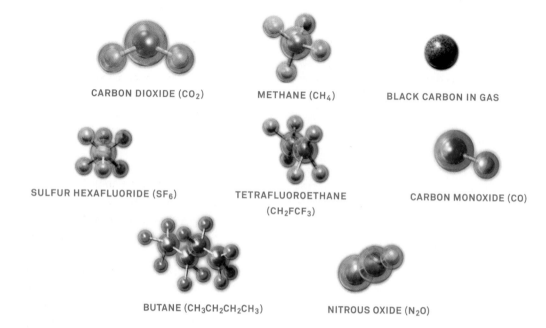

CARBON DIOXIDE (CO_2)   METHANE (CH_4)   BLACK CARBON IN GAS

SULFUR HEXAFLUORIDE (SF_6)   TETRAFLUOROETHANE   CARBON MONOXIDE (CO)
                                 (CH_2FCF_3)

BUTANE ($CH_3CH_2CH_2CH_3$)   NITROUS OXIDE (N_2O)

A GUIDE TO GLOBAL WARMING POLLUTANTS

All global warming comes directly or indirectly from the effects of six families of pollutants (see "The Sources of Global Warming," right). The largest role is played by carbon dioxide (CO_2), the most abundant and most rapidly increasing greenhouse gas. Methane (CH_4), also a greenhouse gas, is the second worst cause, followed by black carbon (soot). Important roles are also played by industrial chemicals invented in the 20th century—chlorofluorocarbons; halocarbons, such as tetrafluoroethane (CH_2FCF_3); and sulfur hexafluoride (SF_6). All of these chemicals trap heat in the atmosphere. Carbon monoxide (CO) and volatile organic compounds (VOCs)—such as butane—do not trap heat directly but interact with other pollutants to create compounds that do trap heat. Finally, nitrous oxide (N_2O)—which is mainly a by-product of nitrogen-intensive agriculture—plays a smaller but still significant role in trapping heat in the earth's atmosphere.

Protocol) that was adopted worldwide in response to the first global atmospheric crisis, the hole in the stratospheric ozone layer. As an added benefit of that treaty, this category of global warming pollution is now slowly but steadily declining. It still represents roughly 13 percent of the total problem—a significant number—so efforts to further strengthen the Montreal Protocol that are already under way will help. For example, many scientists are critical of the U.S. insistence in 2006 that the phaseout of methyl bromide be delayed indefinitely for certain agricultural uses. In addition, there is growing concern among scientists that some of the chemicals used as substitutes for halocarbons—particularly chemicals known as hydrofluorocarbons—should also be controlled under the Montreal Protocol because they are potent global warming pollutants and their volume is growing rapidly.

Three other chemical compounds in the halocarbon family that do not destroy stratospheric ozone (and thus were not covered in the earlier treaty) are also potent greenhouse gases. These are controlled under the Kyoto Protocol (which the United States didn't ratify). Some halocarbons stay in the atmosphere for thousands of years. (One of them, carbon tetrafluoride, lingers in the atmosphere for an incredible 50,000 years—though, thankfully, it is produced in small volumes.)

It is important to note that the world's efforts to protect the stratospheric ozone layer represent an historic success. Even though the affected industries initially fought acceptance of the science that alerted us to the gravity of the threat, political leaders in country after country wasted very little time in coming together across ideological lines to secure an effective treaty in spite of some residual uncertainty in the science. Three years after the treaty was signed, they revisited the subject and toughened the original standards. In the years since, it has been strengthened again several times. Significantly, some of the same corporations that had opposed the original treaty worked in favor of strengthening it after their experience in finding substitutes for the offending chemicals. As a result, the world is now well on the way to solving this particular problem. Scientists say it may take another 50 to 100 years before the stratospheric ozone layer is fully healed, but we are now moving in the right direction. They caution us, however, that the one thing that could reverse this trend is failure to solve global warming, which according to some scientists could threaten to make the ozone hole above Antarctica start growing again. Continued heating of the atmosphere (and cooling of the stratosphere) could threaten to restart the destruction of stratospheric ozone and thin the ozone layer to the point where it could once again become a dangerous threat to human life.

The next family of air pollutants contributing to global warming includes carbon monoxide and volatile organic compounds (VOCs). Carbon monoxide is mostly produced by cars in the U.S., but is also produced in large quantities in the rest of the world by the burning of biomass. VOCs are produced mainly in industrial processes around the world, but in the U.S., a quarter of these emissions comes from cars and trucks. These pollutants actually do not trap heat themselves, but they lead to the production of low-level ozone, which is a potent greenhouse gas and unhealthy air pollutant.

These pollutants are not included in the official list of chemical compounds controlled under the Kyoto Protocol—just as black carbon is not yet included—but scientific experts include them among the causes of global warming because they interact with other chemicals in the atmosphere (including methane, sulfates, and, to a lesser extent, CO_2) in ways that further trap significant amounts of heat and contribute to global warming.

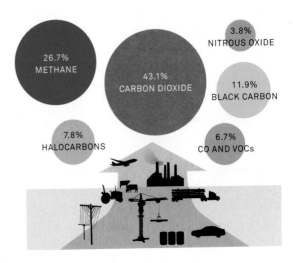

THE SOURCES OF GLOBAL WARMING
Global warming can be traced to six families of pollutants, whose proportional role in the problem is seen above. These gases and black carbon are emitted from many human endeavors, from transportation to farming to heating.

SOURCE: Drew T. Shindell, et al., *Science*, in press, 2009

A WORKER SPREADS FERTILIZER PELLETS ON A FARM IN NORTHERN MEXICO. THE MAJORITY OF NITROUS-OXIDE EMISSIONS COME FROM AGRICULTURAL PRACTICES, INCLUDING RELIANCE ON FOSSIL FUEL–INTENSIVE, SYNTHETIC NITROGEN–BASED FERTILIZER PELLETS.

Any comprehensive strategy for solving the climate crisis should, therefore, focus on these pollutants, along with the other five causes of global warming.

The details of such a strategy must also include attention to other chemicals in the atmosphere that add complexity to the problem, such as sulfur dioxide (which leads to the formation of sulfate particles), nitrogen oxides (which contribute to the formation of ozone), sulfates, nitrates, and organic carbon. All of these have a net cooling effect on their own, but they also interact with global warming pollution and impact public health and ecosystems in ways that affect problem-solving strategies.

The last cause of global warming is nitrous oxide. The vast majority of nitrous-oxide emissions comes from agricultural practices that rely heavily on nitrogen fertilizers, greatly magnifying the natural emissions resulting from the bacterial breakdown of nitrogen in the soil. In the past 100 years—since two German chemists discovered a new process for combining hydrogen with atmospheric nitrogen to create ammonia—we have doubled the amount of available nitrogen in the environment. Traditionally, farmers rotated crops to replenish nitrogen depleted from the soil after several years of growing the same crop. By planting legumes and applying animal manure, farmers found they could restore fertility to their land. However, modern agriculture has come to rely heavily on vast quantities of synthetic ammonia fertilizers that continually add nitrogen to soils otherwise too depleted to grow crops. This Faustian bargain has greatly increased crop yields. The trade-off has been nitrous-oxide emissions into the atmosphere and nitrogen runoffs into rivers and creeks, where it stimulates the rapid and unsustainable growth of algae blooms. When these algae blooms die and decompose, the oxygen in the water is depleted, forming "dead zones" where fish and many other species cannot survive. Moreover, since these synthetic ammonia fertilizers require large amounts of fossil fuel to produce, the manufacturing process adds significant amounts of CO_2 to the atmosphere.

Smaller amounts of nitrous oxide are also emitted from burning fossil fuels, from a variety of industrial processes, and from poor management of livestock manure and human sewage.

Although nitrous oxide is the smallest contributor among the six causes of global warming, it is nevertheless significant and can be reduced if we change the way we use nitrogen.

Finally, it is important to note the role played in the atmosphere by water vapor. Some commentators like to point out that water vapor traps more heat than CO_2. While this is technically correct, the extent to which water vapor traps more heat than normal in the earth's atmosphere is determined by the extent to which global warming pollutants raise the air and ocean temperatures, increasing the amount of water vapor the atmosphere can hold. The amount of water vapor in the air is responsive to its temperature and to atmospheric circulation patterns that help determine the relative humidity. Because changes in these variables are being driven by the emission of CO_2 and other global warming pollutants, human activities are really controlling the change in atmospheric water vapor. Consequently, the only way to reduce the role of water vapor is to solve the climate crisis.

So there it is: the solution to global warming is as easy to describe as it is difficult to put into practice. Emissions of the six kinds of air pollutants causing the problem—CO_2, methane, black carbon, halocarbons, nitrous oxide, and carbon monoxide, plus VOCs—must all be reduced dramatically. And we must simultaneously increase the rate at which they are removed from the air and reabsorbed by the earth's oceans and biosphere.

OUR SOURCES OF ENERGY

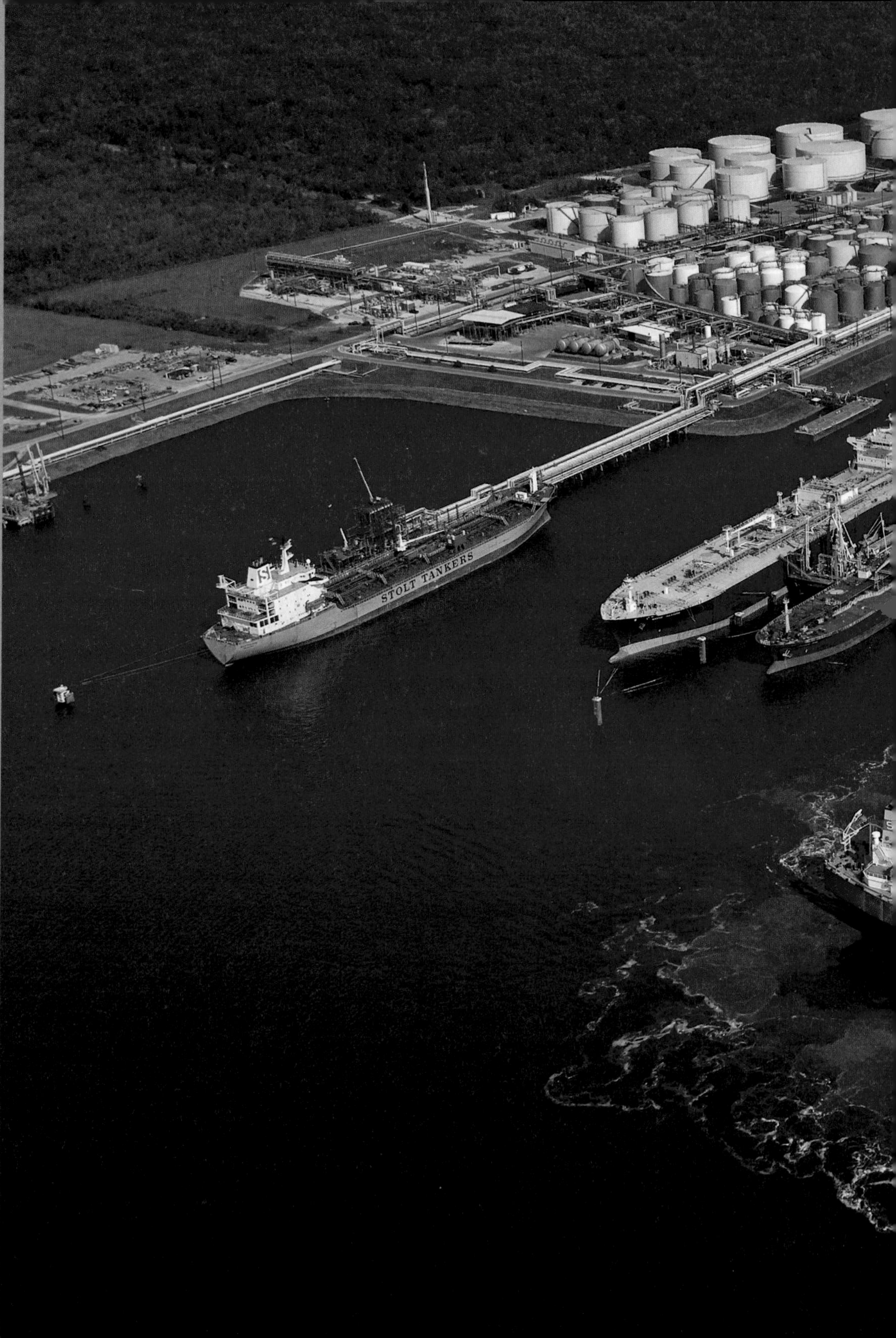

CHAPTER TWO

WHERE OUR ENERGY COMES FROM AND WHERE IT GOES

AN EMPTY SUPERTANKER LEAVES A TANK FARM IN HOUSTON ON ITS WAY TO GET MORE OIL.

The single largest source of man-made global warming pollution is the production of energy from fossil fuels—coal, oil, and natural gas. Consequently, the most important solutions for the climate crisis require the accelerated development and deployment of low-CO_2 substitutes for producing the energy needed for the global economy. Our present dependence on fossil fuels is relatively new in human history. Although both coal and oil were known in antiquity, they were used only in very small quantities at locations where they were easily accessible on the earth's surface.

Wood remained the principal source of energy until the latter part of the 18th century. The slow but steady growth in human population throughout the Middle Ages (which accelerated after Europeans colonized the Americas) led to a dramatic reduction of Europe's tree cover, from 95 percent at the time of the Roman Empire's collapse in 476 to only 20 percent at the beginning of the Scientific Revolution in the early 17th century.

The combination of widespread wood shortages and newly developed, energy-using machines (as the Scientific Revolution gained momentum) led to wider use of coal, which had always been considered inferior to wood as a home heating source because of the air pollution it produced in houses and cities. But coal's higher energy-to-weight ratio made it preferable—first in metallurgy, where higher temperatures were necessary for producing iron and steel, and then in a variety of emerging manufacturing processes that set the stage for the Industrial Revolution.

As the search for coal pushed underground, coal mining and the key technologies of the early Industrial Revolution developed in tandem. The steam engine and the use of steel wheels on steel rails both emerged first in coal mines. As these technologies improved, steam-powered locomotives on aboveground rail lines and use of the steam engine as a replacement for sails in powering mid-19th-century oceangoing vessels contributed to the surge in coal demand. Soon, the Industrial Revolution led to widespread use of coal as the principal source of energy for factories. But it was the harnessing of electricity for commercial use in the late 1800s that set the stage for today's dominant use of coal.

When considering the best choices we can make for producing low-CO_2 energy from new sources rather than fossil fuels, it is important to think differently about each of the three discrete forms of energy that are moved from one place to another: liquid fuels, gaseous fuels, and electricity. Each of these three sectors of the energy marketplace has its own special characteristics.

Liquid forms of energy come almost entirely from oil and have very different characteristics from those of either electricity or gas. Because they are easily stored and contain more energy, pound for pound, than coal, liquid fuels derived from oil account for almost all of the energy used in transportation throughout the world. In the U.S., more than half of all oil use is in cars and trucks,

A TRAIN FILLED WITH COAL DESTINED FOR POWER GENERATION PREPARES TO LEAVE A MINE NEAR WRIGHT, WYOMING.

with most of the rest used in industry for stationary diesel engines and as petrochemical feedstock. A little less than 10 percent is used for heating homes and businesses, and less than 6 percent is used for generating electricity.

Overall, oil is now the largest source of energy we use, providing considerably more energy than either coal or gas. However, the burning of petroleum products produces about 30 percent less CO_2 than the burning of coal, per unit of energy generated.

(It's ironic that the first successful oil well—in Pennsylvania—was drilled in 1859, the same year the great Irish scientist John Tyndall determined that CO_2 molecules intercept infrared radiation, a discovery that led to the science of global warming.)

Liquid fuels pose very different challenges than do the other two forms of energy, electricity and natural gas. Most nations, including the United

States, are dependent on foreign sources for oil, and the global market is dominated by the huge reserves in the Middle East, the flows of which have been interrupted several times in the past 35 years for geopolitical reasons. Moreover, wildly oscillating oil prices have periodically wreaked havoc with efforts by the U.S. and other large oil consumers to sustain national investment strategies to accelerate the development of renewable forms of energy. (The lower-CO_2 alternatives for oil-based liquid fuels are discussed in Chapter 6.)

Natural gas, the second primary form of energy we use, is principally methane. Gas now provides for approximately 23 percent of the world's energy consumption. Almost 40 percent of all natural gas is used in industry as a chemical feedstock and as a source of heat for boilers. In recent decades, almost one third of natural gas has gone to the production of electricity, and almost 20 percent is used directly

CARBON DIOXIDE FROM CARBON-BASED FUELS

Our carbon-based fuels have very different characteristics. Oil and natural gas have more energy, pound for pound, than coal. But oil produces 40 percent more CO_2 than gas, and coal creates 40 percent more than oil. Wood, the only renewable carbon-based fuel, contains the least energy by weight.

POUNDS OF CO_2 PRODUCED PER 1 MILLION BTU

SOURCE: University of California, Irvine; Oak Ridge National Laboratory; gas conversion: Tulsa Gas Technologies

in homes for heating and cooking.

Gaseous carbon fuels were first produced from coal in an expensive conversion process, but later, oil drillers began discovering large amounts of natural gas in the same underground formations from which they were pumping petroleum. At first they simply burned the gas off in order to get at the oil, but now they collect both oil and gas. As the value of natural gas became more apparent, geologists learned to locate wells that produce only gas.

Because methane has many more hydrogen atoms per carbon atom than either coal or oil, it produces only 70 percent of the CO_2 produced by oil—and roughly half of the CO_2 produced by coal—for the same amount of energy. That is why many consider natural gas an important transition fuel in the shift away from coal and oil. Over the next few decades, however, the reductions in CO_2 needed to halt destruction of our climate will require shifting away from gas, as well as from coal and oil. After all, one fifth of the CO_2 produced in the energy marketplace already comes from the burning of gas.

Gaseous fuels have special characteristics. The best new gas-fired power plants are roughly twice as efficient as a typical coal plant.

Electricity is a primary energy carrier made from sources as diverse as coal, sunlight, and wind. It is responsible for delivering a large part of the world's energy, growing faster than any other energy sector. The invention of incandescent lighting by Thomas Edison in 1879 and the subsequent invention of alternating current by Nikola Tesla nine years later led to the progressive electrification of streetcars, factories, and homes, and a rapidly proliferating array of electrical devices whose progeny are still ubiquitous in our lives.

In spite of its widespread use and convenience, however, electricity also has important disadvantages. First, a large percentage of the energy contained in the fuels burned to produce electricity is lost in the relatively inefficient conversion to electrical current. Second, electricity has been expensive to store.

Finally, of course, as the demand for electricity has grown rapidly, and the early dependence on hydroelectric generators has given way to coal-fired steam generators, the production of man-made CO_2 from coal has grown exponentially. More than 40 percent of all the electricity in the world is still produced by burning coal, and roughly half that much electricity (20 percent) comes from burning natural gas.

The remainder of the world's electricity comes from sources that do not produce large amounts of CO_2. Hydroelectric dams provide 18 percent, and less still—15 percent—comes from nuclear power, once considered the natural successor to fossil fuels. A small amount comes from solar, wind, and geothermal, sources that are all expected to grow rapidly in the next quarter-century.

The growing popularity of electricity compared with other forms of energy is due not only to the ease with which it can be used but also to the versatility of the electricity transmission and distribution infrastructure to handle a variety of different energy sources to power electric turbines. That is because, with some exceptions—solar photovoltaic energy and fuel cells, for example—most forms of electricity production utilize energy to rapidly turn the blades of a turbine in order to spin copper coils or magnets in relation to one another, thereby producing electrical current (see "How a Turbine Works," page 60). Coal, gas, and oil are all burned to make steam that is pressurized to turn the turbines. Windmills and hydroelectric dams turn the blades directly.

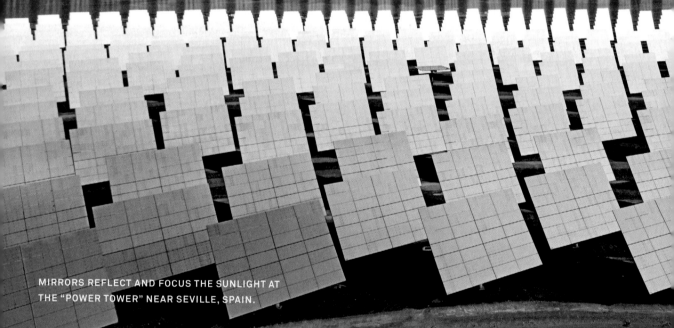

"I'd put my money on the sun and solar energy. What a source of power!"

THOMAS EDISON

MIRRORS REFLECT AND FOCUS THE SUNLIGHT AT THE "POWER TOWER" NEAR SEVILLE, SPAIN.

Altogether, oil, coal, and natural gas still provide 86.5 percent of the primary energy we now use on Earth. (Oil is responsible for 36.5 percent; coal for 27 percent; and natural gas for 23 percent.) Taken together, these three fossil fuels make up the single largest cause of global warming. That is why the world has begun to focus with such intensity on new alternatives for producing energy without emitting large quantities of CO_2.

The exciting news about renewable, low-CO_2 sources of energy—especially for producing electricity—is that there is an almost limitless amount available. Indeed, all of the oil, coal, and natural gas in the world contain the same amount of energy as the earth receives in only 50 days from the sun. The earth is bathed in so much energy from the sun that the amount falling on the surface of our planet every hour is theoretically equal to the entire world's energy use for a full year. Even taking into account all of the technical difficulties in capturing and using solar energy, it would take only seven days' worth of sunlight hitting the earth to meet the annual energy needs of the planet.

Almost a hundred years ago, Thomas Edison said, in a conversation with Henry Ford and Harvey Firestone: "I'd put my money on the sun and solar energy. What a source of power! I hope we don't have to wait until oil and coal run out before we tackle that."

Similarly, one month's worth of the energy that could be captured from wind and from the geothermal energy emitted from the earth itself could each supply civilization's entire energy use for a year. Add to that the energy contained in flowing rivers and the waves and tides of the ocean, and it's obvious that renewable sources of energy—if developed—could completely replace CO_2-rich fossil fuels.

The difficulty, of course, lies in the focused

effort and large investments needed to develop and build cost-effective renewable-energy systems that allow us to capture and efficiently use these enormous natural flows of energy.

All forms of energy are expensive, but over time, renewable energy gets cheaper, while carbon-based energy gets more expensive. Renewable energy will decline in price for three reasons.

First, once the renewable infrastructure is built, the fuel is free forever. Unlike carbon-based fuels, the wind and the sun and the earth itself provide fuel that is free, in amounts that are effectively limitless.

Second, while fossil fuel technologies are more mature, renewable-energy technologies are still being rapidly improved. So innovation and ingenuity give us the ability to constantly increase the efficiency of renewable energy and continually reduce its cost.

Third, once the world makes a clear commitment to shifting toward renewable energy, the volume of production will itself sharply reduce the cost of each windmill and each solar panel, while adding yet more incentives for additional research and development to further speed up the innovation process.

Consider, for example, what has happened to the cost and effectiveness of computers during the past 20 years. The growing demand for inexpensive computers has led to larger budgets allocated by manufacturers of computer chips to research and develop less expensive and more powerful ways to process information.

The well-known phenomenon known as Moore's Law (which predicts that the number of transistors on a computer chip will double every 18 to 24 months) has led to a 50 percent reduction in price for the same information processing power every year and a half or so for the last several decades. It is, instead, a self-fulfilling expectation by competing companies, all of whom assume their competitors will do what's necessary to continue this breakneck pace of advance.

Since each company anticipates that the industry as a whole will continue to follow this pathway, each company competes vigorously to make sure it is not left behind. And since the demand for computers around the world has continued to grow dramatically, the incentive for staying ahead of the competition is large enough to justify spending more money to build the best and most powerful computers at the lowest possible cost.

Moreover, as the most powerful computers enhance the ability of these scientists and engineers to explore ever newer ways to handle bits of information—by using new materials, subatomic processes, designing ever more advanced fabrication tools, and exploring new designs by simulating them on computers without having to actually build them—there is a positive feedback loop that continues to drive exciting new breakthroughs. These breakthroughs are expected to continue this extraordinary pattern of radical improvement every 18 to 24 months for at least another two decades.

In much the same way, the explosion in demand for innovative new approaches to the production of energy from renewable sources is fueling ever larger research and development budgets to find innovative new approaches at lower costs. In other words, what has been done with bits of information is beginning to be done with electrons. And as the cost comes down, the demand goes up, reinforcing this pattern of constant improvement, just as in the computer industry. A global commitment to a massive shift toward renewable energy will greatly accelerate this trend.

TURBINES: THE WAY MOST ELECTRICITY IS CREATED TODAY

Most of the electricity we use is produced by turbines. The power station converts a primary energy source—such as coal, natural gas, or uranium—into heat. The heat turns water into steam, which spins the blades of a turbine. The blades of the turbine power a generator to produce electricity, which can then be transmitted into homes and businesses for everyday use. (Hydropower uses falling water to turn the blades of a turbine. A windmill is a turbine, converting the wind into motion. The blades of a windmill are really the blades of a turbine.)

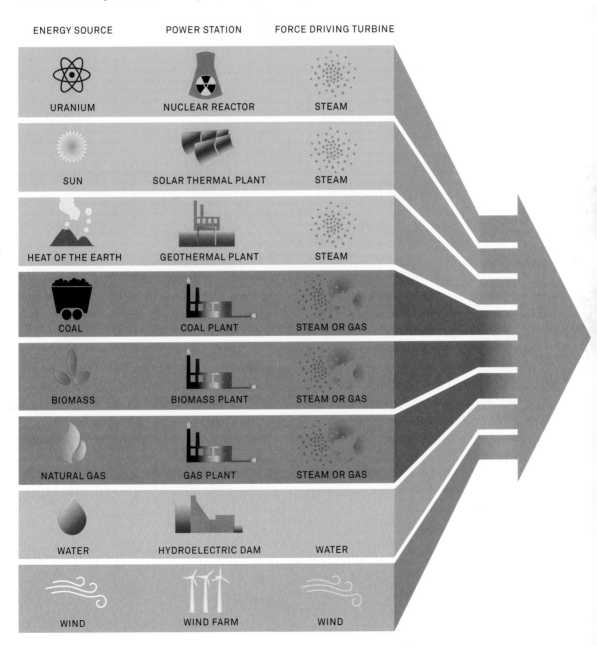

ENERGY SOURCE	POWER STATION	FORCE DRIVING TURBINE
URANIUM	NUCLEAR REACTOR	STEAM
SUN	SOLAR THERMAL PLANT	STEAM
HEAT OF THE EARTH	GEOTHERMAL PLANT	STEAM
COAL	COAL PLANT	STEAM OR GAS
BIOMASS	BIOMASS PLANT	STEAM OR GAS
NATURAL GAS	GAS PLANT	STEAM OR GAS
WATER	HYDROELECTRIC DAM	WATER
WIND	WIND FARM	WIND

HOW A TURBINE WORKS

In a typical generator, steam, gas, water, or wind spins turbine blades. The turbine then spins a shaft that is connected to a cylinder of insulated wire coils inside magnets or to magnets inside of wire coils. Either way, the magnets and coils spin in relation to each other. In the design shown here, the coils spin inside two opposite poles of a magnet.

Each section of the generator's wire, when spun in the magnetic field, becomes a small electromagnetic conductor; together, they create one large current. The current is sent to a transformer, which converts the electricity into a form that can be transmitted over high-voltage lines.

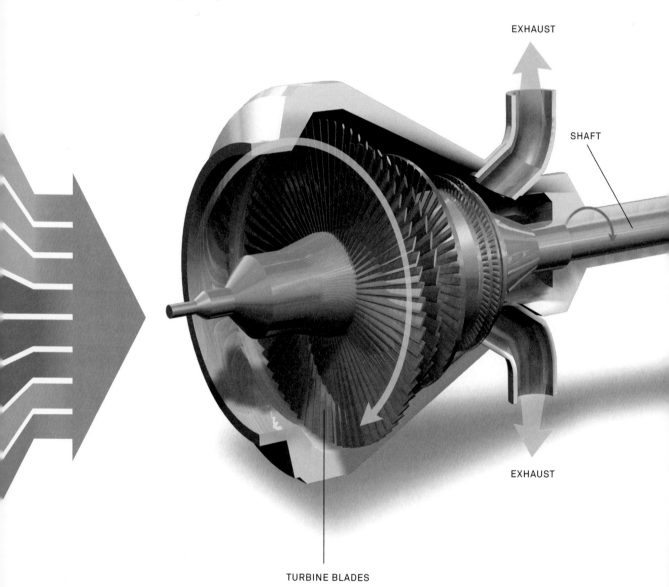

EXHAUST

SHAFT

EXHAUST

TURBINE BLADES

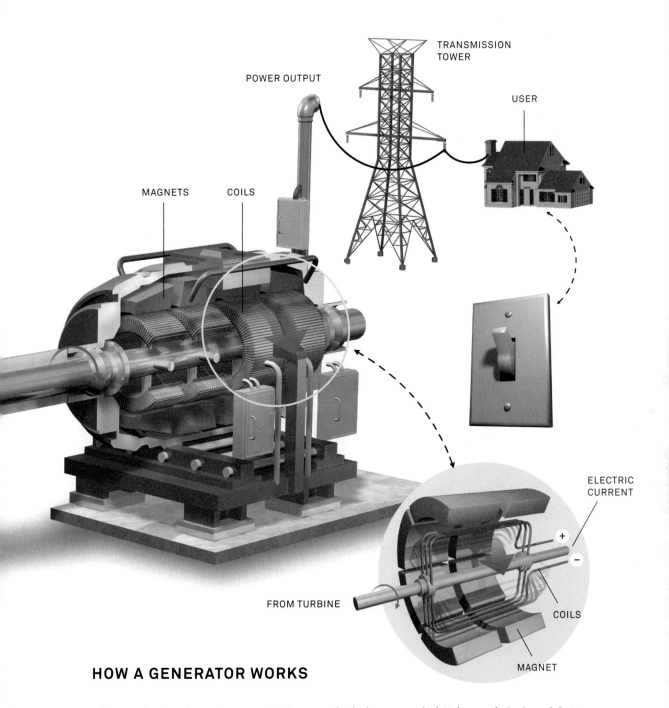

TRANSMISSION
TOWER

POWER OUTPUT

USER

MAGNETS

COILS

ELECTRIC
CURRENT

+

−

FROM TURBINE

COILS

MAGNET

HOW A GENERATOR WORKS

When a conductive wire, such as one made of copper, spins inside a magnetic field, its negatively charged electrons are thrown out of their orbits and can move from one atom's orbit to that of another. The movement of these "free" electrons is an electric charge; when many coils are spun quickly, they produce a continuous current of electric charge, or electricity.

OUR SOURCES OF ENERGY

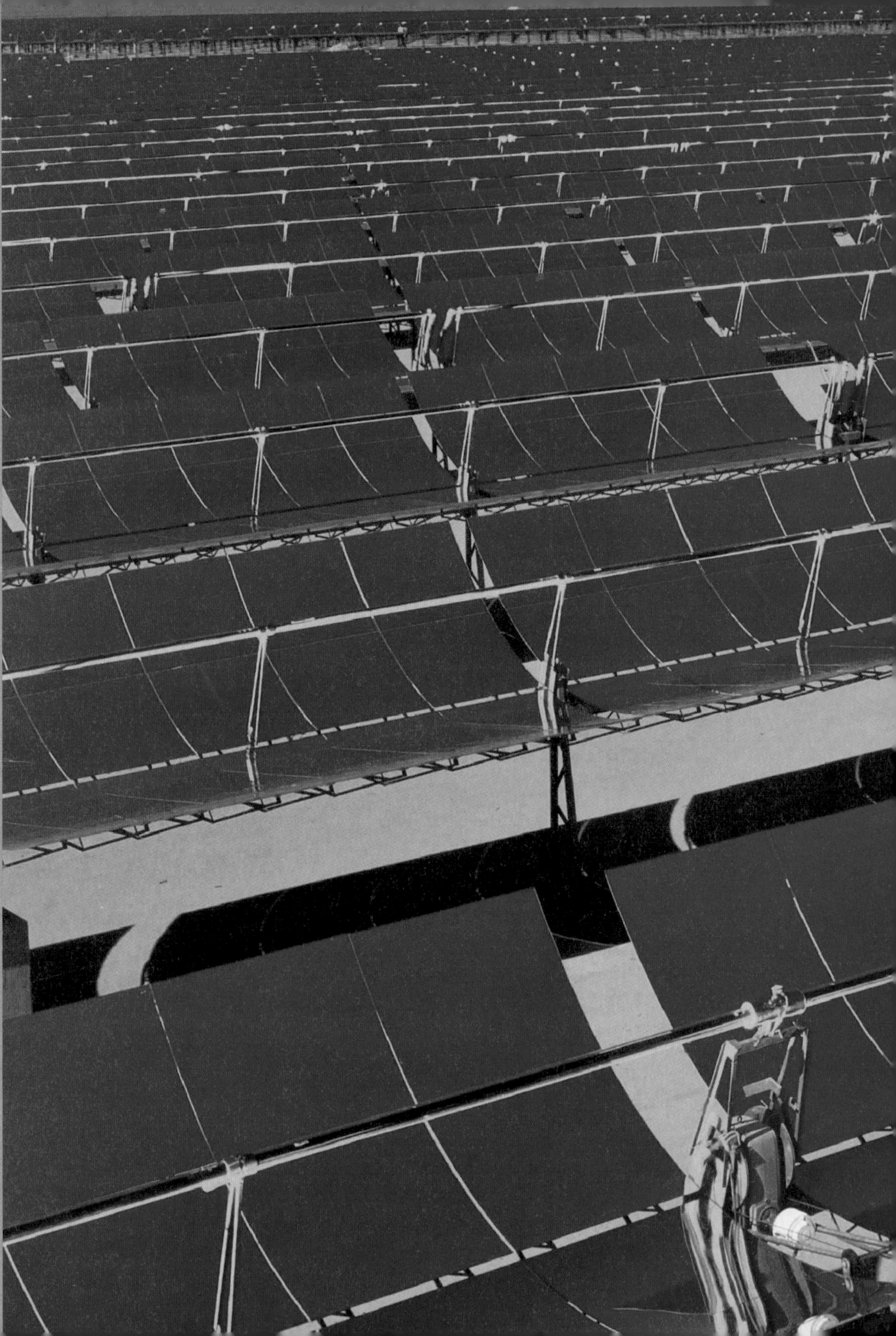

CHAPTER THREE

ELECTRICITY FROM THE SUN

PARABOLIC MIRRORS CONCENTRATE THE SUN'S ENERGY AT ONE OF SEVERAL SOLAR ELECTRIC GENERATING PLANTS IN CALIFORNIA'S MOJAVE DESERT. THE FACILITIES PRODUCE 150 MEGA-WATTS OF POWER.

Electricity can be produced from sunlight in two main ways—by producing heat that powers an electricity generator or by converting sunlight directly to electricity using solar cells.

The first approach, known as concentrated solar thermal (CST), uses sunlight to heat liquids that are used to power electricity generators. This requires using mirrors to concentrate sunlight.

There are several ways to do this. Some CST plants use curved mirrors (called parabolic trough mirrors) that track the movement of the sun across the sky and focus sunlight on a pipe, heating water or other liquid. Other designs use long rows of nearly flat, rotating mirrors to accomplish basically the same result with lower-cost materials.

Another innovative and visually exciting design uses a large array of flat sun-tracking mirrors spread out in a semicircle around a centrally located "Power Tower." This structure supports a large container of liquid heated to very high temperatures by the solar energy focused by all the mirrors simultaneously. As in the other designs, the heat collected is then used to boil water and produce steam for electric turbines. Most experts believe this technology involves higher risk—but potentially provides a higher reward—than the first two designs. Among other things, the higher temperatures make it easier to store heat energy for longer periods when sunlight is not available. It is not clear, however, that there is a cost advantage to the Power Tower approach.

Yet another design for producing thermal solar electricity utilizes a high-efficiency Stirling engine suspended in front of each parabolic mirror. The intense heat from the concentrated sunlight drives the engine, which is connected to a small electricity generator. While this is an innovative approach, most experts believe it will continue to be more expensive than other forms of CST, because it costs more money to build a separate Stirling engine for each array of mirrors.

Regardless of which design is chosen, CST plants are large, utility-scale installations that require substantial capital investments. Because they use steel, glass, and concrete—and do not require any precious or rare materials—they are usually not vulnerable to strategic bottlenecks in the supply of materials needed to scale up their size. As a result, CST plants could be used immediately to provide larger amounts of electricity.

These plants are connected to the existing transmission and distribution grids. However, this technology is most efficient in areas with the most direct sunlight, and the desert areas that qualify are far away from the population centers where most of the electricity is needed. As a result, full utilization of this potential will require the building of new high-technology transmission lines that connect to a smart grid distribution system.

SPAIN'S OLMEDILLA PHOTOVOLTAIC PARK USES
MORE THAN 160,000 PV PANELS TO GENERATE
60 MEGAWATTS OF ELECTRICITY.

AT THE POWER TOWER NEAR SEVILLE, SPAIN,
MIRRORS FOCUS THE SUN'S ENERGY ON THE
300-FOOT STRUCTURE, HEATING WATER INTO
STEAM THAT TURNS AN ELECTRIC TURBINE.

In addition, some designs for CST require almost as much water as do conventional fossil fuel generating plants, although the industry is moving toward other designs that require far less water.

The second way to produce electricity from sunlight is by using solar cells made of materials with special properties to convert the energy of the photons in the sunlight directly into electricity. No steam turbines are involved.

These photovoltaic (PV) cells are semiconductor devices like transistors. The energy in the photons of sunlight frees electrons from atoms in the PV cells so they can flow out of the cell as an electrical current.

Until recently, most experts have assumed that photovoltaic electricity was likely to remain somewhat more expensive than the electricity produced through concentrated solar thermal. However, continuing improvements in the efficiency of all forms of photovoltaics have led some experts to conclude that we are at or near a threshold beyond which photovoltaics will actually have a cost advantage over CST. Moreover, if the cost-reduction curve continues to drive the cost of photovoltaics down at the same rate, they may soon have a cost advantage over electricity generated by fossil fuels.

That is not to say that concentrated solar thermal will not have a robust future even if the cost of photovoltaics continues to come down sharply. In fact, the ability to construct CST installations out of readily available commodity materials and link them to the transmission and distribution grid will make them an attractive option in many areas—especially areas with high sunlight or areas that connect to high-efficiency transmission lines.

Advanced PV cells can benefit from an innovation-and-cost-reduction curve similar to—though not as powerful as—the phenomenon known as

Moore's Law, which has regularly produced sharp reductions every 18 to 24 months in the cost of computer chips. By contrast, CST plants utilize materials that are commodities and, as a result, they will almost certainly not come down in their cost as rapidly as photovoltaics.

Moreover, solar PV cells also appear to benefit much more than CST from economies of scale in their production. Thus far, each doubling of the cumulative production volume has produced a 20 percent cost reduction. And they are not as dependent for efficient performance on high, direct sunlight in the same way CST is.

In addition, PV cells can be deployed in a distributed manner. That is, small installations are just as efficient as a utility-scale deployment. For example, PV can be deployed on rooftops, while CST obviously cannot. PV will eventually be competitive in many areas of the world—including virtually all of the United States—while CST will probably be feasible only in areas with high levels of direct sunlight.

The use of "net metering," which allows homeowners and business owners to install photovoltaic cells on their rooftops and sell electricity back into the grid at times when they don't need it, will also

HOW CONCENTRATED SOLAR THERMAL POWER WORKS

At concentrated solar thermal (CST) plants, solar energy is concentrated by mirrors, which focus the sun's rays on a pipe filled with a synthetic oil called Therminol or water, heating it to several hundred degrees Fahrenheit. The liquid is pumped though the entire system, and a heat exchanger uses that heat energy to generate steam, which turns an electricity-generating turbine.

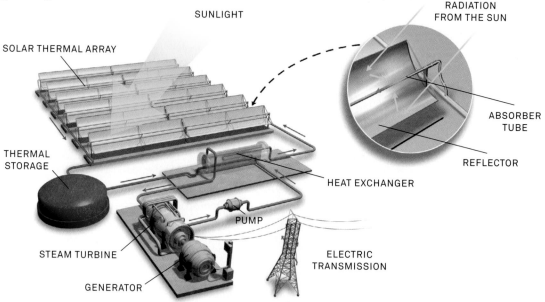

provide powerful incentives for the rapid spread of photovoltaics.

More than 90 percent of the photovoltaic cells currently in use are made from silicon. First-generation photovoltaic cells require the processing and transformation of silicon into one of two forms, either single-crystal polysilicon (many small crystals) or amorphous silicon.

Recent increases in the demand for solar electricity have led to a dramatic expansion of the world's capacity for making silicon-based photovoltaic cells and have subsequently driven prices down. Because silicon is the second most abundant substance in the crust of the earth (after oxygen), there is no shortage of raw material. Moreover, as the cost has come down, demand has increased further, leading to a "virtuous cycle" of cost efficiency. Until recently, PV electricity was still more expensive than that made by concentrated solar thermal plants, but this situation is now changing rapidly as new forms of photovoltaic energy are developed.

Single-crystal silicon cells are coming down in cost and are steadily being made more efficient. These cells still require the use of relatively expensive glass with metal frames, and that has led

HOW PHOTOVOLTAIC POWER WORKS

Solar photovoltaic (PV) cells make electricity directly, without a turbine. When sunlight hits the panel, usually made of semiconducting silicon, the photons in the sunlight free electrons from the atoms in the photovoltaic material so they can flow out of the cell as an electrical current. When the electrons are forced to move in one direction, they become electric current. An inverter is needed to convert the direct current into the alternating current we use in our homes.

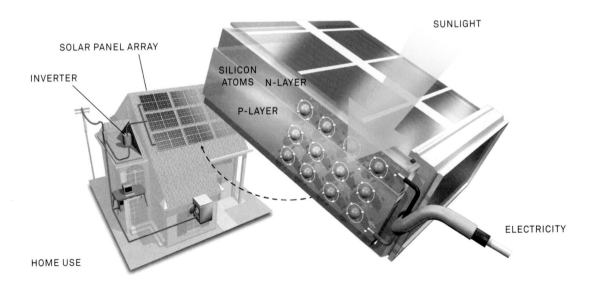

SUNLIGHT

SOLAR PANEL ARRAY

INVERTER

SILICON ATOMS   N-LAYER

P-LAYER

ELECTRICITY

HOME USE

scientists to explore the use of innovative, cost-saving materials and designs.

New types of "thin film" cells are cheaper to manufacture and install, although at present the efficiency with which they convert photons to electrons is less than that of the single-crystal silicon cells. The most common materials for these thin-film cells are amorphous silicon; a cadmium-and-tellurium combination; and a combination of copper, indium, gallium, and selenium.

The trade-off between first-generation silicon cells and the newer thin-film cells is between higher cost/higher efficiency and lower cost/lower efficiency. Even so, both types are coming down in cost while increasing in efficiency (although progress has been greater of late with thin-film cells).

The emergence of a new generation of photovoltaic cells currently sparks the greatest excitement among researchers and engineers. New molecular structures made possible by newly developed chemical processes and fabrication technologies offer the promise of much higher levels of efficiency at much lower cost. These cells may use sophisticated nanotechnology to achieve dramatic increases in efficiency by capturing and using more photons and displacing more electrons. They may also be able to avoid some costly features of the current generation of cells. Since photovoltaic cells are relatively expensive to manufacture, it may be possible to cut costs by using inexpensive lenses or mirrors to concentrate sunlight on a small, highly efficient but expensive cell. Indeed, some next-generation cells can actually use nanostructures to perform the concentrating task within the surface of the cell itself.

However, some designs for new-generation PV thin-film cells may require the use of precious materials that are not abundant (selenium, for example), raising the possibility of periodic supply bottlenecks and potentially higher costs if those designs are manufactured in high volumes.

Unlike fossil fuel or nuclear plants, solar electric plants can produce a constant electrical current only when the sun is shining. Even though the fuel—sunlight—is free, there is obviously no sunshine at night, and passing clouds during the day affect the flows of solar electricity. This so-called intermittency problem requires utilities to think differently about the role played by solar electricity.

One offsetting advantage of solar electric plants is that they generate their peak output when the sun is hottest. Happily, this coincides with the normal peak demand for electricity, which is usually driven by peak air-conditioning usage.

CST plants contribute another partial solution to the intermittency problem: the heat collected in the liquid tanks can be stored efficiently (usually in molten salt) when clouds block the sunlight. At present, heat can be stored for about one hour, but soon for five to six hours. One Power Tower design is expected to store heat for 15 hours! Since PV cells do not generate heat in the production of electricity, they cannot use thermal storage. Most photovoltaic plants have gas-fired generators on standby to solve their intermittency problem.

Another partial solution to the intermittency problem is the completion of a unified national smart grid, because when it is spread out over a large geographic area, the intermittency challenges in one location will usually be offset by the availability of sunlight elsewhere.

The ability of the electricity grid to integrate intermittent flows of electricity is a manageable challenge so long as the overall percentage of intermittent sources does not exceed roughly 20 percent of the total. Beyond that, however, the challenge becomes considerably more difficult. One innovative solution involves a side benefit of shifting the auto fleet to plug-in hybrid electric vehicles (PHEVs) and all-electric vehicles (EVs), because these millions of PHEVs and EVs can serve as a widely distributed battery. Other emerging forms of electric energy storage can also smooth out the intermittent flows of electricity.

Today, all forms of solar electricity are still priced higher than electricity from the burning of coal or gas, largely because the enormous costs associated with greenhouse gas pollution are excluded from calculations of the true value of electricity from those latter sources. Innovations in solar energy are quickly driving costs down, however. Many experts predict that we are only a few years

SPACE-BASED SOLAR POWER

There is an exotic proposal for space-based solar energy that has been discussed and debated for decades. In space, there is no intermittency problem, and solar energy is stronger. Sunlight that hits rooftop solar panels in North America carries around 125 to 375 watts per square meter, delivering about one kilowatt-hour of electricity a day. In theory, it is possible to put into orbit a large array of photovoltaic cells on a geostationary satellite 22,300 miles above the earth, where solar radiation is more than eight times stronger.

Scientists propose placing several satellites in a kilometer-wide fixed orbit. Each satellite would use reflector arrays collecting constant sunlight and directing it onto photovoltaic cells. A single photovoltaic array in space could collect six to eight times the daily power collection as one on Earth, and the orbit could support thousands of arrays.

The satellites would beam the power back to ground receivers through microwave frequencies. Advocates say the beams would be a highly efficient system for moving power and insist these beams pose no threat to humans, birds, and other life-forms. However, there is still great skepticism about the public's acceptance of beams from space and any system requiring multiple rocket launches as part of its installation.

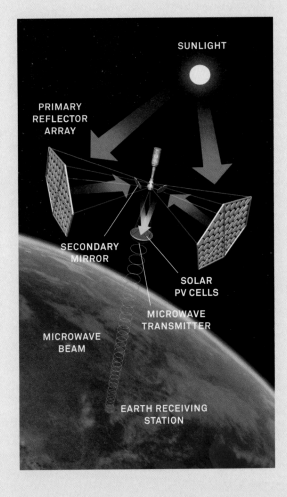

Large panels would focus the sun's energy on PV panels. The electricity generated would be transmitted to collecting facilities on Earth via microwaves.

MORE THAN 1,000 PV PANELS ATOP THE
PAUL VI AUDIENCE HALL AT THE VATICAN
PROVIDE THE BUILDING'S HEATING, COOLING,
AND LIGHTING ENERGY.

away from being able to produce PV electricity at rates competitive with the current cost of coal-fired generators. At present, however, the higher cost of solar electricity means that financing for these technologies depends upon government policies to compensate for the artificial difference in cost between solar electricity and fossil fuel electricity.

In many countries—especially in the United States—government policies have been fickle, following a frustrating stop-and-go pattern of development that has periodically dried up the capital needed for the industry to stay on a consistent development pathway.

For example, nine early CST plants were built between 1984 and 1991 in the Mojave Desert of Southern California (with a total of two million square meters of mirrors) and have operated continuously and very successfully for a quarter-century. Although none of these store energy, they utilize gas-fired generators to produce electricity during periods when clouds are a problem. Their proven track record has encouraged many to believe that similar plants employing newly available improvements to the basic technology could be built quickly to provide large amounts of electricity. However, in the period after these plants were built, market conditions and changes in government policy led to the drying up of the capital needed to build additional plants.

In general, there have been two primary reasons for erratic shifts in U.S. energy policy: the roller-coaster pattern of world oil prices and the periodic change in political control of the White House and Congress. Whenever oil prices reach a peak, there is a surge of public support for alternative sources of energy. But when oil prices come back down again, that support has tended to dissipate fairly quickly. As for the second factor, oil and coal companies, as well as coal-burning utilities, have worked hard to build opposition to solar electricity in both of America's political parties.

Nevertheless, within the past decade, new and improved CST plants have been built in Arizona and Nevada, and many more are now under development and/or construction in the American West and in numerous other countries. The recent establishment by the U.S. government of new incentives for solar electricity—and new laws in California and several other states requiring utilities to obtain a certain percentage of their electricity from renewable sources—has led to a resurgence in the construction of new solar electricity plants. For example, Florida Power and Light recently began construction of a utility-scale photovoltaic plant to produce electricity for its customers.

In some other countries—Spain and Germany, for example—innovative government policies have stimulated demand for and dramatically expanded the use of solar electricity technology. China and Taiwan have made major commitments to become the world's leaders in the production of photovoltaics. By contrast, even though photovoltaics were developed in the United States, only one of the 10 leading manufacturers is now based in the United States.

New, more robust and consistent government policies are necessary in the U.S. and elsewhere to accelerate the development of markets for solar electricity and to create millions of new jobs in this 21st century industry.

Once the world chooses to set ambitious goals for scaling up solar electricity development and commits to the investments necessary to further improve the technologies involved, there is no question that solar energy will provide a major percentage of the world's electricity.

DESIGNING FOR THE SUN: PASSIVE SOLAR HOMES

There are other valuable forms of solar energy that do not produce electricity. So-called passive solar energy can play a significant role in reducing energy consumption in homes and commercial buildings. (Approximately 40 percent of all U.S. greenhouse gas emissions come from buildings' energy use.)

Passive solar structures take advantage of sunlight as a direct source of heat. Designs rely on the physics of how heat moves, so that buildings themselves become solar collectors, heat absorbers, and heat distribution systems. Heat absorption is maximized in winter and minimized in summer. Houses are oriented so that windows face the sun's path, with overhangs angled to let more of the sun's energy in during winter, and keep it out during summer. Thermal mass—such as stone walls—absorbs and stores solar heat; proper ventilation allows heat to circulate within the building; and walls and windows are insulated to prevent heat (or cool air) from leaking out. Solar hot-water heaters on the roof also utilize the sun to reduce energy use.

Passive solar design can greatly lower a building's energy requirements. Combined with solar PV panels or other technologies, the result can be a "net zero energy" building, requiring no outside power.

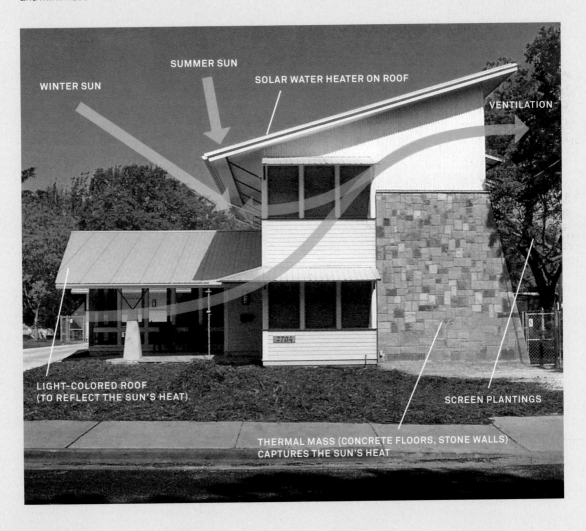

WINTER SUN

SUMMER SUN

SOLAR WATER HEATER ON ROOF

VENTILATION

LIGHT-COLORED ROOF
(TO REFLECT THE SUN'S HEAT)

SCREEN PLANTINGS

THERMAL MASS (CONCRETE FLOORS, STONE WALLS)
CAPTURES THE SUN'S HEAT

CHAPTER FOUR

HARVESTING THE WIND

WIND TURBINES ON A FARM IN SHERMAN
COUNTY, OREGON

Wind is actually another form of the sun's energy. Because some parts of our planet receive more direct sunlight than others—the tropics, for example, obviously get more than the polar regions—the resulting difference in air temperature creates large planetary flows of wind. In addition, air above land heats up faster during the day and cools more quickly at night than does the air above the ocean. The differing surface characteristics of the land itself also affect the heating and cooling of the air above it. Mountains have an impact on the altitude—and thus the temperature—of air. Deserts are hotter during the day and cooler during the night, while forests don't heat up as much during the day or cool down as much at night. Whenever air near the surface is hot, it expands and rises, creating a pressure vacuum into which cooler air rushes. That's the wind.

The permanence of geographical features—oceans and continents, mountains and the passes between them, hills and valleys, deserts and forests—make most wind patterns predictable. If air were as visible as water, we could see "lakes" of air where the wind is quiet, flowing "streams" where it blows gently, and raging "rivers" in those places where winds average more than the 15 miles per hour necessary for efficient production of electricity. That last is where the windmills are sited.

Overall, the earth's wind resource is so large that it could technically provide five times the total energy consumed by the entire world from all sources. The amount of wind power available in the United States could technically provide 10 times the amount of electricity the country consumes annually.

In each of the past two years, wind power has been the greatest source of increased electricity-generating capacity in the United States. Not only is it the fastest-growing source of renewable electricity, it is the fastest-growing source of any form of energy, surpassing the net additional capacity from coal-fired, gas-fired, and nuclear power plants combined. In fact, in 2008, the United States led the world in the installation of new wind capacity.

In terms of the total volume of electricity already being produced from wind, the United States is in first place. Germany, with a much smaller population, is a close second. Spain, with a population half that of Germany, has more than two thirds of Germany's wind capacity. China is fourth in overall installed wind capacity, but second only to the United States in new capacity added in 2008, and is expected to climb to second place overall in 2010. India is in fifth place in installed capacity, but third in the number of new windmills added in 2008.

Denmark, the leader among all nations in the share of its overall energy received from wind, now gets more than 21 percent of its electricity from

A TECHNICIAN INSPECTS A TURBINE BLADE AT
THE WETHERSFIELD WIND FARM IN WESTERN
NEW YORK.

windmills. (Denmark's first modern commercial-scale windmill, by the way, was built after World War II with assistance from the Marshall Plan.) Germany and Spain both get more than 5 percent of their power from wind, and some regions get significantly more.

The widespread availability of the wind resource is one of the main reasons it has become the most popular new source of electricity in the world. For the time being, wind power also has the lowest cost of any form of renewable energy other than geothermal. Even though other sources—particularly solar photovoltaics—will come

GLOBAL WIND ENERGY PRODUCTION

Wind energy is experiencing rapid growth and now totals more than 120,000 megawatts worldwide. In 2008 the United States added 8,500 megawatts of wind power, a 50 percent increase in a single year.

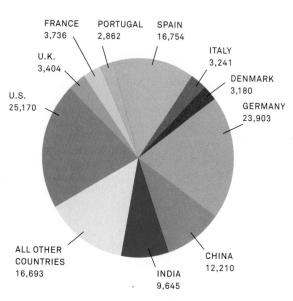

FRANCE 3,736
PORTUGAL 2,862
SPAIN 16,754
ITALY 3,241
U.K. 3,404
DENMARK 3,180
U.S. 25,170
GERMANY 23,903
ALL OTHER COUNTRIES 16,693
CHINA 12,210
INDIA 9,645

TOTAL INSTALLED WIND-POWER CAPACITY, 2008
(in megawatts)

SOURCE: Global Wind Energy Council

down in cost very rapidly over the next few years, wind power is already a mature technology and competitive even without additional break-throughs. As the world moves toward putting a price on carbon to establish a more realistic valuation of fossil fuel–generated electricity, wind is positioned to continue its rapid growth as a major source of electricity for the world.

Most windmills today have a similar appearance, with three large blades mounted on top of a tall tower. The typical commercial-size windmill has blades 89 to 147 feet (27 to 45 meters) long mounted atop a tower 147 to 344 feet (45 to 105 meters tall). The most popular turbine engine produces on average 1.5 megawatts of electricity—enough to supply all of the electricity needed by 400 average American homes. These large windmills are usually installed in clusters of dozens or hundreds in utility-scale "wind farms" connected to the transmission and distribution grid.

The blades of modern windmills are designed according to a principle similar to the one used in airplane wings. The curved shape of the top of the blade causes the wind to speed up relative to its slower flow across the bottom. The resulting difference in air pressure makes the blades turn with great force. The rotation of the blades spins a drive-shaft that turns an electric generator to produce electricity in much the same way that steam drives the blades of the much larger turbines in fossil fuel, nuclear, and concentrated solar thermal plants.

Over time, the construction of larger rotors and taller towers has improved efficiencies. The difficulty of transporting the largest blades on highways is one of the factors that have led to predictions that we are at or near a practical limit in the length of blades. However, many similar predictions in the past about limits on the size of blades have proved to be wrong.

HOW A WIND TURBINE WORKS

Most modern wind farms have towers with three blades on a horizontal axis. The rotor and electrical generator are at the top of the tower, and a computer-controlled motor points the blades into the wind. The curved shape of each blade increases wind velocity above it, which decreases air density. This low pressure (compared with the pressure along the bottom of the blade) creates aerodynamic lift, which pulls the blade, causing it to rotate. The rotation of the blades spins a driveshaft, which turns the electric generator, located directly behind the blades.

HIGHER VELOCITY,
LOWER AIR DENSITY = LIFT

LOWER VELOCITY,
HIGHER AIR DENSITY

VERTICAL WINDMILL

Vertical-axis turbines are less efficient, but because they can collect wind in all directions and spin faster under slower winds, they have niche applications on a smaller scale.

GENERATOR

THE WELSH ISLAND OF ANGLESEY HOSTS
SEVERAL WIND FARMS THAT TAKE ADVANTAGE
OF THE LOCATION'S SUSTAINED WINDS.

Since towers, blades, and turbines cost money to transport over long distances, there is an emerging cost advantage to building windmills in the country where they are used. Tens of thousands of new jobs have already been created in the United States in the manufacturing and installation of windmills. For example, Cardinal Fastener in Bedford Heights, Ohio, the company that made the giant bolts used in both the Golden Gate Bridge and the Statue of Liberty, now expects half of its revenue to come from making large bolts for new windmills. Many more such jobs are now in prospect.

Among their other advantages, wind farms are scalable: when more electricity is needed for the marketplace, more windmills can be added. In fact, windmills can be built and installed within two months. (Modular designs have made assembly quick and easy.) This makes them more attractive than many other sources of electricity that can take a decade or more to build. Maintenance costs are low, and experience shows that windmills are durable over a long period of time. Moreover, wind power—unlike many other technologies—does not require water, an increasingly important advantage in dry regions. Wind power uses less land than any other renewable-energy option, but it is the most visible on the horizon. I am among those who believe windmills are usually a beautiful and appealing addition to the landscape, but there are some who don't like their appearance and have

ARE WINDMILLS A THREAT TO BIRDS?

The fact that birds are sometimes killed by flying into the blades of windmills has created controversy over the siting of some new wind farms.

The total number of bird deaths from windmills in the United States each year is only 0.5 percent of the number of birds killed by communications towers. It is 0.005 percent of the number killed by buildings in the United States and less than 0.03 percent of the number killed by housecats. (Statistically, this means that the average housecat in the United States kills approximately the same number of birds each year as does the average windmill.)

Moreover, the CO_2 produced by fossil fuel power plants is one of the leading causes of global warming—which, according to one study, could contribute to the extinction of more than a quarter of all bird species in the world. Still, improvements in windmill design that would avoid further bird kills are desirable, and engineers are hard at work on the issue.

Testing is under way on sensors capable of detecting the approach of large bird flocks in order to shut down operations as necessary.

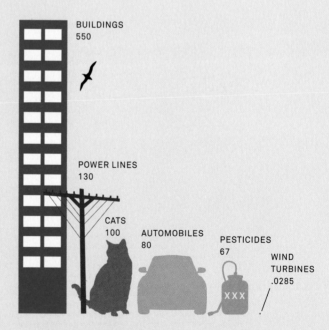

BUILDINGS 550

POWER LINES 130

CATS 100

AUTOMOBILES 80

PESTICIDES 67

WIND TURBINES .0285

CAUSES OF BIRD DEATHS (in millions annually, in U.S.)

SOURCE: Wallace P. Erickson, et al., in *Bird Conservation Implementation and Integration in the Americas, 2002*

MUCH OF THE UNITED KINGDOM'S NEW WIND POWER IS BASED OFFSHORE. THIS FIVE-MEGAWATT TURBINE STANDS NEAR INVERNESS, SCOTLAND.

mounted campaigns opposing new installations. On the other hand, enthusiasm for wind power should not be allowed to override truly legitimate objections to specific sites. For example, the proposed location of a large wind farm in a carbon-rich peat bog on the Shetland Islands would appear to negate the CO_2-reducing benefits of that particular project.

While most windmills today are located on land, use of offshore wind is growing rapidly for several reasons. For one thing, there are usually fewer objections to the siting of windmills offshore. Winds are generally stronger, more predictable, and less turbulent over the ocean, because the surface is flat with few obstructions. Experience gained in construction of offshore oil-drilling platforms on continental shelves has created the expertise and confidence needed to build windmill platforms that rise from the ocean floor at a manageable cost. Indeed, some wind entrepreneurs believe that many oil platforms are located in areas where it makes economic sense to actually place windmills on top of the oil platforms themselves. Two companies, StatoilHydro in Norway and Siemens in Germany, announced in 2008 that they have begun work on a floating platform for offshore windmills that can be located in deeper waters and tethered by three anchor cables to the ocean bottom.

In addition, transmission cables buried under the ocean bottom to carry the electricity to shore are relatively inexpensive—even if the platforms are tens or hundreds of miles from land. Denmark, Sweden, the United Kingdom, Ireland, the Netherlands, and China are all beginning to use offshore wind as an attractive source of renewable, pollution-free electricity. Multiple projects are in the development stage in the United States, though none have yet begun to operate.

At present, the world's largest offshore wind farm is in the North Sea near Skegness, England. The Lynn and Inner Dowsing wind farm has 54 huge windmills, each with blades more than 175 feet long and a turbine that sits 265 feet above sea level. Together, these machines produce almost 200 megawatts of electricity at peak generation. The United Kingdom is also constructing a much larger, 270-turbine, 1,000-megawatt offshore wind farm in the outer Thames Estuary, which will be the largest offshore installation in the world when it is completed in 2012. In early 2010, Denmark will start operating a 209-megawatt array. The Horns Rev 2 contains 91 turbines located between 19 and 25 miles west of Denmark in the North Sea.

The predictability and reliability of wind in certain areas has led to the exploitation of this natural flow of energy since ancient times. Wind has been used as a source of energy at least since the invention of sails more than 5,000 years ago.

Most historians give Persia credit for inventing windmills in the first millennium. From there, the technology spread to China and later was taken to Europe by returning Crusaders. The technology was improved upon in the Netherlands and Great Britain and used to grind grain, pump water, and saw wood, among other purposes. Before the widespread use of underground coal reserves began in the early 17th century, there were hundreds of thousands of windmills in Europe and as many as 500,000 in China.

The widespread use of coal led to declining interest in windmills, even before the production of electricity as a source of power. However, windmills continued to be used in areas where fossil fuel was hard to come by. In the early 20th century, they were used to generate electricity in some rural areas not connected to large power plants by transmission lines.

THE BOY WHO HARNESSED THE WIND

William Kamkwamba salvaged his father's old broken bicycle for a frame, took a rusted shock absorber for a shaft, a tractor fan for a rotor, and melted down PVC pipes from an old bathhouse for the blades. He scoured his town's scrap yard for ball bearings, finally finding the precious components in an old groundnut-grinding machine. With a nail through a maize-cob handle for a drill and bicycle spokes fashioned into a makeshift screwdriver, he turned these motley parts into a windmill, precariously set atop a ladder made of blue gum tree wood. When it was finished, the wind blew, the blades spun, and the then-14-year-old Malawian held a glowing electric bulb in his hand.

William had been forced to abandon his secondary-school studies when his family could no longer pay his tuition: they had narrowly survived one of Malawi's most severe famines. Heartbroken that his schooling had come to an early end, William followed his friends' class notes: history, English, geography, and science. When he stumbled upon the English textbook *Explaining Physics* at the library one day, his independent study was infused with the passion of an inventor. A friend of William's father had a bicycle with a dynamo attached that powered a lightbulb when he pedaled. With this basic concept in mind—that the mechanical spinning coils of wire inside a magnetic field could create an electrical current—and the knowledge that electricity could bring an easier life to Wimbe, his hometown of about 60 families, he began to teach himself basic electrical engineering from the book's illustrations.

Photographs of a windmill in another library book inspired him to start building. "We have enough wind in Malawi," he said. "[I thought] maybe I can make one so that I can have electricity in my home." Soon after his first success with the lightbulb, William got to work to increase his machine's power, adding a car battery to store power, a circuit breaker made from nails and speaker magnets, and handmade light switches. He installed lightbulbs in every room in his home and two outside, then supplemented that power with solar panels on the roof. Now every home in Wimbe has a solar panel and a battery to store power.

William's goals go beyond lighting. In a normal year, two thirds of Malawian households are unable to produce enough maize for their own use. In 2001

William Kamkwamba stands atop one of the windmills he has built from salvaged parts in his village in Malawi.

and 2002, the drought-prone country's annual hungry season stretched into a famine that starved many of its 11 million inhabitants. William's newest windmills are meant to prevent his village from suffering. One pumps water to irrigate his family's vegetable garden, and a solar-powered pump at the public well fills water tanks for all the people in Wimbe.

In the years since he built his first windmill, in 2003, William has become a symbol of grassroots innovation in Africa. Through his book (*The Boy Who Harnessed the Wind*), his blog (williamkamkwamba.typepad.com), and appearances at conferences around the world, he has transmitted a message of hope and human potential through clean energy. And he has returned to school, at the African Leadership Academy near Johannesburg, where he studies so that one day he can start a company that builds windmills in Africa. "People want technology, but they cannot use it without electricity," he said. "I'm planning to [bring] reliable electricity."

Modern interest in windmills for the production of electricity took off after the Mideast oil embargoes of 1973 and 1979 led to steep price hikes for fossil fuels. The resulting search for alternative sources of energy led immediately to wind.

New policies were developed in many countries, particularly in the U.S., where many of the key technologies now in use around the world were first developed in response to incentives passed under President Jimmy Carter.

Early generations of electricity-producing windmills were often based on modified airplane propellers and were noisy, generating complaints from those who lived nearby. However, modern designs have all but eliminated noise complaints. Any noise from the new, highly efficient turbines is difficult to hear. And noise from the turbulence created in the air has also been minimized to the point where complaints are now few and far between.

Engineers have redesigned windmills to make them far more efficient. By using new, lighter-weight materials and better designs, they were able to start building the predecessors of today's modern windmills, all the while improving windmill efficiency. Larger blades and larger turbines were placed on taller towers to take advantage of stronger winds at higher altitudes.

Within a few years after the Carter program began, 85 percent of the world's wind energy was being used in the United States. Moreover, these and other policies promoting renewable energy—in combination with the impact of higher oil prices—led to a dramatic reduction in America's dependence on imported oil.

Unfortunately, the political transition from Jimmy Carter to Ronald Reagan in 1981 was accompanied by an 80 percent reduction in renewable-energy programs, and the U.S. industry stopped moving forward. As oil prices fell, U.S. dependence on foreign oil once again began to grow rapidly.

This unfortunate history illustrates one challenge that wind energy has in common with solar energy: it requires innovative and consistent policies that can make up for the artificial advantage that subsidies and distorted cost-accounting currently give to oil and coal. Those nations that have made the most rapid advances in wind power are those that have persistent, long-term policies designed to increase the market demand for renewable energy and boost the incentives for manufacturing and production.

The Clinton-Gore administration renewed incentives for wind (and solar and other renewables), but the change in control of the U.S. Congress in the 1994 elections stopped the funding once again. Several times, the tax credit that helped produce the initial expansion of wind power has been allowed to expire. And several times it has been renewed for a duration of only two years, which discourages the flow of investment needed to expand the use of wind-generated electricity. As a direct result, the United States has lost its early lead in wind technology. Even though the U.S. uses more windmills than any other nation, half of them are now purchased from foreign manufacturers. More important, of the 10 leading wind-technology companies in the world, only one of them—General Electric—is now located in the United States.

Fortunately, in the absence of consistent federal policy, several states—including the early leaders, Texas and California—filled the policy vacuum with state incentives that encouraged development of this important industry. Texas—which most people associate with the production of energy from oil—is actually the leader in installed wind capacity, with more than twice as much as any other state. Iowa and California are second and

THE WIND FARMS AT ALTAMONT PASS IN CALIFORNIA WERE FIRST BUILT IN THE 1970S AND TOGETHER FORM ONE OF THE LARGEST INSTALLATIONS IN THE U.S.

third, respectively. Minnesota, in fourth place, leads all states in the percentage of its overall electricity that comes from wind, with Iowa a close second.

Twenty-two states now have more than 100 megawatts of installed electricity production from windmills (enough to supply 30,000 homes), and more are being built every day. The world's largest onshore wind farm is in Texas, west of Dallas. The Horse Hollow Wind Energy Center (owned by Florida Power and Light) has 421 giant turbines delivering 735 megawatts of peak power. Even larger wind farms are presently under construction. Three states—Minnesota, Iowa, and Colorado—already get more than 5 percent of their electricity from windmills.

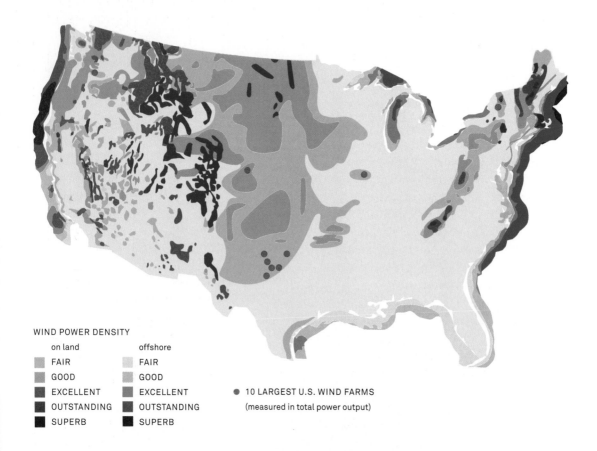

WIND POWER DENSITY

on land	offshore	
FAIR	FAIR	
GOOD	GOOD	
EXCELLENT	EXCELLENT	● 10 LARGEST U.S. WIND FARMS
OUTSTANDING	OUTSTANDING	(measured in total power output)
SUPERB	SUPERB	

WIND RESOURCES IN THE UNITED STATES

Just as water power, geothermal, and solar potential vary according to geography, the amount of energy we can harness from the wind is dependent on local conditions. The darkest green and blue areas indicate locations with wind speeds high and predictable enough to consistently generate power. However, until the United States builds a super grid, many of these areas remain too far from electricity transmission lines for the wind to be harnessable.

SOURCE: American Wind Energy Association; National Renewable Energy Laboratory

With both wind and solar power, once the systems are in place, the fuel is free. However, limitations on the use of wind power are similar in many respects to limitations on the use of solar power. Electricity can be produced only when the wind is blowing sufficiently to turn the blades of the windmill. If the wind stops blowing, the windmill stops generating electricity.

Intermittency is a particular problem for both wind and photovoltaic energy because no heat is produced in either process, only electricity. And electricity—unlike heat energy—is still difficult to store efficiently. That will change with the construction of a unified national smart grid and with the widespread use of plug-in hybrid electric vehicles, because a large fleet of PHEVs can serve as a widely distributed and highly efficient battery.

Another limitation that wind electricity shares with solar electricity—particularly solar electricity produced by concentrated solar thermal plants—is that in both cases the best resources for sun and wind are generally found in areas distant from population centers. As a result, new high-technology, long-distance transmission lines are needed to maximize the use of both resources.

Electrical grids are currently designed in ways that make it difficult to integrate intermittent sources of electricity if they provide more than 20 percent of the total. But smart grids over larger areas will be able to exceed that percentage. Moreover, the predictability of wind downtime makes unplanned outages much less of a risk, enabling wind-farm operators to lessen the impact of their downtime. When several wind farms utilize the same transmission lines, downtime in one farm can usually be covered by electricity from another.

There is growing enthusiasm for smaller windmills that can provide electricity for homes, farms, and ranches. In the United States, there are an estimated 13 million homes that are suitable for these small windmills. Virtually all of them are in rural areas, because urban environments do not have wind patterns suitable for windmills.

In this respect, wind power, like photovoltaic solar cells, is adaptable to what is called a "distributed energy" approach, allowing those who use it to sharply reduce the electricity they purchase from utilities. However, because each windmill must have its own electric turbine, and because the cost for each kilowatt-hour of electricity increases as the windmill gets smaller, the expense of small windmills is likely to remain higher than the expense of home-scale solar photovoltaic cells—especially as the cost of photovoltaic energy continues to rapidly decline.

The typical small windmill—from 35 to 140 feet tall for the larger residential versions—produces enough electricity to pay for itself in six-plus years to 30 years. Continuing improvements in efficiency and reductions in cost will continue to shorten that payback period. Experts in renewable energy, however, caution that experience indicates that for all renewable sources of residential energy, the payback period must be only a few years before there is widespread consumer interest. Of course, any increases in the cost of electricity from utilities would also have the effect of shortening the payback period for small windmill–generated electricity and thus would encourage more widespread use of these systems.

At present, 10,000 small windmills are being installed each year in the United States, and that number is rapidly increasing. Some experts believe that the market for small-scale windmills could be expanded significantly by optimizing them for lower wind speeds in appropriate locations.

OUR SOURCES OF ENERGY

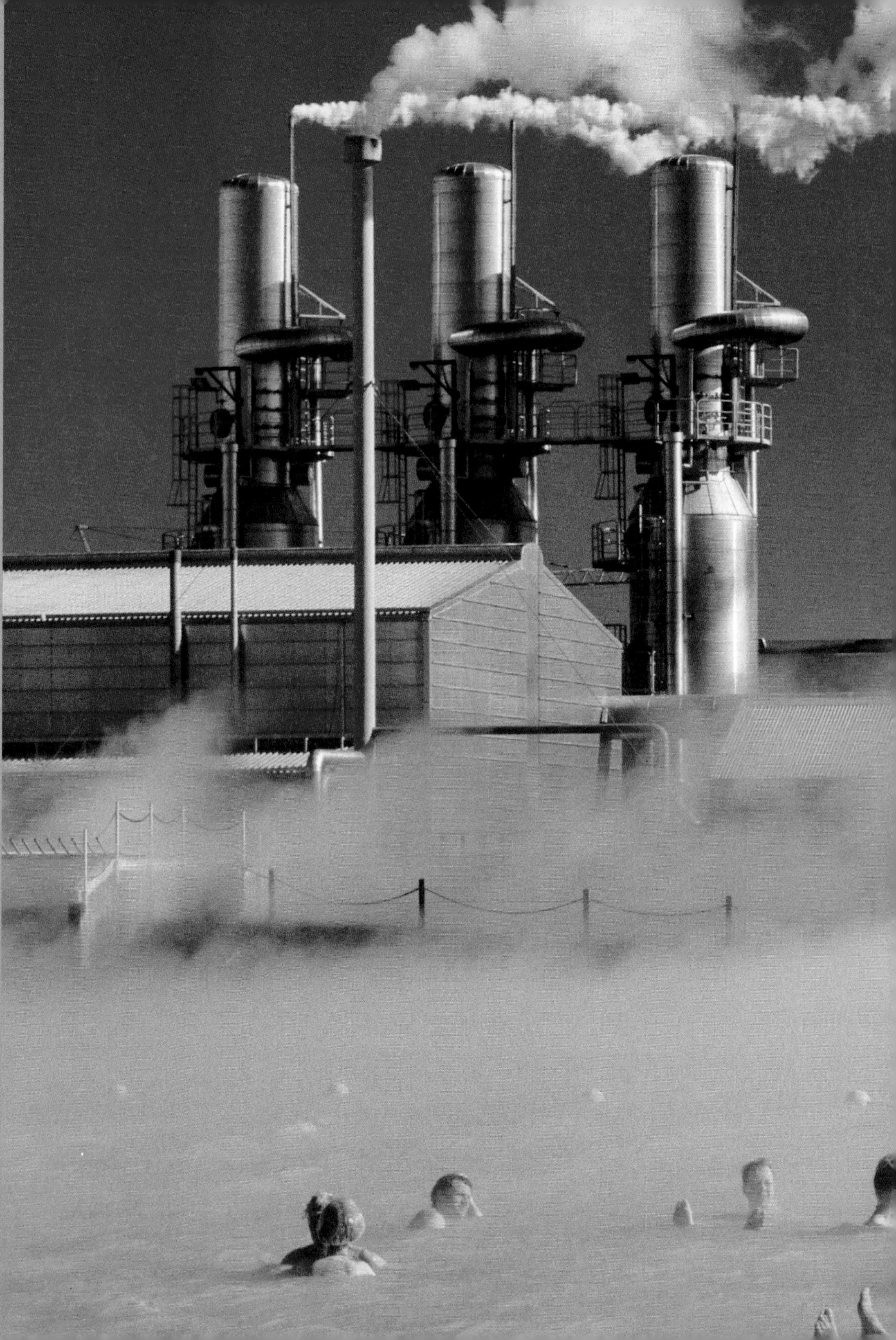

CHAPTER FIVE

SOAKING UP GEOTHERMAL ENERGY

THE BLUE LAGOON SPA IN ICELAND IS FED BY HOT WATER FROM THE ADJACENT GEOTHERMAL POWER PLANT.

Geothermal energy is potentially the largest—and presently the most misunderstood—source of energy in the United States and the world today. Unlike solar energy, which comes to the earth from the sun, geothermal energy comes from deep within the planet itself.

Like solar energy and wind power, geothermal energy could—if properly developed—match all of the energy available from coal, oil, and gas combined. Indeed, the amount of geothermal energy potentially available is, in the words of U.S. Secretary of Energy Steven Chu, "effectively unlimited." Yet even renewable-energy enthusiasts typically speak of "wind and solar" without including in their litany one of the most promising potential sources of all: geothermal.

The amount of geothermal power available is so frequently underestimated primarily because its use as a source of electricity has long been associated with the few locations where hot water bubbles or spouts to the surface. In some of those places, heat-recovery systems that tap into steam or hot water reserves contained underground are used to drive turbines to generate electricity, and because our first experiences with geothermal electricity have been limited to these "hydrothermal" sites, many people are still encumbered with the false impression that that's all there is.

In this respect, the common view of geothermal power—except in the expert community—is similar to the view people had long ago of coal and oil, when the only supplies were the small amounts found exposed at the surface of the earth. Before the first underground coal mines were dug in the 17th century, and before the first oil well was drilled in 1859, those who knew about coal and oil at all thought of both resources as finite novelties, useful only in those few locations where these resources were visible.

However, when engineers developed new mining and drilling technologies to search for and access underground deposits of coal and oil, they began to find the vast reserves that quickly became the dominant energy sources of the last 150 years.

In the same way, what we have long thought of as geothermal energy is only the tiniest fraction of what is actually accessible to us by using newer technologies to drill deep into the earth to tap into this enormous source of energy.

The global geothermal resource base of stored thermal energy is very large. According to the U.N. World Energy Assessment report, the geothermal resource is roughly 280,000 times the annual consumption of primary energy in the world. In the United States alone, according to two other experts, Bruce Green and Gerald Nix, "the energy content of domestic geothermal resources to a depth of three kilometers [1.86 miles] is estimated to be three million quads—equivalent to a 30,000-year supply of energy at our current rate for the United States." The Massachusetts Institute of Technology, in a major assessment of geothermal power in 2006, estimated that the "technically extractable portion" of the U.S. geothermal resource is "about 2,000 times the annual consumption of primary energy in the United States." As a consequence, assuming appropriate improvements in technology over time, geothermal could provide a significant fraction of U.S. primary energy needs in a sustainable manner for electricity generation and for the heating and cooling of buildings.

Moreover, geothermal power has numerous advantages over any other form of energy. Unlike coal, oil, and gas, geothermal energy has virtually no CO_2 emissions. Geothermal plants are modular and scalable—and have the smallest environmental "footprint" on the surface. Like solar power, geothermal is available virtually everywhere on Earth: underneath developing countries as well as wealthier countries. But unlike solar and wind power, it is not intermittent. Once in place, a geothermal electricity plant provides power 24 hours a day.

There are two kinds of geological areas where geothermal resources have traditionally been most easily located. First, the parts of the earth where the temperature beneath the surface gets hottest are usually located at the boundaries where tectonic plates come together and where active volcanoes are often found. Perhaps the best known example is the so-called Ring of Fire that marks the perimeter of the Pacific Ocean from the eastern coast of New Zealand northward to Samoa, then west through Papua New Guinea and Indonesia, and then north through the Philippines and Taiwan and along Japan, then east through the Aleutian Islands and south along the western coast of North America, Central America, and South America.

The other category of hot spot—not related to tectonic-plate boundaries—is where the heat from the magma deep in the earth's mantle finds

THE EARTH'S HOT ZONES
The earth's crust is hottest where tectonic plates meet, as in the "Ring of Fire" encircling the Pacific Ocean. There are also dozens of smaller hot spots around the globe, caused by volcanic activity, mantle plumes, and other geological events.

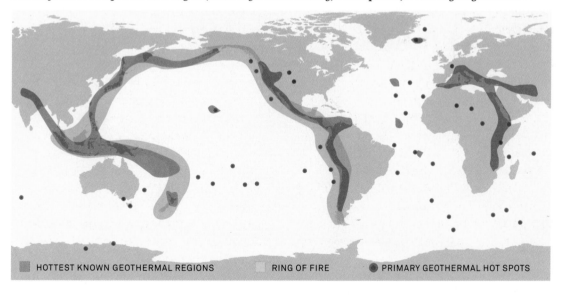

HOTTEST KNOWN GEOTHERMAL REGIONS   RING OF FIRE   ● PRIMARY GEOTHERMAL HOT SPOTS

SOURCE: National Oceanic and Atmospheric Administration; Jonathan T. Hagstrum, *Earth and Planetary Science Letters* 236, 2005

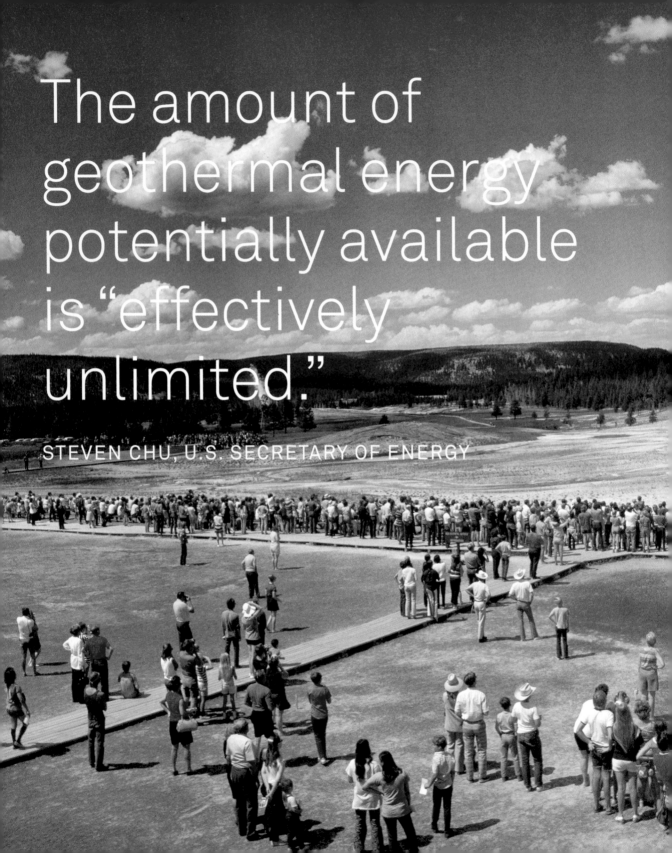

The amount of geothermal energy potentially available is "effectively unlimited."

STEVEN CHU, U.S. SECRETARY OF ENERGY

THE OLD FAITHFUL GEYSER IN YELLOWSTONE
NATIONAL PARK IS A SYMBOL OF THE EARTH'S
AWESOME GEOTHERMAL POWER.

its way naturally to the surface. The bubbling hot springs and famous geysers (like Old Faithful at Yellowstone National Park) are produced by this second geological phenomenon. Scientists have located approximately 45 of these primary hot spots on the surface of the earth. What causes them is less well understood, but science is beginning to find intriguing explanations for the majority of them. (See "The Origin of Hot Spots," below.)

Conventional hydrothermal power was the initial form of geothermal power used to produce electricity—first in 1904 at a site near Larderello, Italy, that was interrupted during World War I and II. Even today, the Larderello field produces about 400 megawatts of baseload electricity. The technology for drilling into conventional

THE ORIGIN OF HOT SPOTS

New discoveries strongly indicate that at least half of the earth's primary hot spots may be the result of large asteroids colliding with the earth in ocean areas at points exactly on the other side of the world, 180 degrees away. If you draw a straight line from Yellowstone through the center of the earth to the opposite side of the globe, you will find the Kerguelen Islands, 1,300 miles north of Antarctica and 2,600 miles southeast of Africa. Evidence suggests an asteroid struck here, in the Southern Ocean—leaving a hot spot on the ocean floor and transmitting shock waves that traveled through the earth, bouncing off the inside of the rocky mantle and reconverging on a focal point at the other side of the sphere. Continental drift has moved the hot spot created there over the last 15 million years or so to the exact location where Yellowstone is now found. During the same period, the Kerguelen Islands were formed by volcanic activity on the ocean floor at the point of the asteroid's impact.

Statisticians analyzing these antipodal pairs of hot spots are 99 percent confident that at least half of the Yellowstone category of hot spots are linked to hot spots on the ocean floor on the other side of the earth, which some evidence suggests were caused by "large-body impacts" from space. According to this theory, ancient asteroid strikes that hit continents did not have the same effect on the opposite side of the planet because the land mass buffered and shielded the shock waves, preventing them from propagating in the same way as ocean impacts. For those who are

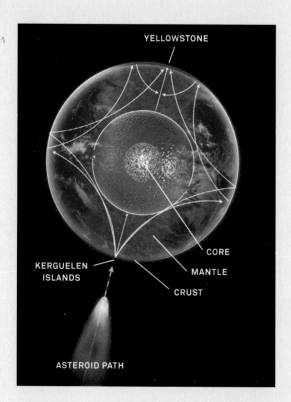

skeptical that such a newly recognized, and seemingly novel, theory could actually be responsible for these phenomena, it is worth remembering that the theory of continental drift and tectonic plates was not widely accepted by the scientific community until the 1960s.

THE 22 GEOTHERMAL POWER PLANTS OF THE GEYSERS IN NORTHERN CALIFORNIA MAKE UP THE LARGEST GEOTHERMAL INSTALLATION IN THE WORLD.

hydrothermal sites is mature, and the cost of the electricity produced is generally competitive with other sources. As a result, hydrothermal electricity is generated at a utility scale, and new sites are continuously being brought online in both developed and developing nations throughout the world. Today, the total global generating capacity from geothermal energy resources is about 10,000 megawatts of electricity.

For example, 60 percent of the average electricity used in California's North Coast region (from the Golden Gate Bridge to the Oregon border) is provided by electric turbines powered by steam from hot springs in an area north of San Francisco known as The Geysers. The Geysers had a peak production of more than 2,000 megawatts in 1987, but now produce only half as much. There are actually 22 separate sites that make up The Geysers. Taken together, they still constitute the largest such system in the world, but a new hydrothermal plant now proposed for Sarulla in North Sumatra, Indonesia, will dwarf each of the individual plants at The Geysers when it is completed.

The best hydrothermal sites have a natural source of water underground, high temperatures, and permeable rocks through which the water circulates and absorbs heat. This heat energy is withdrawn either in the form of steam or very hot water. However, these sites require a lot of hot natural fluids at relatively shallow depths in rocks that are highly porous and permeable. And because some of the water is lost during the process, there is frequently a requirement to add additional volumes of water to maintain the initial productivity of the site. This is now under way in Northern California at The Geysers.

Conventional hydrothermal plants are built according to one of three different designs. The steam can be taken directly through the turbine and then recondensed into water for recirculation. Very hot water can be depressurized and "flashed" into steam. Or the hot water can be put through a heat exchanger to transfer its heat to another liquid—like isopentane—that boils at a higher pressure, producing steam to run the turbine. Because more of the resource is available as hot water over a range of temperatures rather than as "dry steam," the last of the three categories is believed to have the greatest potential for use, particularly for resources below 400°F.

The new and growing excitement about geothermal electricity is based on new technologies that make it possible to exploit sites deep in the earth that have enormous amounts of heat but lack one or more of the characteristics found in hydrothermal reservoirs. For example, many of these sites do not have water.

Capturing energy from these new nonhydrothermal resources has led to new approaches. This new technology for making geothermal power is called enhanced (or engineered) geothermal systems (EGS). By using new technology that exploits the advances in drilling and reservoir stimulation developed in part as a result of the frenzy of oil and gas exploration in the 20th century to create active reservoirs that emulate the properties of hydrothermal systems, geologists and engineers believe they have found ways to produce extremely large sources of geothermal power from regions several miles deep in the crust of the earth.

Instead of searching for conventional hydrothermal sites, geologists look for areas where hot, dry rocks or hot rocks of lower permeability and porosity with minimal amounts of water or brine have temperatures in excess of 300°F to 400°F (150°C to 200°C) and are close enough to the surface to allow cost-efficient drilling. By pumping high-pressure water into these rocks, permeable

HOW ENHANCED GEOTHERMAL SYSTEMS WORK

In the new generation of geothermal power, wells are drilled several kilometers below the surface to reach hot, often dry, rock. Pressurized water is pumped down into the well, "enhancing" the site by cracking apart seams in the rock through which the added water can flow. The now-heated water is pumped back to the surface, where it is converted to steam to drive a turbine, generating electricity.

POWER TRANSMISSION

POWER PLANT

TURBINE

HEAT EXCHANGER

GENERATOR

INJECTION PUMP

SEDIMENTARY LAYER

COOL WATER

HOT WATER

(1.9–3.7 MILES)

HOT ROCK

PERMEABLE ZONE

GEOTHERMAL POWER IS PERVASIVE IN ICELAND, WHERE THIS GREENHOUSE IS WARMED WITH GEOTHERMAL HEAT AND LIT WITH ELECTRICITY FROM A GEOTHERMAL PLANT.

flow paths are created by opening existing sealed fractures. To extract energy, water is pumped from the surface down through injection wells into the fractured region and returned to the surface in production wells as steam or very hot water—either of which can be used as a source of energy to drive an electric turbine generator. These basic techniques for "hydrofracturing and stimulation" have long been used in the development of some oil and gas reserves.

As the injected water circulates through the newly opened cracks in the fractured region, steam (or superheated liquid) is withdrawn from a second set of "production wells" located some distance away from the injection wells. In this process, the heat withdrawn from the fracture zone is produced in essentially the same way it is collected from conventional hydrothermal formations. Essentially, this involves adding the "hydro" to the "thermal" and then constantly recirculating the water to engineer a brand-new, highly productive geothermal system.

In some locations—particularly in the western United States—special attention is being given by the industry to reducing loss of water during the operation of an EGS power plant. Based on early field tests at several EGS sites, around 5 percent of the water is lost to permeation (fracturing the rocks), but the industry's goal is to reduce that percentage to less than 1 percent. Some operators now claim to have reduced the loss to zero.

Ironically, one of the alternative working fluids under consideration for use in EGS reservoirs is supercritical CO_2. In this state, CO_2 has high, liquidlike densities and low, gaslike viscosities that can make it more effective than water in transferring the heat from the hot rocks in an EGS reservoir to the surface. Some of the circulating CO_2 would be "sequestered" inside the rock formation, but the amounts involved—even with a massive expansion of EGS—would be trivial in the context of what would be needed for full-scale carbon capture and sequestration to have a global impact. More important, EGS-generated electricity and heat could replace existing coal-, oil-, and gas-fired generation to remove their carbon footprint directly.

The largest amount of stored thermal energy in the crust of the earth is found in rocks that do not have much natural permeability. The principal challenge for EGS is to artificially create just the right amount of permeability to allow the flow of the injected water to transfer heat on an even, continuous basis. EGS projects need to ensure that there is enough connectivity created within the region being stimulated to permit the maximum heat to be withdrawn from each well without, in

GEOTHERMAL RESOURCES IN THE UNITED STATES

Enhanced geothermal systems (EGS) open the majority of regions in the United States to potential use as a power source. This map indicates where and at what depth the earth's temperature exceeds 300°F—considered by many scientists to be the minimum temperature for cost-effective use of EGS to generate electricity.

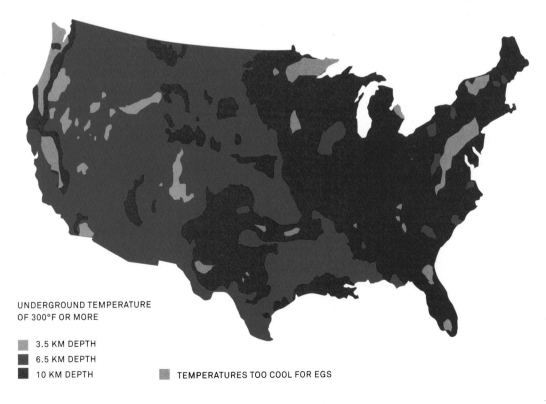

UNDERGROUND TEMPERATURE
OF 300°F OR MORE

- 3.5 KM DEPTH
- 6.5 KM DEPTH
- 10 KM DEPTH

TEMPERATURES TOO COOL FOR EGS

SOURCE: Massachusetts Institute of Technology, *The Future of Geothermal Energy*, 2006

AT THE COOPER BASIN PROJECT IN SOUTH
AUSTRALIA, WORKERS PREPARE TO DRILL
NEARLY THREE MILES INTO THE EARTH TO
TAP THE GEOTHERMAL HEAT.

the process, diminishing the lifetime of the reservoir by cooling it off too quickly during the production of energy.

As is true with enhanced oil recovery and carbon capture and sequestration, the injection of pressurized water into rock formations deep in the ground can, under some conditions, produce minor microseismic events—which requires proper assessment of seismic risks as part of the production process. Areas of high seismic risk where large faults and dangerous earthquakes are possible should be avoided, while risks at other sites must be measured and managed. For example, this risk can be controlled by reducing subsurface pressures if the monitoring picks up any heightened seismic activity. In any case, seismic risk assessment at

INSIDE THE EARTH

Scientists believe the core of the earth is made principally of iron at temperatures comparable to those at the surface of the sun. Everywhere in the world, the crust of the earth constantly absorbs huge amounts of heat transmitted through convection and conduction outward from the white-hot center of the planet.

The superheated solid iron core is 1,500 miles in diameter, about 70 percent as large as the moon. Scientists do not know with precision how hot this inner core really is, but estimates range from 7,800°F to 12,600°F. (Indeed, science doesn't know from any experimental evidence that the core is solid iron; they infer that from the behavior of seismic waves passing through it.) Surrounding this inner core is a molten metal (mostly iron) outer core with temperatures estimated to be 6,700°F to 7,800°F. The thickness of this outer core is approximately 1,400 miles, which means that the overall diameter of the inner and outer core combined is approximately 4,300 miles—the size of Mars.

The next spherical shell—1,800 miles thick—is the mantle, which extends to within 60 to 125 miles beneath the surface of the planet. The temperature of the mantle ranges from 6,700°F in the inner mantle to 1,800°F for the outer mantle. The final layer is the relatively thin crust of the earth, which ranges from less than two miles thick at the bottom of the oceans to more than 60 miles thick below mountain ranges. The average depth under continents is less than 20 miles.

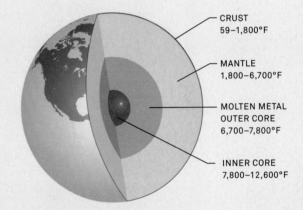

CRUST
59–1,800°F

MANTLE
1,800–6,700°F

MOLTEN METAL
OUTER CORE
6,700–7,800°F

INNER CORE
7,800–12,600°F

In addition to the convective and conductive heat coming from the mantle into the crust, the crust itself has its own independent source of heat: it contains significant amounts of uranium, thorium, potassium, and other radioactive elements that naturally give off heat as they decay. Measuring rock temperatures from the surface with a thermometer placed in a well, scientists find that beyond a depth of a few hundred feet, the temperature begins to increase steadily the deeper they go. Indeed, the amount of heat within the outer six miles of the earth's crust is estimated to contain 50,000 times as much energy as all of the petroleum and natural gas reserves in the world.

THE WAIREKEI POWER STATION IN NEW ZEALAND WAS ONE OF THE FIRST GEOTHERMAL PLANTS IN THE WORLD. TODAY IT CAN PRODUCE 181 MEGA-WATTS OF ELECTRICITY.

specific sites is an integral part of any geothermal project to evaluate the probability that larger induced events might occur. These precautions were not adequately taken a few years ago by a Swiss company that drilled into a known faulted region three miles below Basel, Switzerland, and triggered a magnitude 3.4 earthquake (still quite small) during pressurization. The event was characterized by an audible air shockwave with barely detectable ground motion and led to the project's being shut down. Clearly there are lessons to be learned from this unfortunate experience, but they need to be taken in the proper context of dealing with early deployment of a new technology that carries some level of risk that must be understood and managed.

Soon after the 1973 oil embargo, when the United States and several other nations began to search in earnest for alternative sources of energy, U.S. government scientists at Los Alamos launched the first EGS experiment at Fenton Hill, New Mexico. Since then, progress in research and development internationally has led to the steady maturation of this technology, to the point where most experts are highly confident that it will become a major new source of pollution-free baseload electricity.

The major up-front expense with EGS is the cost of drilling. A single drill hole to a depth of three kilometers (10,000 feet) can cost $5 million or more, and the willingness of investors to continue financing operations is sometimes dependent on signs of success with the initial drill hole. The current economics of EGS put a premium on finding hot rocks at regions between three and six kilometers deep. This capital constraint pushes companies to drill in higher-temperature gradient zones that have more potential for electricity with less risky and less costly drilling.

THE CITY OF BEPPU, JAPAN, TAPS GEOTHERMAL
HEAT FOR THE DIRECT HEATING OF BUILDINGS AS
WELL AS ELECTRICITY GENERATION.

When advances in drill bits, drilling methods, pipe casings, and related technologies allow more economical drilling to depths of greater than six kilometers, the size of the geothermal resource will expand considerably yet again—and will help alleviate financial pressure on the industry while making it progressively easier to utilize the geothermal resource in more and more areas.

Although the basic technology of EGS has been demonstrated in several field experiments—and is creating enormous excitement among energy experts familiar with its potential—reservoir testing at a commercial scale remains as a major barrier that must be crossed before

private investors will view large financial risks as acceptable. At a comparable stage in the oil business, many drillers went bankrupt for every one who developed a gusher. Yet the amount of money necessary to fully develop this exciting technology is trivial in the context of governmental energy R&D budgets. But because of the widespread misunderstanding of the potential for geothermal energy, the necessary R&D funding has been frequently cut to the bone, effectively slowing the development of this incredible potential energy source. Most recently, the Bush-Cheney administration cut the budget for geothermal energy to zero and allowed the Production Tax Credit, which had been driving a surge of private investment, to expire.

The absence of a coherent, well-funded federal program for the development of geothermal electricity has, in turn, reinforced the general impression that this technology must not be very promising. Otherwise, we would be supporting it. That is unfortunate, because energy experts who have examined the data on this incredible resource have argued forcefully for several years that the United States has been making a serious mistake by not developing geothermal energy at top speed.

Advances are currently being sought not only in drilling technology but also in the exploration processes used to locate the best hot rock resources, in reservoir-stimulation techniques, and in improving the efficiency of energy conversion.

But the major study of EGS at M.I.T. concluded, "It is important to emphasize that while further advances are needed, none of the known technical and economic barriers limiting widespread development of EGS as a domestic energy source are considered to be insurmountable."

Since 2008, both research-and-development funding and the geothermal tax credit have been belatedly renewed in the United States, and other obstacles to the development of this resource are being removed. This has led to expanded activity to deploy hydrothermal power plants as well as to develop EGS. For example, a small EGS demonstration project is now under way at Desert Peak, Nevada. When completed, it will expand an existing 11-megawatt hydrothermal facility into a 50-megawatt EGS facility. Geothermal energy can also benefit greatly from the adoption of a national Renewable Portfolio Standard, which gives a powerful mandate to utilities to search for ways to use more low-polluting, renewable forms of energy.

Scientists and engineers are also working on more cost-effective ways to derive energy from areas with lower rock temperatures. Since these hot rocks are found throughout the crust of the earth at varying depths, the size of the resource is truly enormous. If they can do so, the size of the resource will be greatly expanded. In fact, if EGS technology could be perfected to depths of six kilometers, then the entire United States would be available for geothermal power generation (see "Geothermal Resources in the United States," page 103).

Indeed, somewhat lower-temperature rocks (from less than 176°F to 248°F, or 80°C to 120°C) can be used for geothermal power applied to the heating of buildings directly without incurring the energy losses associated with conversion of heat into electricity. The United States installed the first geothermal district heating system in Boise, Idaho, over 100 years ago. Even today, the state capitol and many other buildings in Boise are heated with geothermal hot water. In Klamath Falls, Oregon, heat from geothermal wells has been used for direct space heating for more than a century. Similarly, since Iceland responded to the oil shocks of the 1970s by converting to domestic resources, virtually every building in the entire country is heated by the hot water resources close to the surface of that tectonically active land.

While the United States was asleep at the switch, other nations began to aggressively research and develop EGS. The European Union has an EGS project under way in Soultz-sous-Forêts, France, near the border with Germany. Other projects are under way in Germany, Switzerland, the United Kingdom, the Czech Republic, and elsewhere. Government policy has been most supportive in Australia, where seven publicly traded companies are now actively pursuing and developing EGS opportunities.

The Philippines, El Salvador, and Costa Rica have all recently achieved the production of more than 15 percent of their electricity from geothermal generation. So have Kenya and Iceland, both located in different tectonically active boundary zones. New Zealand, Indonesia, Nicaragua, and the Caribbean island of Guadeloupe all get between 5 and 10 percent of their electricity from geothermal generation.

There are two other kinds of energy outputs being developed in the geothermal power industry. First, the "coproduction" of hot water from some oil and gas wells can profitably be converted to yet another source of electricity. In many cases, all of the heat energy contained in hot water coming out of oil and gas wells is now simply discarded, but the development by geothermal engineers of new heat exchangers and other techniques for transforming hot water from the earth into electricity have opened up the potential for oil and gas drillers to exploit what is essentially a free source of electricity as they drill.

Second, deep reservoirs of high-pressure hot water sometimes contain dissolved methane that can be reclaimed as the hot water is produced, rather than venting the methane into the atmosphere. In the U.S., this resource is heavily concentrated along the Gulf Coast, in the Appalachian Mountain region, and on the West Coast.

Finally, there is yet another form of geothermal energy that is quite different: geothermal heat pumps, sometimes called ground source heat pumps. Even though it is the weakest form of geothermal energy, it is the most easily accessible and could provide an estimated 1,000 gigawatts of energy in the United States alone—enough to provide energy equivalent to more than a third of America's annual electricity use.

Geothermal heat pumps provide a widely distributed, highly economical way to sharply reduce the cost of heating and cooling buildings. While EGS depends upon drilling technologies developed for oil and gas, geothermal heat pumps make use of much more widely available drilling techniques commonly used to drill water wells. Anywhere from a dozen to a few hundred feet below the surface, the earth's temperature averages approximately 59°F (15°C).

More precisely, the temperature at that depth is equivalent to the year-round average air temperature at the surface where the site is located. By circulating water or a freeze-resistant fluid through a ground source closed loop or to the bottom of a geothermal well, and then pumping it back to the surface, thermal energy can be either extracted (during the winter) or deposited (during the summer). These geothermal heat pumps use conventional, refrigerant-based vapor-compression units to transfer heat at a rate four times more efficient than that of air-to-air heat pumps—thereby lessening the need for conventional electricity.

It is much easier to raise the temperature from 59°F to 68°F or 70°F than to start with the normal outdoor temperatures in most places in the temperate zones. In summer, the process is reversed to make the bottom of the geothermal heat pump a sink and to lift the cooler temperatures for space cooling. Once again, it is far easier to cool the air if you have a readily available source of 59°F fluid.

Geothermal heat pumps are usually integrated with traditional heating and air-conditioning equipment in homes. There are federal tax breaks (and some state tax benefits) for purchasing and using geothermal heat pumps.

At present, the principal disadvantage of geothermal heat pumps is the capital cost involved in installing them. Since most builders and developers pay less attention to the annual operating cost of the building after they have sold it or leased it to the first owner or tenant, they are less likely to make even marginally higher investments whose value will be recovered by those owning or using the building and who will pay the heating and cooling bills (see Chapter 15).

The economics of geothermal heat pumps often produce savings of more than 60 percent in heating and cooling bills, which makes them economically attractive in every part of the United States—especially for new construction. Retrofitting old structures usually involves greater expense, but is often still economically attractive.

GEOTHERMAL HOME HEATING AND COOLING

A few years ago, my wife and I decided to install a new geothermal, or "ground source," heat pump system for providing heating, cooling, and hot water to our house. There are a number of different types of home geothermal systems: some use water in open loops or closed loops; some use a refrigerant; some use a deeply drilled well. Newer systems are called *direct exchange systems* (DX) because they exchange the heat directly between the ground and a refrigerant fluid. As there is no water pump required, the DX system is generally more efficient, and DX systems are easier to accommodate on smaller lots of land. At our house, a local company installed a DX system that circulates a nonhazardous refrigerant in copper tubing underground. The system was installed at a depth of up to 300 feet by boring holes underneath our driveway.

Underground, the earth's naturally occurring heat is transferred to the colder refrigerant circulating in the copper tubes. The refrigerant (in the form of a vapor when inside the underground tubes) is then compressed by an exchange unit in the basement. Compression raises the pressure and temperature. In winter, cooler inside air absorbs the heat, warming the house. In summer, the system is reversed. Heat energy from the warm interior is absorbed by the fluid, which circulates it underground and releases heat into the cooler ground.

This geothermal system provides heating and cooling with almost no reliance on fossil fuels. But there are additional advantages. First, reducing the use of elec-

SUMMER COOLING MODE   WINTER HEATING MODE

The earth's heat is transferred into the house in winter; in summer, the air's heat is moved into the ground.

tricity during peak hours helps lower utility costs and ultimately lowers consumer electric rates. Second, the system is silent, eliminating the whirring noise of an outdoor air-conditioning unit in summer. Finally, the air in the house is comfortable: warm in the winter and crisp, cool, and dehumidified in the summer.

OUR SOURCES OF ENERGY

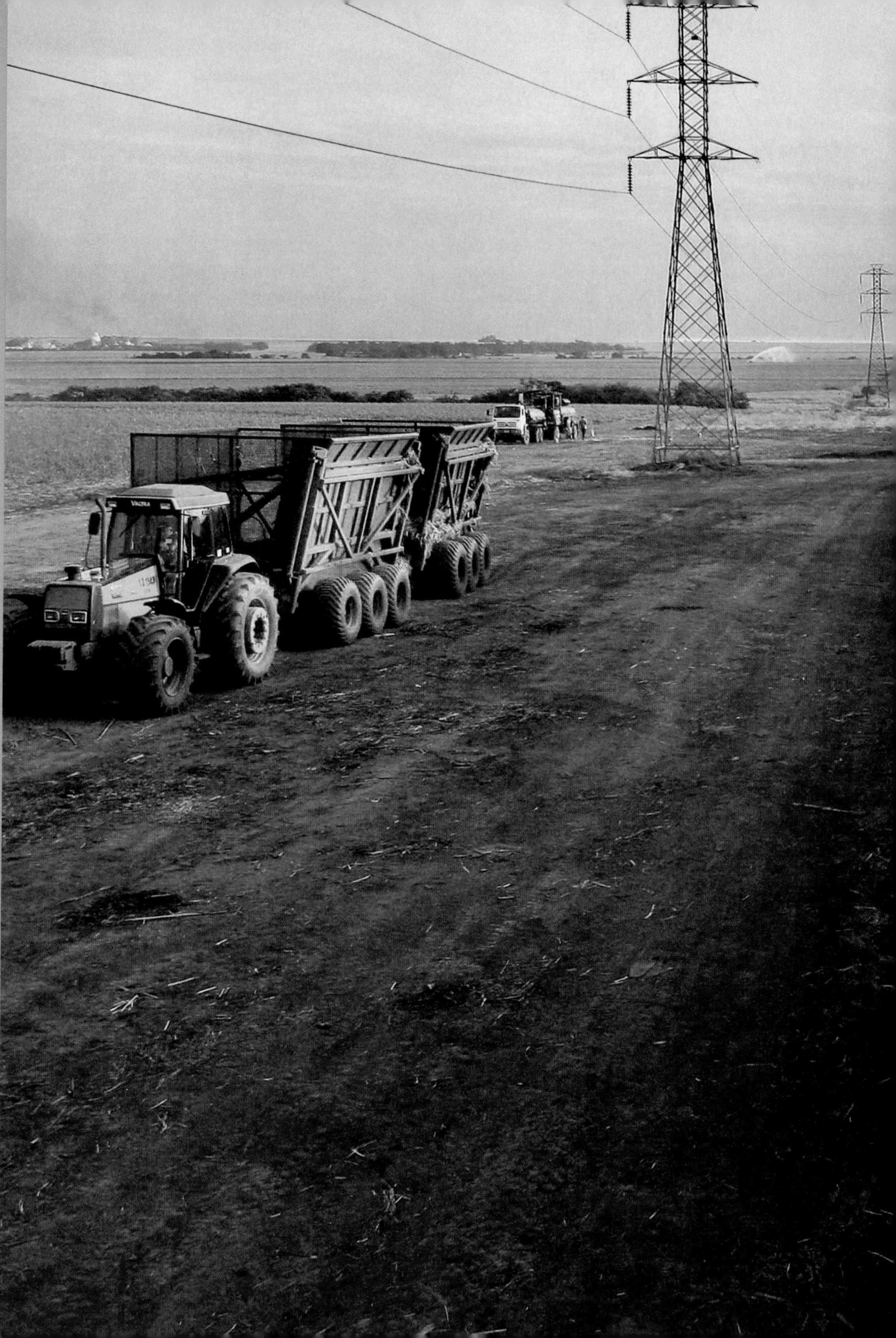

CHAPTER SIX

GROWING FUEL

SUGARCANE HARVEST, SERTÃOZINHO, BRAZIL.
BRAZIL HAS DEVELOPED THE FIRST LARGE-SCALE
BIOFUEL ECONOMY. ABOUT 50 PERCENT OF FUEL
USED IN GASOLINE-POWERED CARS IS ETHANOL.

Biomass energy is one of the most promising ways to reduce significant amounts of CO_2 from the burning of coal and natural gas. It is unfortunate that so many people think of biomass primarily as a way to displace oil as a source of liquid fuels for cars and trucks, because most expert studies have thrown cold water on the early enthusiasm for making alcohol fuels from food crops like corn.

The good news is that nonfood sources of biomass can be burned directly with advanced combustion technologies to produce electricity and heat in ways that deliver large energy savings and reductions in global warming pollution. Moreover, a newer technology for creating liquid fuels from nonfood crops is close to commercialization.

Biomass energy is, in theory, renewable. Since the carbon contained in plants comes ultimately from sunlight—through photosynthesis—any plant used for the production of energy can be replaced with another whose growth is also powered by sunlight.

In reality, however, the energy consumed in transforming the plant material into a usable form of power often comes from nonrenewable fossil fuels. The benefits often still outweigh the costs, but full "life cycle" analyses are necessary to accurately determine which approaches are truly beneficial in solving the climate crisis—and in achieving other desirable goals, such as reducing dependence on foreign oil, conserving water, and biodiversity.

Biomass energy can be produced from a variety of feedstocks: virgin trees and forest waste; food crops (like corn and sugarcane); energy crops (like switchgrass, miscanthus, and sweet sorghum); and from municipal, agricultural, and industrial waste that includes organic materials. These biomass inputs can be used to produce electricity, thermal energy, and liquid fuels for transportation.

In addition, the same processes used to produce liquid fuels from biomass inputs can also be modified to produce biomaterials. Just as fossil fuels are used to produce precursors for plastics, chemicals, and other materials (roughly one fifth of each barrel of oil is used not for fuel but for materials), biorefineries are used to supply a growing market in bioplastics and molecules used for chemical processing, which often have higher profit margins than liquid biofuels.

Excluding the burning of wood for heating and cooking in less-developed countries, more than 90 percent of the biomass energy being produced in the world is used to make thermal energy for industrial processes and the space heating of buildings, and to make electricity in steam-driven generators and combined heat and power generators.

However, most of the policy debates have focused on the production of ethanol and biodiesel fuels in an effort to develop economical substitutes for petroleum-based liquid fuels. In the United States, there has been sustained enthusiasm for ethanol made from corn that can displace some of the gasoline used in automobiles.

The first generation technology now used to

MISCANTHUS GRASS IS A PROMISING BIOFUEL CROP. IT IS LOW MAINTENANCE, CAN BE RAISED ON MARGINAL LAND, AND GROWS VERY QUICKLY.

convert corn, palm oil, soy, and other food crops into liquid fuels for vehicles has created enormous controversy because repeated life-cycle analyses of this process have led to the growing realization that it usually releases almost as much CO_2 into the atmosphere as does the production and use of the petroleum-based fuels being displaced.

First generation ethanol has also been blamed for contributing to increased food prices around the world by diverting prime cropland from food production to fuel production. This diversion of food production capacity has also been blamed for stimulating the additional clearing of tropical and subtropical forests for the production of both food and fuel from biomass in ways that threaten biodiversity and add yet more greenhouse gas pollution to the atmosphere. Finally, critics of this technology have also focused on the large amounts of water required in the process.

Nevertheless, the drive for reduced dependence on global oil markets and the support by farmers and agribusiness for more ethanol production have combined to stimulate rapid growth in these first generation ethanol and biodiesel technologies.

In order to make intelligent policy choices about biomass as a source of renewable energy, we should first of all take steps to ensure that biomass feedstocks are produced in a truly sustainable manner:

▶   The harvesting of biomass should not cause the destruction of virgin forests and the habitat for biodiversity they provide.

▶ CO_2 emissions should be minimized in the growing of biomass.

▶   Plants other than food crops should be used—in order to avoid upward pressure on food prices and additional land clearing.

▶   Water use should be sustainable—in terms of both quantity and quality.

▶   Soil fertility should be preserved and, where possible, enhanced.

▶   The social and economic well-being of stakeholders in the process should be respected and, where possible, improved.

After ensuring that biomass feedstocks are produced and supplied in a sustainable way, policy makers should then ensure that the technologies chosen to aggregate, transport, and convert these feedstocks into energy, biofuels, and biomaterials are environmentally, economically, and socially sustainable.

BIOFUEL FEEDSTOCKS

CORN SILAGE

CHARCOAL

The production of ethanol in first generation biorefineries has been a disappointment. However, it has had the benefit of increasing income for farmers and has led to the emergence of an infrastructure that will prove highly valuable when second generation technologies are available to produce ethanol from nonfood crops.

I feel the disappointment personally because, as vice president of the United States in 1994, I cast the tiebreaking vote in favor of moving forward with a large national commitment to ethanol. In 1978, as a young congressman from a farming district in Middle Tennessee, I organized and hosted a daylong workshop on what was then called "gasohol" for 5,000 constituents, mostly farmers, eager to be a part of the national effort then under way to reduce our dependence on foreign oil. Throughout my 16 years in the House of Representatives and the U.S. Senate, I was a persistent advocate of helping farmers to earn income from the production of alcohol fuels for cars and trucks.

By the mid-1990s, however, there were already ample warnings that the energy and CO_2 balances for first generation ethanol were not nearly as favorable as I would have liked them to be. But my political desire to help the farm economy (in Tennessee and Iowa, for example), coupled with my optimism that improvements in the efficiency with which the crop feedstocks were grown and processed, partly accounted for my desire to go forward with the large-scale development of the technology. Indeed, with no-till farming and even newer "precision farming" techniques to reduce water, fuel, and fertilizer use, the fossil fuel inputs for the growing and harvesting of these feedstocks can be reduced to the point where the life-cycle CO_2 and energy balances are mildly favorable. In practice, however, the results over the last several years have convinced many analysts that producing first generation ethanol from corn is a mistake.

Fast-growing sugarcane in Brazil, with its abundant sunlight and rainfall and cheaper labor and conversion costs, is far more efficient and—for the most part—more environmentally responsible, with roughly one third the greenhouse gas emissions of U.S. corn-based ethanol. Brazil began its National Alcohol Program in 1975, after the oil embargo in late 1973. Sugarcane uses far less petroleum-intensive fertilizer because it is a perennial crop; it produces much more biomass per acre

WOOD CHIPS

WASTE PAPER

than corn; the waste from the sugarcane plants (bagasse) is used as a fuel in the production process; and the leading Brazilian ethanol companies use a closed water cycle and pay close attention to following responsible environmental and labor practices. Moreover, the sugarcane growing areas, currently less than 2 percent of Brazil's arable land, are far from the Amazon and have not caused significant direct or indirect encroachment into the rain forest. In 2003, Brazil began to manufacture and sell flexible-fuel vehicles, and in 2005 it began to sell large volumes of ethanol into foreign markets. Unfortunately, studies have shown that, for a variety of reasons (including climate), sugarcane ethanol made in the U.S. is not competitive with ethanol made in Brazil. As well, high U.S. tariffs sharply limit the importation of Brazilian ethanol.

Where corn is concerned, even with the dramatic growth in yields per acre made possible by hybridization and, more recently, the insertion of new genetic traits, corn yields approximately 400 gallons of ethanol per acre, compared with 650 gallons per acre from sugarcane, according to experts from North Carolina State University. Moreover, because corn yields so much less energy per acre than sugarcane, corn supplies typically must be located no more than 50 miles from the refinery in order to keep feedstock transportation costs within profitable boundaries. And since the refined product—ethanol—cannot be shipped through existing pipeline networks, the distribution of the liquid fuel to wholesalers also relies on heavy trucks.

The emissions from automobiles running on ethanol—whether made from corn, sugarcane, or some other crop—are significantly lower than from those running on gasoline. But the balance shifts when all of the fossil-based energy used in growing and harvesting the plants, then refining and transporting the ethanol, is included in a full analysis.

Unfortunately, largely because modern agriculture is so petroleum intensive, net greenhouse gas emissions from corn-based ethanol turn out to be almost equal to the emissions from gasoline.

Moreover, since the resulting ethanol contains only two thirds of the energy in a gallon of gasoline, vehicles using ethanol-blended fuel get reductions in the number of miles per gallon proportional to the ethanol content in the fuel. At present, approximately five billion gallons of ethanol are blended into nearly half of the 140 billion gallons of gasoline used each year in the United States—at a 10 percent or lower blend in most cases principally because of the U.S. Renewable Fuel Standard.

Finally, there is also an upper limit to how large a role corn-based ethanol could play in the transportation fuels marketplace, even if it were produced and used in a sustainable fashion. *Scientific American* published a study in 2009, finding, "There is simply not enough available farmland to provide more than about 10 percent of developed countries' liquid fuel needs with first generation biofuels." The Congressional Research Service found in a 2007 study that even if the entire U.S. corn crop were completely devoted to the production of first generation ethanol, it would supply only 13.4 percent of current national gasoline usage.

Nevertheless, in the late 1980s, the United States established an incentive program to encourage automakers to manufacture cars and light trucks that are capable of burning blended fuels containing as much as 85 percent ethanol (E85). In return, the automakers are given credit from the government for higher gas mileage than they actually achieve. Since it requires expenditures of only about $100 per vehicle to replace gaskets and fuel

TRUCKS LINE UP WITH SUGARCANE AT AN ETHANOL PLANT NEAR SÃO PAULO, BRAZIL.

lines with ones that are compatible with the more corrosive ethanol blends, the automakers can save money by avoiding the much larger investments necessary to actually increase the average mileage of the cars and light trucks they manufacture.

However, more than a decade after the introduction of this incentive program, studies show that only 6 percent of the flexible-fuel vehicles use E85 fuel as their primary fuel. Part of the reason is that less than one half of 1 percent of U.S. vehicle fueling stations actually sell E85, with a slightly larger number selling biodiesel fuel. Moreover, the alternative fuel stations are largely clustered in the upper Midwest, where ethanol production is also highest. In many other parts of the United States, millions of people own flexible-fuel vehicles without even realizing it. Some notice the logo designating their cars as E85 cars but search in vain for an E85 filling station near where they live. The principal benefit of this program has been to allow automobile manufacturers to evade requirements for higher mileage standards.

Two other large factors have been responsible for the shift in expert opinion toward a negative conclusion on corn-based ethanol. First, the significant increase in world food prices during 2007–08 was partly blamed on the diversion of cultivated land at the margins from the growing of food to the growing of crops for ethanol. Detailed studies later found that other causes—including the historic drought in Australia, which took a large percentage of food grains off the world market—were responsible for the majority of the price increase. Most of the pressure on the world's agricultural base comes from population growth and soaring demand for land-intensive animal protein as a result of rising incomes and changing diets in the developing world. But there is no doubt that further diversions of cropland from food to fuel will

put upward pressure on food prices at a time when many impoverished regions of the world are facing growing concerns about food security.

Moreover, public support for biofuels in developed countries usually peaks when oil prices are high, but because of the oil intensity of food production, these are the periods when food prices are also at their highest—particularly in developing countries. We face repeated cycles of rising food and energy prices, with each cycle renewing the focus on the perceived conflict between food and biofuel.

Second, the U.S. National Research Council found in 2008 that, on an average national basis, each gallon of corn ethanol requires four gallons of water at the refinery (compared with one and a half gallons of water for the refining of each gallon of gasoline) and 142 gallons on average for the growing of the corn—785 gallons in irrigated areas. Even though most of the U.S. corn crop is rain fed, expanded cultivation has pushed corn into areas that require such large amounts of irrigation that the average water use for corn has grown dramatically. In addition, predictions of deeper and more frequent droughts in prime growing areas (a trend in mid-continent regions throughout the world long associated with global warming) have enhanced concerns about the future viability of such a water-intensive process.

Some first generation ethanol plants have already been denied permits because of their requests to withdraw water from underground aquifers in volumes that were calculated by watershed regulators to be unsustainable. Some plants in Minnesota and Wisconsin have faced public opposition from citizens concerned about losing access to their well water. Other new biorefineries have also been proposed for areas that are currently drawing water from underground aquifers like the Ogallala

Producing first generation ethanol from corn is a mistake.

CORN IS GROWN FOR CONVERSION INTO
ETHANOL NEAR THE COTTAM POWER STATION
IN NOTTINGHAMSHIRE, ENGLAND.

at completely unsustainable rates.

First generation biodiesel—mostly made from soybeans in the U.S.—has larger net reductions in greenhouse gas emissions than corn ethanol, but it shares most of the other problems of first generation ethanol. In other parts of the world, the emerging markets for biodiesel have resulted in gross abuses to the environment, such as the clearing of peat forests in Indonesia and Malaysia for the growing of palm oil plantations. Indonesia has now become the third largest source of greenhouse gas pollution, largely as a result of this practice. (Animal fats have also been used, to a limited degree, as feedstock for biodiesel. Occasional news stories highlight enthusiastic entrepreneurs and early adopters running vehicles on discarded french fry oil from fast food restaurants, but larger projects are now proposed that will use waste fat from large beef, chicken, and pork producers.)

The opportunity for those destroying the peat forests to plant oil palms as a feedstock for biodiesel is illustrated by the comparison of current fuel-equivalent yields. Oil palms yield 610 gallons of biodiesel per acre, the most of any source. Coconuts, another productive source, yield 276 gallons per acre, followed by rapeseed (122 gallons), peanuts (109 gallons), sunflowers (98 gallons), and soybeans—which, at 46 gallons per acre, are the biodiesel feedstock of choice in the United States. In 2008, the majority of the U.S.

FUEL YIELDS FOR BIOFUEL CROPS

The figures below show the yields for leading biofuel crops, in gallons of fuel per acre, and projected yields for second generation ethanol crops. Current biofuel crops are processed into either biodiesel or ethanol. The next generation of biofuels will convert cellulosic, nonfood plants such as switchgrass and miscanthus (elephant grass) into ethanol. This technology is not yet in use.

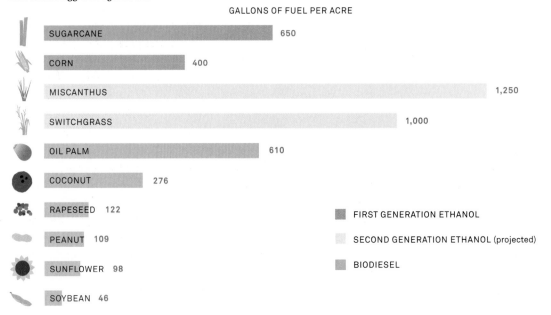

GALLONS OF FUEL PER ACRE

SUGARCANE 650
CORN 400
MISCANTHUS 1,250
SWITCHGRASS 1,000
OIL PALM 610
COCONUT 276
RAPESEED 122
PEANUT 109
SUNFLOWER 98
SOYBEAN 46

FIRST GENERATION ETHANOL
SECOND GENERATION ETHANOL (projected)
BIODIESEL

SOURCE: Ethanol: North Carolina Cooperative Extension Service, October 2007; biodiesel: National Sustainable Agriculture Information Service

biodiesel supply was exported, mainly to Europe.

Policy makers who are aware of the discouraging life-cycle balances for CO_2 and energy with corn-based ethanol have taken comfort in the argument that the expansion of this liquid biofuel market has, at the least, created a production and distribution infrastructure that can soon be supplied with more economically and environmentally responsible biofuels produced by the long-anticipated second generation technology.

"Cellulosic" biofuels can be made from tall grasses (like switchgrass), fast-growing trees, and other plants with a much higher cellulose content than food crops, and with organic waste from agricultural, forestry, and industrial waste streams. By using nonfood crops, this second generation technology can avoid some of the serious problems that have arisen with the production of ethanol from corn. In addition, cellulosic biofuels are expected to have far lower greenhouse gas emissions but will vary in platform and conversion technology. However, the development of profitable versions of this second generation technology has taken more time than many had hoped.

The second generation technology for producing ethanol—when it becomes commercially available—has a significant advantage over the first generation technology: instead of using food crops, it will make liquid fuels from perennial grasses, fast-growing trees, and waste streams with a high cellulose content.

Some of the plants used as feedstocks—like switchgrass—actually help to sequester carbon in the soil as they grow, and research indicates that regular harvesting of switchgrass actually accelerates sequestration of soil carbon. Incidentally, while most studies of cellulosic ethanol in the United States have focused on switchgrass, recent data indicate that miscanthus (a perennial grass

native to Africa and South Asia) has substantially higher yields than switchgrass.

One of the principal benefits of this second generation process is that the feedstock plants flourish on land that is unsuitable for food crops—thereby eliminating the geopolitically and environmentally problematic competition between food crops and fuel crops—and can also be planted on degraded lands that have been abandoned by farmers, where they reinvigorate the fertility of the soil as they grow. Moreover, most of these fuel crops require no petroleum-intensive fertilizers, cultivation, or replanting (they are perennials).

Some have proposed the widespread use of agricultural waste, like corn stover, as a feedstock for cellulosic ethanol, but soil experts caution that most such agricultural waste is best left on the ground to replenish the fertility of the soil. Nevertheless, according to most experts, both purposes can be served if the amount removed from the fields is less than 25 percent of the total.

In a new twist on crop rotation, some farmers are now planting "energy crops" after the fall harvest of food crops and before the following spring's planting. Others are using "mixed cropping" systems, in which both food and energy crops are grown simultaneously. If implemented carefully, both of these approaches can actually enrich the fertility of the farmer's soil for food crops and provide an additional source of income from the energy crops.

Similarly, fast-growing trees—like poplars, hybrid willows, sycamores, sweetgums, and eucalyptus trees—can be harvested annually. Some can be cut off close to the ground (a practice called coppicing), which actually stimulates more rapid regrowth and continues to stabilize, replenish, and recarbonize the soil.

In addition, forest waste from the timber

industry, wood-processing mills, pulp and paper mills, the construction industry, and the clearing of underbrush during fire prevention activities could supply more than 350 million tons of cellulosic feedstock each year, according to the U.S. Departments of Energy and Agriculture. The study by these two departments found that, overall, the United States could process at least 1.3 billion tons of cellulosic biomass per year—enough to produce bioliquid fuels equal to one third of the current U.S. consumption of gasoline and diesel.

It should be noted that many conservationists and environmentalists are wary of assertions that large-scale harvesting of "forest waste" and large-scale "selective" cutting of trees in mature forests will in fact be carried out in a truly sustainable pattern that respects the biodiversity and ecological complexity at risk in forest ecosystems. In many developing countries, there is no institutional capacity for monitoring or enforcing the principles of "sustainable forestry." In the U.S., the history of the occasionally close relationship between timber companies and some officials in the U.S. Forest Service has fed additional concerns about the impact of a large increase in the removal of wood from national forests as a source of biomass energy.

There are two leading families of technologies for the production of second generation biofuels. Whereas first generation ethanol is produced essentially through the ancient process of fermentation, which transforms sugars and starches in food crops into alcohol, second generation technologies are focused on breaking apart the more rigid molecular structures found in plants much higher in cellulose, hemicellulose, and lignin. The National Renewable Energy Laboratory notes that, "Plants have evolved over several hundred million years to be recalcitrant—resistant to attacks from the likes of bacteria, fungi, insects, and extreme weather . . . For cellulosic ethanol production, the primary challenge is breaking down (hydrolyzing) cellulose into its component sugars."

The competition among several different technology pathways has yet to produce a clear winner for the hydrolyzing process, although the use of precisely engineered enzymes may emerge next year as the first commercially available technology for the competitive production of second generation cellulosic ethanol.

The genetic modification of enzymes and microbes in ways that are useful in the conversion of biomass into energy has created excitement in recent years. Many biomass entrepreneurs are also enthusiastic about the dramatically increased yields they can obtain by inserting genetic traits into the plant varieties upon which they are relying as feedstock, though this application has aroused controversy. Some are also inserting traits that change the chemical nature of compounds made by the plant as it grows. These new techniques have generated opposition—especially in Europe— because of fears that the selection of traits by scientists working for businesses could result in genetic modifications that pose hidden risks for people and for ecological systems.

It is worth remembering that the food crops on which humanity now depends were themselves modified through patient trait-selection over many generations by our ancestors in the Stone Age. Yet gene splicing is, of course, a process that is not only much faster but also much more powerful. And both of these differences can increase the risk that an unintended side effect could become widespread before it is fully recognized and understood. Nevertheless, the choice has effectively been made by our civilization to move forward even further into genetic modification. In most of the world, genetically modified (GM) crops are already being grown in significant amounts. Indeed, in 2008, 8 percent of all cultivated land in the world was planted with GM crops. And most scientists have come to the conclusion that the risks from genetic modification are extremely low, while the economic benefits are very high.

However, there have been a few cases where genetic changes have caused unintended consequences. Since there are still some risks of events that are classified as "low probability but high impact," it is important for governments to establish a review process that focuses on the few modifications that may fall into this category.

The second leading candidate in second generation biofuel production relies on a thermochemical process to convert the cellulosic material into a syngas, which can then be transformed in one of several ways to make bioliquid fuel. At present, these techniques are more expensive, but much work is under way to bring down the cost sufficiently to make them competitive.

A much cheaper and much more abundant

HOW BIOMASS BECOMES BIOFUEL

First generation biofuels convert biomass with readily available starches—such as corn, palm, or sugarcane. The plant material's starches are converted into sugar in a process called mashing. Second generation biofuels are made by breaking down the cell structure of nonfood plants such as switchgrass to release their sugars. For both types of biofuel, the sugars go through a fermentation process that produces alcohol, which is then distilled into fuel-grade ethanol.

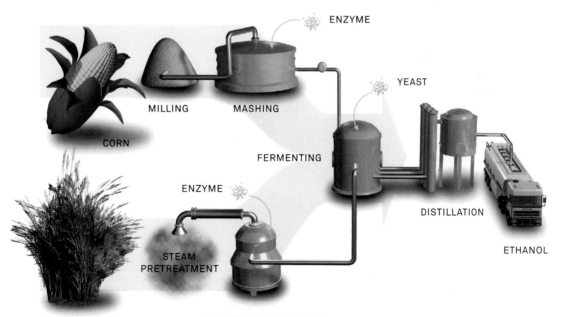

Landfill gas is the second largest source of man-made methane in the United States, partly because the U.S. produces so much landfill waste.

SOME LANDFILLS NOW CAPTURE A PORTION OF
THE METHANE ESCAPING FROM THEIR DECAYING
ORGANIC MATTER. THE GAS FROM THIS LANDFILL IN
NEW JERSEY HELPS POWER 25,000 LOCAL HOMES.

source of biomass energy in gaseous form lies buried in municipal landfills all over the world— sites that are currently responsible for approximately 14 percent of global methane emissions. Since methane (the primary component of natural gas) is a valuable resource, there are business opportunities in the capture and productive use of the enormous quantities produced in landfills.

Methane accounts for the second largest volume of global warming pollution after CO_2. Each molecule is more than 20 times as powerful as a CO_2 molecule in trapping heat. Landfills are the third largest man-made source of methane emissions globally (14 percent, compared with more than 50 percent from agriculture—mainly livestock and rice cultivation, and 16 percent from natural gas systems, mostly from leaks).

Landfill gas, which is typically half methane and half CO_2, is the second largest source of man-made methane in the United States, partly because the United States produces so much landfill waste; China, with four and a half times as many people, did not surpass the U.S. in household garbage until early 2009. The U.S. emits as much landfill gas as the entire continent of Africa and more than two and a half times the amount generated in the next largest national source, China. The amount of landfill gas collected and burned is increasing every year, but the amount of additional methane released from growing landfill volumes is increasing more rapidly, according to the U.S. Environmental Protection Agency (EPA).

The technology for capturing methane from landfills is well developed, mature, and cost-effective. In the most commonly used technique, a series of vertical extraction wells are drilled into the landfill or horizontal gas-collection trenches are used to pipe the gas to a central location, where it is collected and processed. Many landfills with

such collection systems simply flare the landfill gas, which at least reduces the global warming potential of the methane by emitting CO_2 instead. However, this approach is not optimal, and it wastes the opportunity to use the gas as a productive source of energy. It can be burned to produce electricity or sold to businesses for use in boilers as a source of space heating. Some businesses and municipalities also have fleets of vehicles modified to run on landfill gas.

Many developing countries dispose of municipal waste in open dumps, within which the waste decomposes aerobically to produce mainly CO_2, with much smaller quantities of methane. The flow of rainwater through these dumps also produces liquid leachate that often contaminates underground water aquifers. Waste disposed in landfills, by contrast, is usually layered periodically with dirt and covered in ways that cause anaerobic decay (in the absence of oxygen), which begins after one to two years to produce methane. The leachate from older, unlined landfills can be particularly toxic to underground water supplies.

In 1996, the United States enacted a Landfill Rule requiring all new municipal landfills to capture the methane they produce and either simply flare it or use it productively. Approximately 37 percent of U.S. landfills are now covered by these regulations, but roughly half of them flare the gas they collect. That leaves approximately 63 percent of landfills that are not required to collect their gas. And while about 20 percent of these landfills do so on their own, only one third of them put the gas they collect to productive use instead of flaring it.

Many companies have partnered with landfill operators to generate electricity and heat from landfill gas. For example, the BMW plant in Greer, South Carolina, meets 70 percent of its energy needs with landfill gas, which it uses to generate both electricity and heat. The company estimates that in six years of operation, it has saved $5 million in annual energy costs by using landfill gas.

The EPA has organized an international partnership, which now includes 30 countries, to focus on exploiting the opportunities to recover methane from landfills. This Methane to Markets partnership is designed to facilitate the development of projects to recover landfill gas.

Recently, entrepreneurs have devised an innovative new technique for capturing much more methane from municipal solid waste by placing it in large concrete buildings with hermetically sealed doors and efficient gas-collection systems that harvest the methane on an ongoing basis. The gas can be completely evacuated before the doors are unsealed to permit the addition of more waste. And the process is made more efficient still with the introduction of more anaerobic bacteria that accelerate the production of methane.

With all of the focus on producing liquid fuels for transportation from biomass, many have

A NEW INDUSTRIAL REVOLUTION

"For 21 years I never gave a thought to what we were doing to the earth in the making of our products," admitted Interface Flooring founder Ray Anderson. But in the 1990s, customers began asking what Interface was doing for the environment. "The real answer was not very much," he remembered.

Anderson was panicked when Paul Hawken's book *The Ecology of Commerce* fell on his desk. In reading it, his epiphany was simple: "The way I had been running Interface was the way of the plunderer. Someday people like me will end up in jail."

In the following 15 years, Anderson has pursued Mission Zero, a commitment to eliminate Interface's negative impact on the earth by 2020. For the carpet industry, with an estimated 5 billion pounds of its product ending in landfills each year, reducing waste lies at the heart of a sustainable future.

Interface began sourcing renewable materials, replacing oil-based polymers with corn-based ones. Then, through its product reclamation program, it began using post-consumer nylon and vinyl for new carpets. Over the past 23 years, more than 100 million pounds of materials have been diverted from landfills.

In 2003 Interface applied the same goals to its energy demand. The company launched a program with the EPA to convert methane gas from the landfill in LaGrange, Georgia, into a renewable-energy source.

Ray Anderson stands near machines that bind together waste fiber scraps at the Interface plant in West Point, Georgia.

The system captures and burns the landfill methane, converting it to heat and power.

Company-wide, Interface's net greenhouse gas emissions are down 83 percent in absolute tonnage. And the change has been good for business. Interface sales have increased by two thirds and profits have doubled. Anderson disagrees with those who claim we have to choose between the environment and the economy.

begun to ask whether it makes sense to spend so much money, time, and energy in scaling up the transformation of biosolids into bioliquids. After all, while more than 90 percent of the controversy over biomass involves bioliquids, 90 percent of the biomass energy actually produced in the world is in the form of heat and electricity. Researchers at the John F. Kennedy School of Government and the Stockholm Environment Institute concluded in 2007, "If greenhouse gas mitigation is a major objective, then a more effective strategy may be to prioritize the use of biomass to displace coal-based power over the use of biomass to displace transport fuels."

A team of researchers from Michigan State University, the University of Minnesota, and the Swedish University of Agricultural Sciences published a study last spring detailing the argument that the use of biomass for electricity generation can reduce greenhouse gas emissions far more effectively than through the expensive transformation of biomass into liquid fuels. They pointed out that more than 30 percent of the energy contained in cellulosic biomass is presently lost during the recovery of fermentable sugars. Then, another 27 percent of the energy contained in the sugars is lost during fermentation. Worst of all, 75 percent of the energy remaining in the liquid fuels produced is then lost when it is burned in inefficient internal-combustion engines.

Consequently, even though conventional electricity-generating plants lose, on average, 65 percent of the energy in the fuels to wasted heat in the process by which they burn fuels, the much higher efficiency of electric vehicles means that the use of biomass for electricity displaces twice as much petroleum with biomass than when it is converted into liquid fuels. In addition, if the electricity-generating plants burning cellulosic biomass are equipped with combined heat and power plants (cogeneration), more than 60 percent of the energy contained in the biomass can be efficiently used.

Moreover, we should anticipate and accelerate the progressive conversion of the world's automobile fleet to electricity, which will allow increased energy efficiency in the transportation sector, less dependence on unstable global oil markets, and the most efficient use of biomass—which is in the production of electricity and thermal energy.

It should be noted, however, that even if all cars were shifted to electricity over time, in many developing countries, the building of electrical smart grids capable of powering electric vehicles will take considerably more time than in developed nations. And even then, there would still be a need for lower-carbon liquid fuels to power heavy trucks and airplanes. Luckily, in addition to producing ethanol and biodiesel, biorefineries can also produce jet fuel; several airline companies are aggressively exploring the use of jet fuel made from biomass.

Meanwhile, scientists are hard at work on a third generation process. While much of the second generation cellulosic effort is aimed at producing ethanol from nonfood crops, the main focus of the third generation is end products that are superior to ethanol, including new molecules (like biobutanol) that can be mixed directly with gasoline and diesel, eliminating blending problems. The production of transportation fuel from algae is also frequently classified as a third generation technology. Several major oil companies have focused on the potential to create biofuel from algae. In 2007, Royal Dutch Shell announced that it will build a pilot facility in Hawaii and, a month later, Chevron announced a partnership to study algal fuels. In the summer of 2009, ExxonMobil announced a partnership with a company founded by genetics entrepreneur Craig

Venter. One argument in favor of algae is that it can be grown in arid environments using brackish water, salt water, or polluted water. However, in spite of the enthusiasm of algae entrepreneurs—and the large bets placed by energy companies—some experts are still skeptical that competitive fuels from algae will be available soon.

Nevertheless, whatever the technology and whatever the feedstock, the laws of physics may inherently limit the relative energy efficiency of making liquid fuels from biomass—compared with using advanced combustion technologies to harvest much larger percentages of the energy contained in biomass in the form of thermal energy and electricity produced by steam-driven and gas-fired generators.

The use of biomass in combined heat and power (CHP) generators is, at present, a cheaper source of renewable energy than solar electricity, and only slightly more expensive than onshore wind electricity. Indeed, many energy experts have come to believe that, in most cases, the most efficient use of biomass is as fuel for direct combustion to create thermal energy for the space heating and cooling of buildings and for the making of steam to power electricity generators. Moreover, biomass generation can serve as baseload power, while wind and solar cannot, so it is an effective way to reduce the amounts of carbon-intensive coal and natural gas that are burned to produce electricity.

The Natural Resources Defense Council concluded in a 2004 study: "Given the current mix of fuels used to generate electricity in the United States, using a ton of biomass to generate electricity provides a moderately larger reduction in greenhouse gases than any of our fuel-producing options. This situation will change over time, though, especially if we make a concerted effort to reduce overall greenhouse gas emissions."

Biomass used for the production of electricity and thermal energy comes mainly from wood—principally forest waste. Slightly more than half of all biomass energy consumed is derived from wood, and 65 percent of wood consumption is in the industrial sector, where it is used on-site to produce heat for industrial processes, as well as smaller amounts of electricity; residential and commercial use accounts for 25 percent, and the electric power sector consumes only 9 percent. In 2006, electricity generation from biomass (from electric utilities and from industry cogeneration) accounted for approximately 7 percent of global production of renewable electricity.

Biomass CHP generators can also support small-scale, distributed applications—thus avoiding the worst of the logistical nightmares confronting the collection of large supplies of biomass from a wide geographic area in volumes sufficient to make large biorefineries profitable. Since wood has a much lower energy density than coal, the logistics of supplying adequate quantities of feedstock for very large facilities in a sustainable pattern is often a serious challenge. After all, the high energy density of coal and oil, and their widespread availability from underground reserves, are the principal reasons why coal and oil became our dominant fuels in the first place.

Wood for factories and utilities is frequently processed into pellets that are easy to transport, store, and "co-fire" along with coal in existing boilers—displacing as much as 20 percent of the coal and reducing CO_2 emissions in the process. According to a study by the Worldwatch Institute in 2009, "Co-firing holds the most potential out of all renewables for reducing a significant amount of emissions in the near term."

Modern dedicated generators optimized for wood pellets and equipped with state-of-the-art

ALGAE GROW IN PLASTIC BAGS AT A TEST
PROJECT NEAR PHOENIX, ARIZONA. IN THE
RIGHT CONDITIONS, AN ACRE OF ALGAE COULD
PRODUCE 5,000 GALLONS OF BIOFUEL A YEAR.

emissions controls can produce electricity far more efficiently than coal-fired steam generators and can reduce CO_2-equivalent emissions by as much as 94 percent. For example, the U.K. has announced plans to build a dedicated 295-megawatt advanced biomass generator, and other larger models are now being planned. Significantly, the U.K. government has announced its intention to require that these dedicated biomass plants use carbon capture and sequestration starting in 2030.

The biomass industry has also developed integrated biorefineries that use a variety of feedstocks and several different conversion technologies in order to produce biofuels, biomaterials, thermal energy, and electrical power. Although the concept makes sense and can allow the biorefineries to become close to self-sustaining in their energy consumption, most integrated biorefineries built thus far have focused primarily on the production of ethanol or biodiesel.

At present, the largest market for biomass production of electricity and heat from wood pellets is in Europe, with the pellets coming mainly from suppliers in the northwest forests of North America through the Panama Canal. The second largest flow of wood pellets to Europe is from Australia, with growing supplies planned from the southeastern United States.

In the European Union, the burning of biomass—principally wood and wood wastes—accounts for two thirds of the renewable energy being produced, although solar, wind, and geothermal electricity are poised to grow dramatically. According to the consulting group New Energy Finance, there are presently 3.2 gigawatts of biomass power generation announced, permitted, financed, or commissioned in Europe, the Middle East, and Africa. Many other biomass generators are under construction or have been announced.

U.S. markets for thermal and electric energy from wood pellets are expected to grow rapidly after the adoption of the proposed Renewable Electricity Standard—which will lead to the development of regional biomass-trading relationships, the emergence of new biomass logistics providers, and new innovations in the supply chain. For example, a process called *torrefaction,* which has traditionally been used in the coffee industry, is now being introduced into the biomass industry to heat and dry the wood pellets in ways that allow them to be stored outside without absorbing rainwater. There is also growing interest in the agricultural community in a process by which biomass can be burned in the absence of oxygen (a process known as *pyrolysis*) in order to create biochar, which, as discussed in Chapter 10, is an extremely effective way to regenerate soil fertility while sequestering large amounts of carbon in agricultural soils.

In 2009, a remarkable analysis published in *Science* magazine by 11 experts on energy policy and biofuels—including both advocates and skeptics of biofuel potential—proposed a methodology for resolving controversies over alternative biofuels by applying "two simple principles": "In a world seeking solutions to its energy, environmental, and food challenges, society cannot afford to miss out on the global greenhouse gas emissions reductions and the local, environmental, and societal benefits when biofuels are done right. However, society also cannot accept the undesirable impacts of biofuels done wrong."

These scholars also concluded that "the recent biofuels policy dialogue in the United States is troubling. It has become increasingly polarized, and political influence seems to be trumping science."

They called for the adoption of "meaningful science-based environmental safeguards," government support for a "robust biofuels industry," and the assurance of a "viable path forward" for investors in first generation biofuels.

One positive sign for the development of a sensible and coherent public policy on biomass came in 2007, with the formation in the United States of the Council on Sustainable Biomass Production. Made up of farmers, producers, refiners, oil companies, biotech companies, federal officials, and academic researchers, this council has earned widespread respect in a short period of time and is developing a comprehensive, voluntary certification program and education and training programs—all based on standards designed to address "the full complement of sustainability issues through principles, criteria, and indicators applicable to both agriculture and silviculture." They plan to begin implementing their standards-based program in the spring of 2010, "well in advance of full-scale production of cellulosic bioenergy."

Among governments, the European Union has taken the lead in establishing sustainability standards for biomass. The U.K., for example, now requires producers to apply for Renewable Obligation Certificates that, in the case of generating electricity from biomass, include provisions for the careful review and oversight of the nature, production methods, and origin of all biomass feedstocks they plan to use.

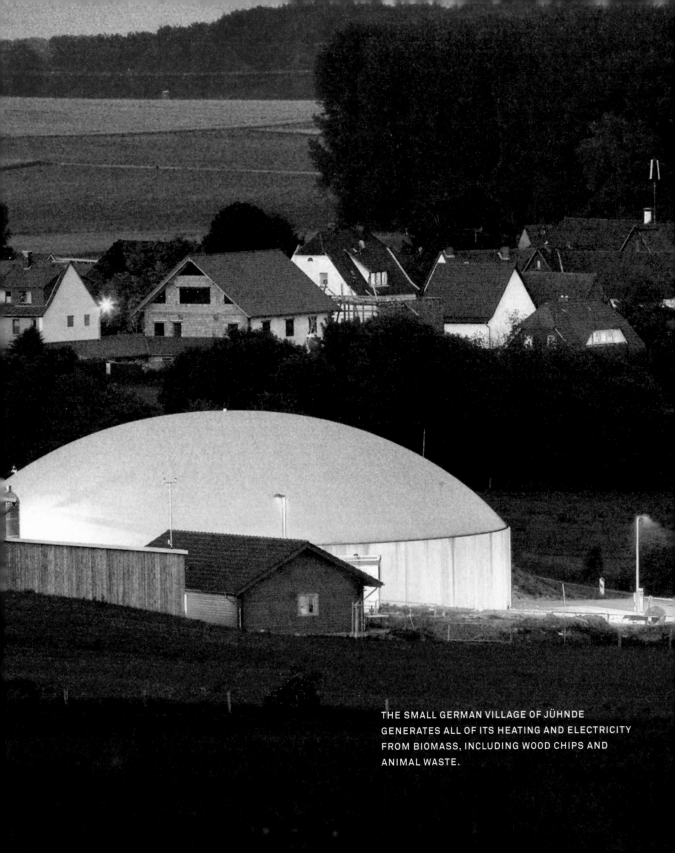

THE SMALL GERMAN VILLAGE OF JÜHNDE
GENERATES ALL OF ITS HEATING AND ELECTRICITY
FROM BIOMASS, INCLUDING WOOD CHIPS AND
ANIMAL WASTE.

OUR SOURCES OF ENERGY

CHAPTER SEVEN

CARBON CAPTURE AND SEQUES- TRATION

THE IN SALAH CCS PROJECT AT THE KRECHBA GAS FIELD IN ALGERIA INJECTS ABOUT 1 MILLION TONS OF CO_2 UNDERGROUND EACH YEAR.

The idea of "carbon capture and sequestration" (CCS) is compelling. In theory, the world could capture all of the CO_2 that is presently emitted into the atmosphere by fossil fuel electricity plants and sequester it safely in repositories located deep underground and beneath the bottom of the ocean. We could then continue to use coal as a primary source of electricity without contributing to the destruction of human civilization in the process.

The reality, however, is that decades after CCS was first proposed, no government or company in the world has built a single commercial-scale demonstration project capturing and sequestering large amounts of CO_2 from a power plant.

All the technologies for capturing, compressing, transporting, and sequestering CO_2 have been developed and tested on a small scale. All of them work. But the components have never been integrated and implemented on a large-enough scale to build the degree of confidence necessary for the truly massive commitment the world would have to make were this option to be chosen as one of civilization's main strategies for solving the climate crisis.

Why?

T.S. Eliot once wrote, "Between the idea and the reality, between the motion and the act, falls the shadow."

The shadow of implausibility that falls between the idea of CCS and the failure to make it a reality is cast by two huge obstacles looming on the horizon. First, the exorbitant energy penalty for capturing CO_2 would require the coal industry to increase the amount of coal it now burns by 25 to 35 percent in order to produce the same amount of electricity it presently generates. If companies did not build new coal plants, then the industry would produce 25 to 35 percent less electricity while burning the same amount of coal. And second, extraordinarily

complex and time-consuming questions about the specific location of appropriate, large-scale, underground repositories—and how much CO_2 could be safely stored at each one—have yet to be answered.

The sheer volume of the CO_2 currently being emitted from coal-fired and gas-fired generators is itself the underlying reason that so many industry experts regard CCS as an intriguing idea that still strains credulity. If all of the CO_2 now being vented into the atmosphere by U.S. coal electricity plants were captured and converted to its liquid form, the volume would be equivalent to 30 million barrels of oil per day—three times the volume of all the oil imported by the United States each day. If the CO_2 were then transported by pipeline to repositories, as proposed, the amount (by volume) would be one third that of all natural gas now being transported in pipelines throughout the U.S.

There is no shortcut to ensuring truly safe storage of huge amounts of CO_2—or for determining how much CO_2 might be stored safely in underground sites—even though the basic science gives cause for optimism on both counts. Underground geological storage may eventually be workable and safe, and geologists already know many of the general areas in which appropriate potential repositories almost certainly can be found.

Still, even as work begins to locate and characterize potential repositories, many energy experts remain skeptical about the practicality of burning a third more coal to produce the same amount of electricity. The Union of Concerned Scientists put it this way: "It is like having to build one new coal plant just to power the carbon capture process for every three to four conventional plants." The Massachusetts Institute of Technology in its recent study "The Future of Coal" concluded that "If carbon capture and sequestration is successfully

also be significant. Coal mining is very destructive to the environment. For example, the obscene practice called "mountaintop mining"—and the dumping of toxic mining waste into streams in the valleys below—is an ongoing environmental atrocity. Not only are the mountain peaks transformed forever into ugly moonscape plateaus but the waste includes arsenic, lead, cadmium, and other dangerous forms of heavy-metal pollution that leach into drinking-water supplies.

Significant progress has been made over the past 20 years in forcing the reduction of sulfur oxides, nitrogen oxides, and particulates from the

> # "It is like having to build one new coal plant just to power the carbon capture process for every three to four conventional plants."
>
> UNION OF CONCERNED SCIENTISTS

adopted, utilization of coal likely will expand even with stabilization of CO_2 emissions." However, Howard Herzog of M.I.T. argues that because CCS would be mandated as part of a climate policy and would raise the price of coal-based electricity, the overall result would be a decrease in the number of coal-burning plants.

The additional CO_2 associated with any increase in the mining and transportation of more coal would not be captured and sequestered. Nor would the CO_2 resulting from the transportation, injection, and sequestration activities.

The associated environmental costs of any expansion of the mining and burning of coal would

burning of coal. Reductions in sulfur dioxide mandated by law have lessened the severity of acid rain, though acid rain continues to be a problem. It would be worsened by a significant increase in the burning of coal, which remains the second largest source of nitrogen oxide (NO_x), one of the components of smog and a contributor to acid rain.

Recent regulations to limit mercury emissions from coal plants are widely considered to be far too weak. Some state governments—with Pennsylvania leading the way—have tightened mercury emissions, but in most of the U.S. and throughout the world, the burning of coal remains the largest man-made source of mercury pollution.

The 130 million tons of coal ash and sludge produced by U.S. coal plants each year is already one of the largest industrial waste streams in the country. In 2008, three days before Christmas, one billion gallons of this toxic sludge burst out of its confinement reservoir and destroyed a neighborhood of homes in Harriman, Tennessee.

Nevertheless, many feel the stakes are so high that no option that might conceivably help solve the climate crisis should be discarded. The expense and risk of CCS, after all, would be far less than what scientists are warning will happen if we continue dumping all of that CO_2 straight into the atmosphere. It's imperative that we quickly find alternatives that stop the ongoing destruction of the habitability of the earth for human civilization. In addition, CCS would, in theory, allow the world to avoid incurring large stranded costs from the retirement of at least some of the existing fossil fuel plants before the end of their original lifetimes.

What should be discarded, however, is any illusion that CCS will be available anytime soon at a scale large enough to make a dent in our CO_2 emissions. We are many years away from understanding the answers to questions that must be resolved before CCS could become one of the viable solutions for global warming.

This last point is crucial, because some coal companies and coal-burning utilities have aggressively promoted the illusion that CCS is actually near at hand. They have a powerful incentive to create this impression, because if the public and policy makers believe that CCS is likely to be available in the near term, they might be persuaded to allow utilities to continue building coal-fired generating plants and simply purchase an empty field adjacent to them outfitted with reassuring billboards reading "Future Site of CCS."

Unfortunately, while the coal industry has a powerful incentive to promote the illusion that CCS is almost ready, it has no incentive whatsoever to actually invest significant sums of money to make CCS a reality—unless and until the nations of the world place a high price on CO_2 (a price avoided only if the CO_2 is captured and sequestered safely). Most of the coal companies, of course, oppose the placing of any price on CO_2 because it will allow other electric generation technologies like natural gas, nuclear, wind, and solar to capture their market share. Thus, they continue to promote the illusion that CCS will soon be available so they can continue to sell coal to dirty old power plants.

Many utilities are now using CCS as an excuse for inaction. Some are arguing that they should be allowed to go forward with construction of new coal-fired generating plants that are—in their phrase—"capture ready." The utilities imply that they can build these plants in ways that prepare them in advance to be retrofitted with CCS technology as soon as it is available for commercial use.

However, the M.I.T. study stated that the idea of a "capture ready" coal plant is "as yet unproven and unlikely to be fruitful." These experts added: "Pre-investment in 'capture ready' features for . . . coal combustion plants designed to operate initially without CCS is unlikely to be economically attractive."

An estimated 75 percent of the cost involved in CCS is in the energy required to capture CO_2 from the exhaust emissions of generating plants. This presents an enormous practical challenge. The low pressure of the gas coming out of the plant's exhaust—and the small percentage of that gaseous mixture made up of CO_2—means that a very high volume of gas must be treated in order to remove most of the CO_2.

Ninety-nine percent of all coal-fired power plants in the United States burn pulverized coal

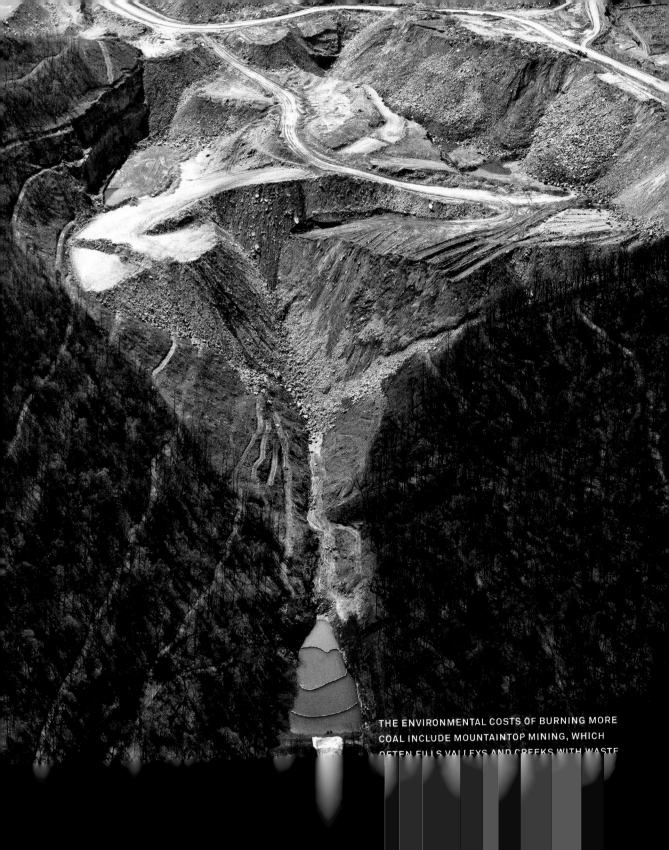

THE ENVIRONMENTAL COSTS OF BURNING MORE
COAL INCLUDE MOUNTAINTOP MINING, WHICH
OFTEN FILLS VALLEYS AND CREEKS WITH WASTE.

(mixed with air) and emit enormous volumes of flue gas, composed of 10 to 15 percent CO_2. The vast majority of the existing plants are based on old, inefficient technology that effectively utilizes, on average, only 32 percent of the heat energy contained in the coal. Large amounts of money are presently required on a regular basis simply to keep these plants operating.

These older, thermally inefficient, coal-fired generating plants would suffer such a large energy penalty that most experts are skeptical that it will be practical to retrofit them with CCS. However, others argue that it will be difficult for the world to meet the necessary CO_2 reduction

substantial retrofitting of boilers, turbines, gas cleanup systems, and other significant components. These expenses would be further increased, because many of the same factors that have produced unsustainable hikes in construction costs for nuclear power plants have also been driving up construction costs for coal-fired generating plants.

After CO_2 is captured, it is compressed to a "supercritical" state, which is neither a gas nor a liquid but has some properties of both. It is then ready to be shipped by pipeline to an appropriate sequestration site. (CO_2 can be transported either as a liquid or in its liquidlike supercritical form.)

Significant amounts of energy are required to

The idea of a "capture ready" coal plant is "as yet unproven and unlikely to be fruitful."

MASSACHUSETTS INSTITUTE OF TECHNOLOGY

goals without dealing somehow with CO_2 emissions from these inefficient plants—particularly the many inefficient *new* plants being opened in China. This accounts for the high interest in a joint U.S.-China program to quickly explore this option.

The newer designs for pulverized-coal plants—supercritical and ultra-supercritical technologies—are capable of utilizing up to 40 percent of the energy in the coal. Another design—fluidized bed combustion—allows the mixing of lower grades of coal or biomass and typically produces less sulfur dioxide and nitrogen oxide. However, even though these new designs are somewhat more efficient, they do not make the task of CCS any easier.

Older and newer plants alike would require

pressurize the CO_2, but overall, the cost for transporting CO_2 by pipeline over reasonable distances is not considered prohibitively expensive. In addition, economic studies indicate that the cost of transporting CO_2 via pipelines can be reduced significantly thanks to economies of scale at volumes above 10 million tons per year. If the CCS option is adopted on a broad scale, the development of pipeline networks would reduce the need for more expensive, dedicated pipelines between each source and each sink.

There are now more than 3,900 miles of CO_2 pipelines in the United States being used for enhanced oil recovery. Thus far, there have been no safety concerns. Although the sudden release of a

HOW CARBON DIOXIDE IS SEQUESTERED

When coal is gasified, the CO_2 can be separated from the resulting gas before it is burned. Post-combustion CCS technology separates CO_2 from a power plant's other flue gases—water vapor, sulfur oxides, and nitrogen oxides. With both techniques, a compressor pushes the captured CO_2 gas through injection pipelines thousands of feet deep into the earth. The CO_2, now pressurized into a denser, liquidlike "supercritical" state, is stored, or sequestered, in rock formations, where it is trapped in the pores of the rocks. The high pressure and temperatures at this depth keep the gas in its supercritical state.

COAL PLANT EQUIPPED
WITH CARBON CAPTURE

COMPRESSOR

CARBON INJECTION
PIPELINES

STORED CARBON

STORED CO_2—TRAPPED
AS A SUPERCRITICAL FLUID
BY ROCK FORMATIONS

SALINE AQUIFER

LIQUID CO$_2$ IS ALREADY BEING TRANSPORTED
BY PIPELINE AND USED FOR ENHANCED OIL
RECOVERY, SEEN HERE IN TEXAS. THE CO$_2$ IS
PUMPED INTO THE GROUND, WHERE IT PUSHES
OIL TOWARD THE SURFACE.

large amount of CO_2 in populated areas would be dangerous if the CO_2 concentrations were higher than 7 to 10 percent in the air, there have been no significant problems with existing CO_2 pipelines. Most engineers believe this risk to be very low. The much longer experience with natural gas pipelines has also produced a high level of public confidence that pipeline safety is a manageable risk.

More than half of all coal-fired power plants operate in areas where geologists have identified nearby underground regions likely to have areas appropriate for CO_2 sequestration. However, scientists cannot be certain of this without extensive work to characterize potential sites. It is also important to note that quite a few coal-fired generating plants are located far from areas considered likely candidates for underground sequestration.

The final stage of the CCS process is the sequestration of the captured CO_2 in a safe location from which it cannot escape back into the atmosphere. Scientists and entrepreneurs are hard at work in pursuit of innovative technological breakthroughs that will capture, stabilize, and "embody" CO_2 in new building and paving materials. These exciting potential breakthroughs could lock significant amounts of CO_2 into the structure of the materials themselves. But at present, virtually all the focus is on geological storage sites deep in the earth.

The most likely candidates for geological sequestration are saline aquifers in rocks sufficiently permeable to absorb the CO_2 and sufficiently stable and isolated to ensure that the CO_2 remains in place indefinitely. The pores of the rocks inside these saline formations trap the CO_2 with what scientists refer to as "capillary forces."

Over long periods of time, the CO_2 dissolves into the liquids in the saline formation and the minerals in the rocks. Even though CO_2 is buoyant and naturally migrates toward the surface if not held in place, when it is injected into the right geochemical environment at a depth of one kilometer or more, the deep-earth pressure and heat will preserve the CO_2 in a supercritical, quasi-liquid form that scientists feel confident will remain in place. Moreover, sites such as these are typically covered by impermeable layers of shale or by minerals and salts left when ancient bodies of water evaporated in an earlier geological age.

The limited experience gained thus far with these saline formations is encouraging. Two major studies—one by the IPCC and another by M.I.T.— have both concluded that once the CO_2 is appropriately sequestered in these formations, virtually all of it is likely to remain there. According to these studies, the highest risk of leakage comes during the injection process, when it is first stored.

Even though there are remaining uncertainties about how these geochemical processes will work in particular locations, independent geologists express a very high level of confidence that the CO_2, once successfully sequestered, will not only remain safely underground but will also become progressively safer as time elapses. Enough is already known about the basic geological and chemical forces at work to produce a very high level of confidence in the safety of this technique.

However, the geological, geochemical, geophysical, and geographical natures of the potential sites differ significantly from one location to another. Moreover, it is difficult and time-consuming to estimate the "pore volume" in particular geological formations that, while generally appropriate for CO_2 sequestration, may present complications that limit the amount of CO_2 they can safely contain.

The earth's crust is described by geologists as a complex, heterogeneous, nonlinear system. In other words, there is such a variety of intermingled geological formations deep underground, it is inherently difficult to map the precise boundaries of reservoirs that initially look promising in order to ensure the CO_2 is not injected into an area with a "fast path" escape route into an adjacent geological area from which it could migrate to the surface.

For example, geologists would have to identify and adequately plug and cap abandoned wells that may have been drilled into parts of these reservoirs and long since forgotten. Abandoned wells, if not identified and plugged, could, under some circumstances, serve as chimneys through which the sequestered CO_2 could find its way up. Geologists must also locate any freshwater aquifers that might be contaminated by large volumes of CO_2.

New fine-grain seismic techniques capable of time-lapse analysis have shown promise in monitoring the behavior of the injected CO_2. They may be usable as a means of detecting any rapid underground migration of the injected CO_2 into areas where it would no longer be safely contained.

Several teams are hard at work improving their understanding of these risks, but this takes time. Moreover, what is learned in the study of one potential reservoir may have only limited relevance to an understanding of the next. Difficult issues related to liability insurance, ownership of tidal to subsurface regions, appropriate ways to monitor safety and regulate safe practices, and others must all be addressed as well.

In spite of these uncertainties, most experts agree it is very likely possible to safely store large amounts of CO_2 in saline aquifers. These same experts, however, stress the importance of conducting studies and large-scale demonstration

THE LAKE NYOS TRAGEDY

Experts agree that the tragedy at Cameroon's Lake Nyos in 1986 is not relevant to the risks associated with carbon capture and sequestration. It does, however, illustrate one reason for public concern about the siting of CCS repositories.

The sudden release of large amounts of CO_2 from deep beneath the bottom of the lake in northwest Cameroon killed more than 1,700 people and 3,000 cattle. The original source of the CO_2 was melted magma 50 miles beneath the bottom of the lake. The gas made its way up through vents in the rocks underneath the lake and saturated the water at its bottom depths. The natural churning of freshwater, from the top down to the bottom, caused a sudden, explosive geyser of CO_2 to spout powerfully to the surface. Because it is heavier than air, the CO_2 spilled over the shores of the lake and down the hillsides and valleys, asphyxiating all who were in the path of the deadly plume. This rare, naturally occurring event is known to have happened before at Lake Nyos, at another lake in Cameroon, and at a

Scientists and workers launch a raft with CO_2-monitoring equipment on Lake Nyos, Cameroon.

lake in neighboring Congo. All three lakes have now been equipped with relatively inexpensive monitoring systems designed to alert people before the next dangerous buildup of CO_2.

projects for many years in order to assure that their tentative conclusion is correct.

And, as is true with any large energy-related project, public opposition is an important factor in the choice of sites. One proposed CO_2 sequestration site in The Netherlands has been blocked—at least temporarily—by strenuous opposition from people who live nearby. Royal Dutch Shell has proposed to locate the site two to three kilometers below the surface in a location near Barendrecht and operate it in a joint venture with ExxonMobil.

The Barendrecht town council has voted to oppose the site, arguing that it is beneath one of the most densely populated areas of the Netherlands. Some environmental groups are concerned about the way in which the safety of sequestration sites will be determined and the effectiveness with which they will be monitored. The Dutch government, however, strongly supports the site. The decision on whether or not to go forward will be heavily influenced by the results of careful study by an independent commission.

Some scientists suggest that deep coal seams that can't be mined—seams that include organic minerals containing brine and other gases— should also be explored as potential storage sites. However, much less is known about these sites than about the saline formations that have received the most attention.

So far, sequestration has been achieved only for small volumes placed in locations known to have ideal conditions. The effort to store vastly larger volumes of CO_2 may lead us to discover weaknesses in the case for geological sequestration at particular sites. For example, much higher volumes of CO_2 could potentially put enough pressure on otherwise stable formations to produce cracks through which the CO_2 could migrate unpredictably.

Seismic risks are also a factor in determining site safety. Such risks include not only naturally occurring earthquakes but also seismic events induced by the injection of very large volumes of CO_2 into some geological formations. This is particularly true, according to the M.I.T. expert study, for "rapid injection of large volumes into moderate-low permeability rock" because "large volume, high rate injections…have a higher chance of exceeding important process thresholds." Most examples of such induced earthquakes have thus far been small, although in the 1960s several somewhat large quakes were induced in Denver—the largest hitting 5.3 on the Richter scale.

Currently, CO_2 in relatively small amounts is sold to oil drillers, who use it for enhanced oil recovery (EOR). By injecting the CO_2 into the bottom of mature oil wells, they use the pressure of the gas to force the remaining oil upward, thereby making it recoverable. In the United States, the technique is used in oil fields in west Texas, southern Louisiana, southwest Oklahoma, and the border region overlapping Utah, Colorado, and Wyoming. In addition, CO_2 from a coal-to-synthetic-gas plant in North Dakota is used for enhanced oil recovery in Saskatchewan, Canada.

Experts on CCS are virtually unanimous in concluding that the use of CO_2 for enhanced oil recovery offers little that is relevant to the long-term, safe storage of vast quantities of CO_2 from electric generating plants. The underground geology of oil wells is almost always fractured and disrupted by the drilling process, the amounts of CO_2 used are minuscule compared with what would have to be stored in large underground secure repositories in order to make CCS possible, and—with one partial exception—the kinds of measurements and studies necessary to gain valuable information about the practicability of CCS at EOR sites have not been conducted.

There are three larger-volume CCS demonstration projects involving the production of natural gas now under way. Two of the three have chosen saline formations for their repositories. Norway is sequestering CO_2 under the ocean bottom in the North Sea between Norway and Scotland at the Sleipner gas field. Statoil, the Norwegian company responsible for the Sleipner project, has also joined with British Petroleum and Sonatrach (of Algeria) to demonstrate the sequestering of CO_2 in natural gas reservoirs at In Salah in Algeria. In addition, a CCS project has been launched at Weyburn in Saskatchewan in connection with coal gasification. It is using CO_2 for the long-established purpose of enhanced oil recovery. Though a commercial venture, it has been designated as an international project to study CCS, and some instruments have been used to monitor potential leakage.

Thus far, none of these three CCS demonstration projects has found any leakage of CO_2. However, the M.I.T. report noted that the projects it studied—Sleipner, In Salah, and Weyburn—"do not address all relevant questions…. Many parameters which would need to be measured to circumscribe the most compelling scientific questions have not yet been collected, including distribution of CO_2 saturation, stress changes, and well-bore leakage detection…. Important nonlinear responses that may depend on a certain pressure, pH, or volume displacement are not reached."

Luckily, more CCS projects are coming soon that may answer some of these questions. Statoil has recently opened a second CCS operation called the Snøhvit Project, located underneath the bottom of the Barents Sea north of the Arctic Circle. The largest incentive motivating Statoil is the CO_2 tax it would otherwise have to pay on each ton of CO_2 that is safely sequestered. A Swedish government-owned company, Vattenfall, which operates a number of coal-fired power plants in Europe, has recently announced the beginning of development work on the first full-scale CCS plant in Europe, to be located in Denmark. If it is successful, Vattenfall plans to build a second plant at the Jänschwalde power plant in Germany.

A large-scale CO_2 sequestration project associated with ethanol production is also nearing completion in Illinois. The plan calls for up to a million metric tons of CO_2 to be injected into a saline formation more than a mile underground. Several other projects at the one-million-ton-per-year scale have been proposed in the United Kingdom, Australia, Germany, Norway, Canada, China, and the U.S. CO_2 capture projects are also under way in Brazil, India, Malaysia, and Germany, and geological storage projects are under way in Australia.

One innovative CCS plant, now in the planning stages, is proposed for Linden, New Jersey, near New York City. It combines electricity generation from the burning of coal with the production of fertilizer. Because the New York City market has very high electricity rates, the plant expects to be profitable by selling electricity during peak-pricing periods, and then—with the flip of a switch—producing fertilizer until the next peak-pricing period. Essentially all of the CO_2 would be captured and transported by pipeline 70 miles offshore, where it would be injected one to two miles beneath the bottom of the Atlantic Ocean into a sequestration site that has all the appropriate geological, geophysical, and geochemical characteristics.

The only large CCS demonstration project planned for the United States—called FutureGen—was announced in 2003. It was canceled in early 2008 because of large cost overruns and what many regarded as muddled objectives. A frequent criticism from wags was "too much future, too little gen." It is currently being restarted after yet more

AT VATTENFALL'S PILOT CCS PROJECT IN GERMANY,
CO_2 IS COMPRESSED INTO ITS LIQUIDLIKE
SUPERCRITICAL STATE IN PREPARATION FOR
UNDERGROUND SEQUESTRATION.

money was appropriated to it by Congress.

Part of the argument for spending enormous sums of taxpayers' money to facilitate continued reliance on coal as the primary source of electricity in the U.S. is based on the widespread assumption that the U.S. has a sufficient supply of coal to last for 250 years, and that the supplies available in several countries elsewhere in the world are similarly large. However, the National Research Council, echoing the assessments of other energy experts, concluded in 2007, "It is not possible to confirm the often-quoted suggestion that there is a sufficient supply of coal for 250 years."

China, which burns twice as much coal as the United States, is already experiencing episodic coal shortages and is importing larger amounts from Australia and elsewhere. India is also experiencing

shortages, although shortages in both China and India are principally problems in the supply chain.

Although these supply constraints on coal will be more significant in the future, they should already be seen as a factor in our decision-making regarding future electricity-generating capacity. Limited supplies will drive up the price, even as the prices for photovoltaic electricity and other renewable sources continue to come down.

Because of the scale of CCS operations necessary to handle the enormous volumes of CO_2 now being emitted from power plants, and the expected difficulties in integrating all phases of the operation while assuring the safety of sequestration sites over periods of time far longer than commercial ventures can typically handle, many—including the expert panel at M.I.T.—have recommended the establishment of a new federal government entity to oversee and monitor all aspects of CCS.

Legislation pending in the U.S. Congress in 2009 provides $10 billion for the study, demonstration, and early deployment of CCS, this on top of an additional $6 billion in subsidies enacted over the previous four years.

Many environmentalists have expressed support of robust and vigorous research and development to determine whether or not CCS can become a practical option as part of civilization's arsenal to combat the climate crisis. What is needed now are large-scale demonstration projects in several different kinds of underground geologies in order to determine how realistic this idea might be for the enormous volumes of CO_2 that would have to be captured and safely stored over the long term.

Most experts who have studied the CCS option have concluded that it is probably impracticable for many years to come, because the technology for capturing CO_2 would either require a dramatic increase in the overall use of coal and gas for the same amount of electricity, or sharply reduce the amount of electricity obtained from burning the same amount of fuel as at present—and because every one of the potential geological repositories presents a unique and extremely difficult challenge in characterizing its geology deep underground and estimating both storage capacity and the safety of storing CO_2 there.

There is actually a fairly simple solution to resolving all the questions and uncertainties about whether CCS is economically plausible and, if so, which techniques are the best ones to use: put a high price on carbon. When the reality of the need to sharply reduce CO_2 emissions is integrated into all market calculations—including the decisions by utilities and their investors—market forces will drive us quickly toward the answers we need.

THE SLEIPNER GAS FIELD IN THE NORTH SEA, NEAR NORWAY, WAS THE FIRST COMMERCIAL CCS PROJECT IN THE WORLD.

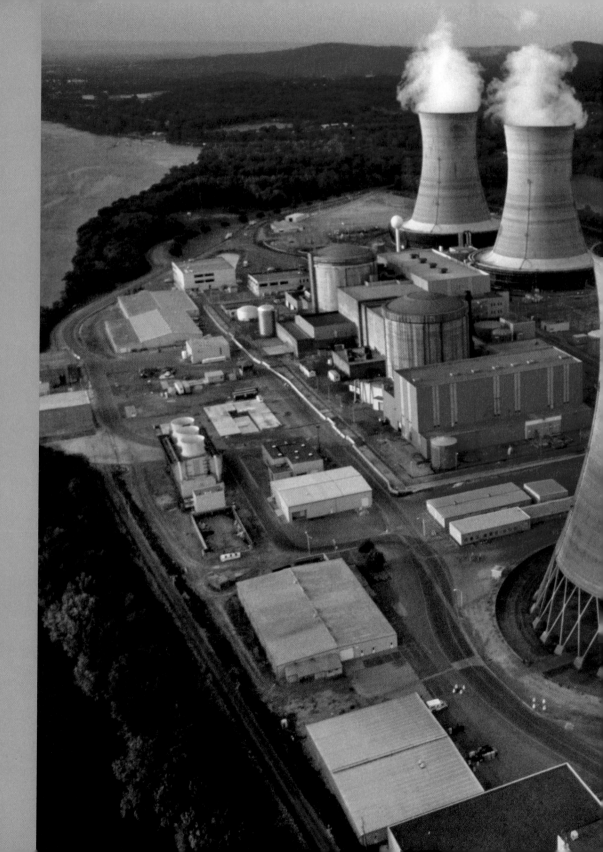

CHAPTER EIGHT

THE NUCLEAR OPTION

THE THREE MILE ISLAND NUCLEAR POWER PLANT IN PENNSYLVANIA. ONE REACTOR IS STILL ACTIVE; THE OTHER SUFFERED A PARTIAL MELTDOWN IN 1979.

In the world's debate over how to produce electricity without generating massive quantities of greenhouse gas pollution, there is a radioactive white elephant in the middle of the room: nuclear power.

The idea of using controlled nuclear fission as a source of heat to turn electric turbines generated enormous enthusiasm during the first quarter-century after World War II. The Atomic Energy Commission predicted at the end of the 1960s that the United States would have more than 1,000 nuclear power plants operating by the year 2000. But only a tenth of that number were ever built. Nuclear power, once expected to provide virtually unlimited supplies of low-cost electricity, has been an energy source in crisis for the last 30 years.

A massive study of the future of nuclear power by the Massachusetts Institute of Technology in 2003 (updated in 2009) concluded that: "Nuclear power could be one option for reducing carbon emissions. At present, however, this is unlikely: nuclear power faces stagnation and decline."

The arguments in favor of nuclear electricity, once very convincing, are still seductive. One pound of uranium contains as much energy as three million pounds of coal. Safety in the operation of nuclear plants has improved, and public acceptance of nuclear power has increased somewhat. The expected lifetime of most older plants has been extended in the licensing process from 40 years to an estimated 60 years. The growing prospect of a price on CO_2 will increase the competitiveness of electricity produced from reactors relative to that produced from fossil fuels. Nuclear power plants offer the promise of modestly reduced dependence on foreign energy sources because of increasingly hopeful prospects that a major portion of the U.S. automobile fleet will shift from oil-based fuel to electricity. The average "capacity factor" of the U.S. fleet of existing nuclear reactors increased from 56 percent in the 1980s to 90 percent during the last seven years. As a result, the amount of electricity produced by the existing fleet of U.S. reactors has increased steadily for the last decade.

Nevertheless, the industry remains moribund in the United States, and its worldwide growth has slowed dramatically, with no new units but instead an actual decline in global capacity and output in 2008. Private investments in new nuclear power plants ordered after 1972 came to a screeching halt in the 1970s, and most of the reactors then on the drawing boards were canceled or delayed indefinitely. In the United States, no nuclear power plants ordered after 1972 have been built to completion.

Nuclear power plants generate heat through a controlled fission chain reaction. Uranium is the heaviest natural element in the earth, and as a result, the strong force that holds together the nucleus of all atoms is attenuated and weaker because its nucleus holds 92 protons (compared with only one in the nucleus of a hydrogen atom). This allows a uranium atom to

WORKERS AT A NUCLEAR PLANT IN SOUTH
CAROLINA SEAL A CONTAINER OF LOW-LEVEL
RADIOACTIVE WASTE.

be split apart more easily when it collides with a neutron.

When the nucleus of a uranium atom splits, it releases a large amount of energy in the form of heat and radiation. It also releases two or three of its neutrons, which collide with the nuclei of other nearby uranium atoms, thus causing continuing releases of heat, radiation, and yet more neutrons in the well-known process called a chain reaction. This process can be modulated by inserting "control rods" (made of cadmium, boron, indium, silver, or hafnium) into a reactor, which absorb some of the flying neutrons that would otherwise split apart yet more uranium atoms.

By using these control rods, engineers can regulate the levels of heat that build up inside the containment vessel at the core of a nuclear reactor. This heat is used to boil water, thereby powering steam-driven electric turbines, which produce electricity. Many reactors first transfer the heat to water held at high pressure and then transfer the heat in this pressurized hot water to a second

HOW A NUCLEAR REACTOR WORKS

In the reactor core, uranium atoms are split in a chain reaction, slowed with control rods. The chain reaction releases gamma rays that create high-energy heat, which heats water. Radioactive hot water is run through pipes past cold water to create steam that drives a turbine, generating electricity. Waste heat, as steam, is released from the cooling tower.

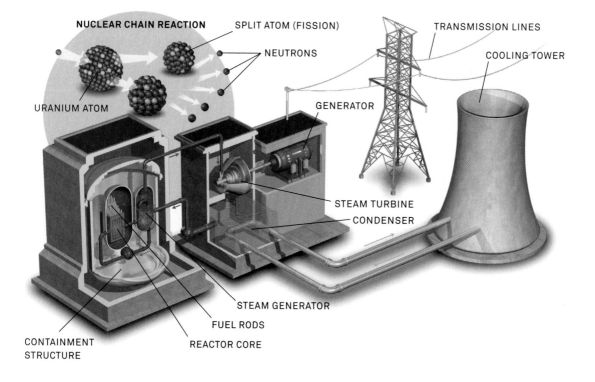

supply of water, which turns into steam without becoming radioactive.

This design (called a pressurized water reactor) is the basis for two thirds of the nuclear power plants in the United States and more than 60 percent worldwide. The remainder of the reactors in the United States are "boiling water" reactors. Other variations on the same basic design are used in several other countries, including "heavy water" reactors and gas-cooled reactors.

In the popular imagination, the blame for all the problems of the nuclear industry is often assigned to two factors: first is the combined effect of the well-publicized accident at Three Mile Island, near Harrisburg, Pennsylvania, in March 1979 and, seven years later, the far worse accident at Chernobyl, near the border between Ukraine and Belarus. The second factor is the long-running and still unresolved dispute over what to do about long-term storage of radioactive waste that remains dangerous for many thousands of years.

Both of these problems are real, and both can almost certainly be solved eventually. Yet neither represents the real cause for the sharp decline of the nuclear power industry. Instead, two other problems—problems that may not be as susceptible to solution—are primarily responsible for blocking the once hoped-for expansion of nuclear power.

First, the driving force that has converted once vibrant nuclear dreams into debilitating nightmares for electric utilities has been the grossly unacceptable economics of the present generation of reactors. To begin with, the cost of constructing nuclear power plants has escalated wildly, to the point where most utilities have long since abandoned any idea of ordering new reactors. By 1985, *Forbes* magazine had concluded: "The failure of the U.S. nuclear power program ranks as the largest managerial disaster in business history, a disaster

on a monumental scale," and added, "For the United States, nuclear power is dead—dead in the near term as a hedge against rising oil prices and dead in the long run as a source of future energy. Nobody really disputes that."

Some in the nuclear industry actually do dispute that conclusion. The industry itself points to other reasons for its decline. Often cited is the

In 1988 I visited Chernobyl and saw the reactor that had melted down two years earlier. I walked through the adjacent ghost town, Pripyat, that is still eerily silent. According to the U.S. Nuclear Regulatory Commission and the International Atomic Energy Association, an estimated 4,000 people will ultimately lose their lives because of the accident at Chernobyl, which released more than 100 times as much radiation as the atomic bombs dropped on Nagasaki and Hiroshima. Three hundred and fifty thousand people were forcibly resettled. More than 1,400 miles away in the United Kingdom, Wales suffered harmful effects from the accident. A quarter-century later, it still is not safe to eat sheep from some areas of Wales.

length of time required for regulatory approval—even though regulatory processes have been streamlined and redesigned to the industry's liking. Continuing safety concerns (complicated by the loss of key expertise within the nuclear construction business) maintain pressure on regulators to avoid dangerous shortcuts.

For a long time, supporters of nuclear power have pointed to perceived successes in France, South Korea, and some other countries as evidence that the technology itself is still attractive and should still be viewed as an option of choice in the United States and elsewhere in the world.

France, which receives more than three quarters of its electricity from nuclear reactors, is often cited as a nuclear power success story. Less known is the fact that the French program is almost entirely owned by the French government, with most of its electricity sold to the government. The degree of government subsidy is difficult to ascertain because of a lack of transparency in the finances of the operation. France is ahead of the United States in solving the problem of where to store long-term nuclear waste—though it relies on the controversial and expensive process of reprocessing—and has thus far compiled an impressive record of safety and reliability. However, it is now apparently facing serious financial difficulties. Moreover, the new modular reactor design it is building in Finland, which was supposed to be faster to construct and both cheaper and safer to operate, is now far behind schedule and way over budget.

The estimated cost of building a nuclear power plant rose from around $400 million in the 1970s to $4 billion in the 1990s, while construction times doubled during the same period. Even before the global economic downturn that began in 2008, cost estimates for constructing nuclear power plants were increasing at a rate of 15 percent per year

(which means the cost of a new power plant at time of completion would potentially increase tenfold yet again in less than 17 years).

Incredibly, it is now difficult to find a single reputable engineering firm in the United States or Europe willing to stand by any estimate of how much it will cost in today's world to build a new nuclear power plant.

As Steve Kidd, the director of strategy and research for the World Nuclear Association, wrote last year in *Nuclear Engineering International,* "What is clear is that it is completely impossible to produce definitive estimates for new nuclear costs at this time." Experience has shown that each year of additional delay in the construction of a nuclear power plant adds another estimated $1 billion to the cost.

As a young congressman from Tennessee in the late 1970s and early '80s, I watched this debacle firsthand as it unfolded in the seven-state area served by the Tennessee Valley Authority. TVA, which had been present at the birth of the nuclear industry in the 1940s (and provided all of the electricity for the enrichment of uranium in Oak Ridge, Tennessee), had been one of the most enthusiastic supporters of nuclear energy. During the 1960s and early '70s, TVA ordered 17 nuclear power reactors at a time when electricity use was increasing at the rate of 7 percent per year.

In the fall of 1973, the OPEC embargo drove oil prices sky-high, leading to sharp increases in the price of coal (the demand for which increased as it was substituted for oil) and, thus, to much higher electricity costs. In response to these increasing costs, electricity use leveled off.

Moreover, President Richard Nixon's Project Independence and President Jimmy Carter's innovative policies promoting energy conservation, efficiency, and renewable energy were both part of

SPAIN STOPPED THE BUILDING OF NEW NUCLEAR POWER PLANTS IN 1984. THIS FACILITY IN ARMINTZA HAS REMAINED UNFINISHED EVER SINCE.

a national shift toward a permanently lower ratio of energy consumption to economic output.

When the smoke cleared, electricity demand settled into a slower growth rate of 1 to 2 percent per year, which eventually led TVA to cancel 10 of the plants it had ordered and to defer another. The overall cost of electricity from TVA continued to rise dramatically, partly because the cost of the unfinished reactors had to be included in the rates charged for electricity from existing generating plants. Other utilities had similar experiences. Overall, 138 nuclear power reactors were canceled in the United States in the 1970s and '80s, although a few older plants ordered prior to 1974 have since been completed after a hiatus in construction.

Of the 253 nuclear power reactors originally ordered in the United States from 1953 to 2008, 48 percent were canceled, 11 percent were prematurely shut down, 14 percent experienced at least a one-year-or-more outage, and 27 percent are operating without having experienced a year-plus outage. Thus, only about one fourth of those ordered, or about half of those completed, are still operating and have proved relatively reliable.

The long hiatus in nuclear construction after the Three Mile Island accident led to the loss of experienced personnel and a deterioration of crucial expertise. Engineers' doubts about the future of the nuclear industry discouraged them from acquiring the training and expertise to make the industry a career option—again reinforcing utilities' doubts that critical personnel would be available for building and operating nuclear power plants. More than one third of the remaining nuclear power workforce in the United States is eligible for retirement by 2012. Thirty years ago, there were 65 academic nuclear-engineering programs in the U.S. Today, there are fewer than 30.

Moreover, doubts about the future of the nuclear power industry have also discouraged large investments in manufacturing capacity critical for expansion. Today, for example, there is only one company in the world able to build the key portion of a nuclear reactor containment vessel. Located in Japan, it is capable of producing no more than four such containment vessels in a year. Although this capacity is being doubled and other companies could build new, specialized foundries, the cost of doing so is very high in both money and time, and other industries compete for similar products from the same suppliers. Doubts that investments in new foundries will be made, in turn, further reinforce doubts by utilities that they will be able to rely on projections concerning how much time and money it might take to build new nuclear reactors.

There are other bottlenecks in almost every part of the manufacturing supply chain. Suppliers of parts are reluctant to make them if they are uncertain of receiving orders, but utilities won't place orders if they're uncertain of receiving funding. And investors won't provide funding if there are supply shortages and bottlenecks that will lead to increased cost and time of construction. The resulting need to order parts from new manufacturers also poses difficulties for quality and safety assurance.

Adding to these problems is the unfortunate fact that, with a few exceptions, each of the 436 nuclear reactors now operating in the world has its own "one-off" design. The lack of standardization has further added to the cost of engineering and construction, while complicating the effectiveness of training and the maintenance of safety protocols, which must be approached in a distinctive way for each distinctive design. The

need for increased standardization has been recognized as an imperative since the cancellation of so many reactors in the early 1980s. South Korea has led in this respect in its national program, while U.S. regulators are pushing hard in the same direction. But even France's strong national control and standardization efforts have not shielded its nuclear program from major escalation in real capital costs and construction times.

There are also doubts about the future of electricity demand in an era characterized by renewed interest in efficiency, conservation, and renewable energy. This uncertainty about demand discourages utilities with limited construction budgets from placing enormous bets on huge, expensive, time-consuming nuclear projects. The reluctance to make large, slow-to-mature investments is further complicated by the fact that, at present, nuclear power plants come in only one size: extra large.

In the early years of the nuclear power industry, most reactors were smaller than the giant ones common today. But early disappointments over the difficulty in producing electricity at a low-enough cost to compete with coal led nuclear power manufacturers to increase the size of their plants to 1,000 megawatts and more—up to 1,600 megawatts—in an effort to reduce costs by maximizing economies of scale. Unfortunately, utilities underestimated the additional cost required to master the new construction complexities of the much larger size.

When utilities took over management of nuclear power plant construction from the specialized contractors who had built turnkey reactors in the industry's early stages, they were unprepared for the extraordinary management challenges they confronted. In practical terms, there is a wide gulf between the culture and practice of nuclear physics, on the one hand, and the culture and practice of welding on the other. Bridging that gulf

NUCLEAR POWER AROUND THE WORLD

There are currently 436 active nuclear reactors around the world, with a capacity to generate approximately 372 gigawatts of electricity. All told, 30 countries have at least one nuclear reactor. The United States, with 104 active reactors, has almost twice as many as France, with 59, and Japan, which has 53. Russia has 31 reactors, and another 35 are in former Soviet Bloc nations, including Ukraine, which has 15. South Korea has 20. The United Kingdom has 19, followed by Canada with 18. Germany and India each have 17.

Within the U.S., 30 of the 50 states have nuclear power plants. Illinois generates the most electricity from nuclear power plants, followed by Pennsylvania, South Carolina, New York, Alabama, Texas, and North Carolina. As a group, U.S. nuclear reactors are responsible for almost 31 percent of all nuclear-generated electricity in the world.

There are 52 more plants now under construction in 14 countries, including 16 in China, nine in Russia, six in India, and five in South Korea.

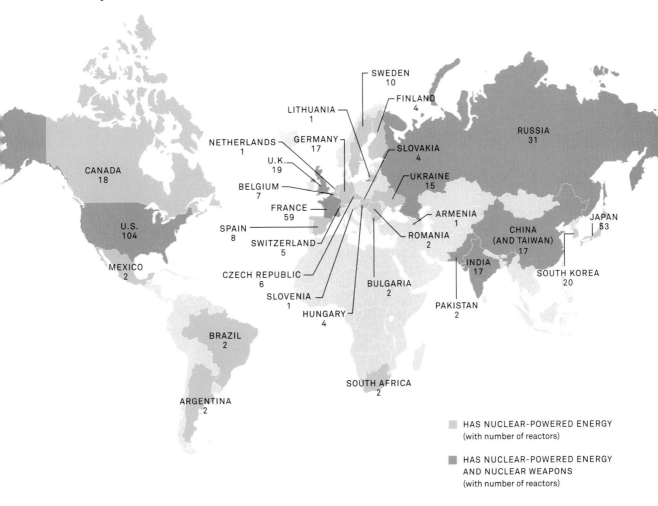

SWEDEN 10
FINLAND 4
LITHUANIA 1
NETHERLANDS 1
GERMANY 17
SLOVAKIA 4
U.K. 19
RUSSIA 31
UKRAINE 15
BELGIUM 7
CANADA 18
FRANCE 59
ARMENIA 1
JAPAN 53
SPAIN 8
ROMANIA 2
CHINA (AND TAIWAN) 17
U.S. 104
SWITZERLAND 5
MEXICO 2
INDIA 17
CZECH REPUBLIC 6
BULGARIA 2
SOUTH KOREA 20
SLOVENIA 1
HUNGARY 4
PAKISTAN 2
BRAZIL 2
SOUTH AFRICA 2
ARGENTINA 2

HAS NUCLEAR-POWERED ENERGY
(with number of reactors)

HAS NUCLEAR-POWERED ENERGY
AND NUCLEAR WEAPONS
(with number of reactors)

SOURCE: World Nuclear Association; Federation of American Scientists

NEXT-GENERATION NUCLEAR

Some experts believe the most promising new approach is a "pebble bed" reactor based on a German design from the 1960s. Rather than using fuel rods, each day 3,000 "pebbles" of uranium oxide are added to the 360,000 pebbles already in the reactor's core, replacing those that are removed from the bottom as spent fuel. Each fuel pebble contains thousands of "kernels" of uranium dioxide, each encased in silicon carbide and a pyrolytic coating. The entire pebble, about the size of a billiard ball, is enclosed in a graphite shell that can withstand temperatures of up to 2,800°C (5,000°F)—far higher than the maximum reaction temperature.

This process would, in theory, make it possible to collect the heat with helium flowing freely through the spaces left between the spheres as they sit, in the description of one physicist, "like gumballs in a giant gumball machine." This elegant combination of the heaviest metal in nature—uranium—and the lightest inert gas—helium—could make the entire process much safer, because helium picks up heat without becoming radioactive; it's this heated helium gas that turns the generating turbine.

One advantage hoped for from this approach is the elimination of any need to stop the reactor for refueling. Moreover, because the pebbles are less likely to catch fire and are more difficult to use in making nuclear weapons, this design is expected to appeal to those worried about accidents and proliferation. Equally important, it is said to be inherently "meltdown-proof" because the pebbles themselves absorb excess neutrons if the temperature begins to climb toward unsafe levels.

One potential problem is a shortage of helium. If many pebble bed reactors were to be built, the supply of helium could well become a bottleneck limiting the design's scalability. China has a small experimental model, and South Africa may soon build a prototype plant. But most experts believe that even successful development of this option would not lead to commercial plants until at least 25 years from now.

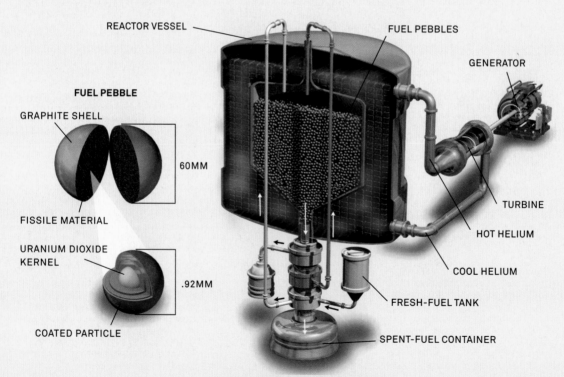

FUEL PEBBLE

GRAPHITE SHELL

60MM

FISSILE MATERIAL

URANIUM DIOXIDE KERNEL

.92MM

COATED PARTICLE

REACTOR VESSEL

FUEL PEBBLES

GENERATOR

TURBINE

HOT HELIUM

COOL HELIUM

FRESH-FUEL TANK

SPENT-FUEL CONTAINER

NUCLEAR PEBBLE BED REACTOR

was easier during the early period of reactor construction, when scientists and engineers from the military and the Atomic Energy Commission maintained seamless control over all steps in the process. The shift in management responsibility to utilities with multiple private suppliers and parts subcontractors opened seams in the culture and ethos of the nuclear design and construction process.

When bottlenecks appeared in response to uncertainty about the scale and continuity of nuclear construction, the search for new contractors and subcontractors put further strains on the integrity and reliability of the process.

Wild swings in the macroeconomic environment also introduced inflationary surges in the price of all of the commodities necessary for the completion of large-scale projects, including not only steel, concrete, and engineering and design services, but capital availability as well. When extended construction times and rising costs destroyed confidence in the reliability of contract terms at the beginning of projects, utilities began to worry about the impact of new reactor purchases on their credit ratings and on their cost for capital generally.

The second major problem slowing the spread of nuclear reactors around the world is the deep concern about nuclear weapons proliferation.

During the eight years I worked in the White House, every nuclear-weapons-proliferation problem we faced was connected to a reactor program. This would have been surprising to nuclear enthusiasts in the 1950s and '60s. They genuinely believed that the scientific and engineering challenges involved in building nuclear weapons were so different from those necessary to build a nuclear power plant that it would be relatively easy to build electricity-generating reactors without increasing the risk of putting nuclear-weapons-construction

capability into the wrong hands.

However, designs for nuclear weapons are now—unfortunately—easily available to those who search for them. And while specialized tools necessary for the fabrication of key components of a nuclear weapon are difficult to come by and are monitored as carefully as possible, the most crucial element of a nuclear bomb is fissionable material. Here again, old assumptions are no longer reassuring.

While it is true that the enrichment of nuclear material for weapons is far more difficult than the enrichment of material for nuclear reactors, advances in enrichment technology have made it much more feasible for nations with a supply of reactor-grade fissionable material to further enrich it to the point where it can be used in a weapon. A team of scientists and engineers capable of managing a nuclear reactor program and at least part of the nuclear fuel cycle can be forced by a dictator to work secretly on a nuclear weapons program. Indeed, that is the principal way nuclear weapons have proliferated in the last 25 years.

The flow of nuclear materials from reactor-fuel use to weapons use can also move in the opposite direction. In 1998, I participated in negotiating an agreement between the United States and Russia to dismantle large numbers of nuclear weapons in the arsenals of both countries. This resulted in a surplus of material that could technically be converted to fuel for civilian nuclear reactors. Unfortunately, this conversion has proved to be difficult in practice, and the surge of potential supply has unsettled and destabilized the market for reactor fuel.

Numerous research-and-development teams are hard at work trying to solve the debilitating problems with the present generation of nuclear reactors by coming up with new designs. They hope

A "GENERATION III+" REACTOR IS CURRENTLY UNDER CONSTRUCTION IN FLAMANVILLE, FRANCE. THE BUILDING OF THE NEXT GENERATION OF REACTORS HAS BEEN PLAGUED BY DELAYS AND COST OVERRUNS.

the next reactors will be cheaper to build, safer and cheaper to operate, far less vulnerable to catastrophic events, less vulnerable to terrorism, and economical in smaller sizes—which will make them more attractive to utilities facing an uncertain future for electricity demand.

There are more than 100 other new reactor designs for so-called Generation IV nuclear plants—including "sodium fast reactors" (or "integral fast reactors"), which use liquid sodium as a coolant. One variation on the South African design, now being explored at the Idaho National Laboratory, is a very high temperature reactor.

At any rate, whatever option the United States and other wealthy, developed nations choose as a strategy for solving the climate crisis will serve as an influential model for the efforts of other nations worldwide. That being the case, it is difficult to imagine the developed nations could get away with saying, "Nuclear power is our choice, but we're not going to let you use it because of our concerns about nuclear weapons proliferation."

Yet, if the world were to decide to make nuclear power the silver-bullet option of choice for electricity production, thousands of additional reactors would be built. And many of them would be placed in countries that most people would agree should not possess nuclear weapons.

One potential and often proposed answer to this problem is the establishment of an international authority under the control of developed nuclear nations that would supply safeguarded nuclear fuel for reactors in less-developed countries. The terms of this transaction would assure that the fuel always remained under the control of the developed nation that supplied it. When spent, the fuel would be removed and replaced with a fresh supply, again on terms that ensured that it was never under the control of the developing nation. However, most developing countries presented with such an option have turned it down, fearing that such an arrangement would put their domestic power programs under the control of other nations.

Another problem with making nuclear power a silver-bullet option is that a massive expansion of the number of reactors in use around the world would stretch the supplies of available fuel. Significantly larger numbers of reactors would place pressure not only on current uranium stockpile reserves but also on the world's limited ability to quickly and safely expand uranium mining and processing operations. Presently, a long lead time is necessary to open new uranium mines. More than half the people employed in the uranium-mining business currently work on the constant cleanup of operations in order to prevent the health and environmental challenges that would otherwise result.

Nuclear enthusiasts offer an answer to this potential shortage of processed uranium: reprocess the spent fuel from reactors in order to extend its useful life. After all, they say, current reactors use only 1 percent of the available energy in the uranium ore, and reprocessing is already being undertaken in Russia, parts of Europe, and—as of 2008—Japan. However, this is a process that separates and recycles plutonium and then transports it for use in reactors.

Many supporters of nuclear power have long argued in favor of reprocessing spent fuel in order to create new supplies that can be put back into reactor cores. This approach has been labeled "recycling" and has sometimes been promoted as an alternative to using long-term repositories. This argument is highly misleading, however, because reprocessing actually increases the overall volume of waste, even though it does reduce the volume of high-level waste. Granted, reprocessing would

extend the uranium resources available in the world to be used in nuclear power plants and would allow the tailoring of waste streams for repositories. But reprocessing also adds significantly to the cost of the nuclear fuel cycle and still requires long-term repositories.

More significantly, reprocessing produces plutonium, which can be used to make nuclear weapons. This controversial option could result in much wider availability of even more dangerous nuclear material, together with the skills, knowledge, and equipment to apply it for both civilian and military use. Under such circumstances, it would be considerably more difficult to limit nuclear proliferation and to keep nuclear weapons out of the hands of terrorists desiring to commit mass murder on a horrific scale.

Currently, flows of plutonium do not have adequate safeguards against diversion. Harvard professor Matthew Bunn, the leading expert on the current set of global safeguards, describes these measures as completely inadequate. He says that the world should place a high priority on dramatically strengthening the global nuclear-security regime. The current international safety regime, Bunn points out, is "entirely voluntary."

The risk is further enhanced by organized efforts to obtain plutonium on the part of al Qaeda, Iran, North Korea, and those who have been identified with the A.Q. Khan network in Pakistan. Professor Graham Allison of Harvard, a leading expert on nuclear proliferation and terrorism, predicts that unless significant additional safeguards are quickly put in place, the chance of a terrorist group detonating a nuclear weapon in a U.S. city during the next 10 years is 50 percent.

In any case, an increasing number of experts have concluded in recent years that reprocessing is a dangerous and poor choice. As the experts at M.I.T. recently pointed out, "We know little about the safety of the overall fuel cycle, beyond reactor operation."

At the other end of the nuclear fuel cycle, the challenge of storing the waste generated by nuclear reactors has all but paralyzed the political process in the United States and several other countries for decades. The well-known "not in my backyard"

issue that is raised whenever controversial projects are sited is a more serious obstacle for long-lived, highly radioactive nuclear waste. There is an international consensus on the advisability of storing nuclear waste in deep underground repositories in locations selected for their long-term geological stability and safety, tectonic stability, and lack of risk posed by groundwater content and flow. These

RELATIVE CO_2 FOOTPRINTS OF ELECTRICITY SOURCES

Much of the new enthusiasm for nuclear power is based on the perception that it is a carbon dioxide–free source of electricity. But that is not entirely correct. The life cycle of a nuclear power plant—from the construction of the plant to the mining and milling of the uranium fuel to the transportation and storage of nuclear waste and the eventual decommissioning of the plant—produces a good deal of CO_2. When all of that CO_2 is allocated among the kilowatts of electricity produced by an average plant during its lifetime, the amount per kilowatt hour is still much less than what is emitted during the generation of electricity produced from coal. However, the CO_2 associated with nuclear plants is many times more than that associated with generating electricity from wind, solar, or hydroelectric power, according to the same life-cycle analyses.

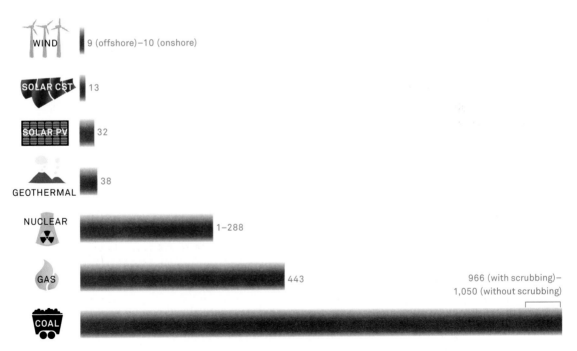

WIND 9 (offshore)–10 (onshore)

SOLAR CST 13

SOLAR PV 32

GEOTHERMAL 38

NUCLEAR 1–288

GAS 443

966 (with scrubbing)–1,050 (without scrubbing)

COAL

GRAMS OF CO_2 PRODUCED PER KILOWATT-HOUR OF ENERGY

SOURCE: Benjamin K. Sovacool, *Energy Policy* 36, 2008

sites must be sufficiently deep, with a perimeter sufficiently far from population centers, yet accessible to transporters of the waste.

In spite of the global consensus on the appropriateness of this storage technique, however, no nation in the world has yet opened such a site. Since the most dangerous nuclear waste products have a half-life of hundreds of thousands of years, the phrase *long term* takes on an entirely different dimension when evaluating each potential disposal site. Sweden and Finland have selected geologic repositories that appear to be appropriate and have secured public approval. France has also selected

a repository site and plans to open it by 2025. All other countries are behind the United States in their planning.

The U.S. Nuclear Regulatory Commission separates nuclear waste into four categories: the first—"high-level waste"—is the used nuclear fuel coming out of reactors. A typical 1,000-megawatt light-water reactor produces approximately 27 tons of high-level waste each year. Advocates of expanding the nuclear option point out that this compares with 400,000 tons of toxic coal ash produced in a typical coal-fired generating plant each year. Around 10,000 cubic meters of high-level waste

UNLOCKING NUCLEAR FUSION

There has been excitement for many years about the potential of a different form of nuclear energy: fusion. While conventional nuclear reactors generate heat by splitting heavy atoms, fusion produces much larger amounts of heat by combining light atoms. The atomic bombs used at the end of World War II were based on fission, while the much more powerful hydrogen bombs built during the Cold War relied on fusion ignited by fission. Fusion is also the underlying process by which the sun produces heat and light. Although enormous sums have been spent in the effort to develop a practical form of fusion power, early enthusiasm has long since given way to a sober assessment that a usable process is still at least several decades away.

Researchers continue to explore two basic approaches: the magnetic confinement of the atoms to be fused, and inertial confinement, which uses high-intensity lasers to trigger fusion. The Tokamak fusion project at the Princeton Plasma Physics Laboratory has made slow progress for several decades. The new National Ignition Facility at Lawrence Livermore National Laboratory, which opened in the spring of 2009, is the leading facility exploring inertial confinement.

The National Ignition Facility at Lawrence Livermore National Laboratory will begin conducting experiments in 2010. At its center is this 10-meter target chamber, where scientists hope to achieve fusion ignition with lasers.

is created globally each year by the nuclear power industry. Though it represents only 3 percent of the total radioactive waste from reactors, it contains 95 percent of the radioactivity. And this is the waste stream that has created political gridlock in the United States.

The second category—"low-level waste"—is produced in much higher volumes, most of which (as the name implies) is far less radioactive. It includes contaminated clothing, filters, rags and tubes, tools, and other items. In some cases, this category includes parts from inside the reactor containment vessel that are very highly radioactive.

The third category—called "waste incidental to reprocessing"—is one created by the Department of Energy for by-products associated with the reprocessing of spent nuclear fuel.

The fourth category—"uranium mill tailings"— includes waste products produced by the processing of uranium ore into reactor fuel. These tailings contain radium, which has a half-life of more than a thousand years. Uranium-238, the most common form of uranium in nature, contains three additional neutrons in each atom compared with uranium-235, the rarer form that is used in the fuel for most reactors.

Though most of the controversy surrounding nuclear waste has involved the selection of a long-term repository, much of the near-term risk is in transportation of large quantities of radioactive waste from reactors to storage sites.

When the U.S. government, responding to the continuing stalemate of its effort to finalize a long-term repository, tried to create a short-term repository where waste could be stored until the long-term repository could be completed, yet another controversy erupted. This effort produced additional opposition in all of the potential short-term sites. Critics observed that the architecture of

WATER DEMANDS OF ENERGY PRODUCTION

Most nuclear plants require large volumes of water, principally for cooling. About 25,000 to 60,000 gallons of water are required for each megawatt-hour of electricity produced by a nuclear power plant using an "open-loop" cooling system, only 445 to 870 of which are actually consumed in the process, depending on the type of plant. Plants using a "closed-loop" system require much less water but still consume just as much. These flows do not pick up radioactivity, but the water—now much hotter— is returned to the rivers, lakes, or seas from which it is withdrawn, sometimes killing fish and creating other problems. Some have proposed ideas for capturing this waste heat.

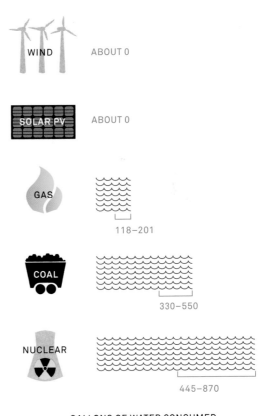

WIND   ABOUT 0

SOLAR PV   ABOUT 0

GAS   118–201

COAL   330–550

NUCLEAR   445–870

GALLONS OF WATER CONSUMED
PER MEGAWATT-HOUR OF ENERGY

SOURCE: U.S. Department of Energy

this plan would essentially double the risk involved by transporting the waste twice: first to the short-term repository and then to the long-term repository. For these and other reasons, spent fuel in the United States is now typically stored in dry casks aboveground at the location of each reactor.

The selection of Nevada's Yucca Mountain as the prime candidate for the long-term repository naturally produced heated opposition in that state. But controversies over the geological safety of Yucca Mountain have led to even more opposition. Upon closer examination, experts found the site to be tectonically active and its geochemical environment to be "oxidizing," in violation of the criteria for long-term storage. Additional opposition has come from communities along the transportation routes that could expect the most traffic in high-level waste.

Although construction of new nuclear power plants has all but stopped in the United States, the Tennessee Valley Authority has refurbished and restarted one of the reactors it shut down more than 20 years ago. It has also resumed construction on another reactor it had previously ordered but mothballed. But no orders for new reactors have been placed by any U.S. utility since 1978.

Even massive new government subsidies and guarantees have failed to attract much private investment back into the construction of new nuclear power plants. These subsidies, estimated by Rocky Mountain Institute cofounder Amory Lovins to total more than $500 billion, include construction-loan guarantees; surcharges on ratepayers' electricity bills to fund the continuing effort to find a long-term storage solution; public guarantees for liability insurance against catastrophic accidents; taxpayer-funded insurance against legal and/or regulatory delays; cost-sharing with taxpayers during the licensing process; federal research-and-development expenditures (more than $150 billion); and public guarantees to help cover "stranded costs," such as unpaid construction debt for old nuclear reactors. In response, some utilities have once again agreed to take a fresh look at the nuclear power option. Seventeen new applications for 26 new reactors have been submitted to the Nuclear Regulatory Commission, but none yet has a commitment for financing and none has begun construction.

Of the 104 reactors operating in the United States, 24 are located in areas that are currently experiencing severe levels of drought (mostly in the southeastern United States). These areas are among those where severe drought is expected to become much more common as a consequence of global warming. Already, a TVA reactor in Alabama was forced to briefly suspend operations when unusually high temperatures reduced stream flow and water levels and limited the ability to withdraw water—and, in some cases, limited the ability to return hot water without causing fish kills.

During the historic European heat wave of 2003, France, Spain, and Germany were forced to shut down numerous nuclear power plants and reduce the power output of others because of low water levels. If, as predicted by scientists, global warming has an even worse impact in the years ahead on water levels and drought conditions, many other nuclear plants on the shores of rivers and lakes may soon face periodic—and expensive—shutdowns, thus making the cost of nuclear-generated electricity even less competitive than it is already.

EACH OF THESE CONCRETE CASKS CONTAINS SPENT FUEL FROM NUCLEAR REACTORS. THE CASKS ARE STILL AWAITING A PERMANENT HOME, FOLLOWING THE CANCELLATION OF THE YUCCA MOUNTAIN STORAGE FACILITY IN NEVADA.

CHAPTER NINE

FORESTS

THE AMAZON IS THE LARGEST RAIN FOREST
ON EARTH, YET EACH YEAR MORE THAN 10,000
SQUARE KILOMETERS (3,900 SQUARE MILES)
OF IT ARE DESTROYED.

The CO_2 emissions from deforestation are second only to the burning of fossil fuels for the production of electricity and heat as the largest source of global warming pollution on the planet. Indeed, an estimated 20 to 23 percent of annual CO_2 emissions—more than that from all the cars and trucks in the world—result from the destruction and burning of forests.

The biggest direct cause of deforestation is the "slash-and-burn" technique used to rapidly clear forests for subsistence farming, plantation agriculture, and cattle ranches—mostly in tropical and subtropical countries. Norman Myers, the distinguished ecologist, recently estimated that 54 percent of current deforestation is due to slash-and-burn agriculture, 22 percent to the spreading of palm oil plantations, 19 percent to "overheavy" logging, and 5 percent to cattle ranching.

The good news is that governments throughout the world have now tentatively agreed to efforts aimed at sharply reducing deforestation. However, they have found that, in order to be successful, the world will have to address the underlying causes of deforestation, which are:

▶ Poverty and population growth in poor countries.

▶ The ravenous appetite in the globalized marketplace for cheap sources of wood and for palm oil, beef, soybeans, sugarcane, and other commodities that large deforesters produce on the land that is cleared.

▶ The failure of market economics to value living forests for anything other than the earnings produced by their destruction.

▶ The failure of the world community to reach a global agreement that puts a price on carbon and assists tropical countries in monetizing the true value of their forests to the world as a whole.

▶ Corruption that undermines the effectiveness of existing laws and regulations designed to prevent irresponsible deforestation.

The biggest change in deforestation patterns in recent years, according to Myers and others, is a significant increase in slash-and-burn farming. Whereas the same small groups of people used to move from one forest area to another, in recent years there has been a large influx of impoverished migrants engaging in the practice—particularly in the Brazilian Amazon and in the African Congo Basin.

Slightly more than one acre of forests is cleared on the earth every second. That amounts to almost 100,000 acres (38,000 hectares) every day, and more than 34 million acres (13.7 million hectares) per year—an area the size of Greece. This is partially offset by new growth and organized tree-planting programs, so the net loss of forests each year amounts to 18 million acres (7.3 million hectares).

This frenzied destruction of forests has a double impact on the climate crisis: first, most of the carbon contained in the trees is emitted into the atmosphere; and second, the planet loses part of its

THE PRODUCTION OF CHARCOAL IS ONE CAUSE
OF DEFORESTATION IN DEVELOPING REGIONS,
WHERE PEOPLE USE IT FOR COOKING. SEEN HERE
IN LIAONING, CHINA, THE PROCESS EMITS CO_2
AND BLACK CARBON.

ability to reabsorb CO_2 because the forests, once destroyed, no longer pull CO_2 from the air.

Most people know by now that the largest contributors to global warming are China and the United States, but many are surprised to learn that the third and fourth nations on the list are Indonesia and Brazil—where the CO_2 being emitted is primarily from deforestation. Astonishingly, satellite data analyzed by the World Resources Institute shows that more than 60 percent of all deforestation in the world today is occurring in Brazil and Indonesia—and is concentrated in the state of Mato Grosso in the Amazon and in Riau

Province and adjacent areas in Indonesia, where large peat forests are located.

The United Nations Food and Agriculture Organization, which maintains statistics on deforestation, notes that in recent years the largest deforesters in the community of nations—after Brazil and Indonesia—are Sudan, Myanmar, Zambia, Tanzania, Nigeria, the Democratic Republic of the Congo, Zimbabwe, and Venezuela. By region, Latin America has lost the most trees to deforestation, while Africa is a close second. Southeast Asia is third, and North America is a distant fourth.

In Asia and Latin America, the single largest cause of deforestation is conversion of forestland to large-scale agricultural use. In Africa, the largest cause is the conversion of forestland to small-scale farms, though large-scale agriculture is increasing in Africa as Chinese interests purchase large tracts for the growing of food they intend to import in the future.

In Brazil—which by itself is responsible for 48 percent of all deforestation in the world—the practice increased yet again in 2008. Almost 20 percent of the Amazon forest has already been destroyed. (The official Brazilian government figure is 17 percent.) After the best wood is removed, the remainder is burned to make way for cattle and crops. According to a recent report by Greenpeace, "80 percent of land deforested in the Amazon from 1996 to 2006 is now used for cattle pasture."

Brazil has been historically sensitive to any efforts by the international community to engage the country in agreements that would give the rest of the world any say in the future of the Amazon. Nevertheless, Brazil has announced a national target of reducing deforestation by 70 percent by 2017. In August of 2008, President Luiz Inácio Lula da Silva announced the creation of a fund and a set of new regulations designed to protect the Amazon. However, these new regulations are not yet being effectively enforced. Carlos Minc, Brazil's environment minister, acknowledged the failure, saying, "We're not content. Deforestation has to fall more, and the conditions for sustainable development have to improve."

Paradoxically, while Brazil is destroying twice as much forestland each year as Indonesia, Indonesia is emitting twice as much CO_2 from deforestation as Brazil—primarily because the

TOP 10 DEFORESTING NATIONS

The problem of deforestation is most severe in developing nations in the tropical latitudes. The map below shows the top 10 deforesting countries, measured in acres of forest lost per year.

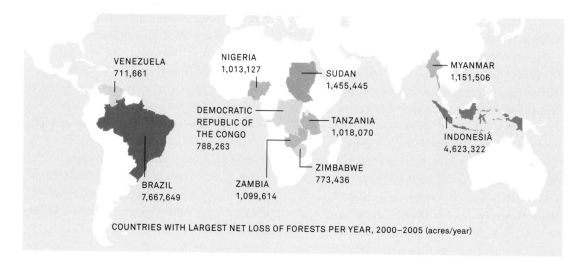

VENEZUELA
711,661

NIGERIA
1,013,127

SUDAN
1,455,445

MYANMAR
1,151,506

DEMOCRATIC
REPUBLIC OF
THE CONGO
788,263

TANZANIA
1,018,070

INDONESIA
4,623,322

BRAZIL
7,667,649

ZAMBIA
1,099,614

ZIMBABWE
773,436

COUNTRIES WITH LARGEST NET LOSS OF FORESTS PER YEAR, 2000–2005 (acres/year)

SOURCE: United Nations Food and Agriculture Organization, *State of the World's Forests 2007*

carbon-rich peatlands from which the Indonesian forests are being cleared dry up when the tree cover is gone and burn much longer when set ablaze, emitting far larger quantities of CO_2 into the atmosphere.

More than 80 percent of the world's palm oil comes from Indonesia and its next-door neighbor Malaysia. (During the past decade, Indonesia surpassed Malaysia as the leading supplier of palm oil.) In both countries, large peatland forests have been cleared of trees and drained to make way for the palm oil plantations. In order to accelerate the drying process, the developers of these plantations burn the peatlands. That is why massive clouds of smoke and soot now cover large portions of the Southeast Asian archipelago every year during the burning season.

Both countries have enacted subsidies and other incentives for the rapid expansion of palm oil plantations. Indonesia's official policy calls for the tripling of palm oil plantations by 2020. The official encouragement to develop new plantations has enabled some loggers to allegedly use their stated intention of growing oil palm trees as a thin excuse for simply clearing the timber in what would otherwise be protected forests.

Indonesia and Malaysia share (along with the kingdom of Brunei) the large island of Borneo. Willie Smits, a conservationist living in Borneo who began an effort to save the endangered orangutan and then expanded his focus to restore as much habitat as possible for the orangutan and the indigenous peoples of the forest, said, "What they're really doing is stealing the timber, because they get to clear it before they plant. But the timber's all they want; hit and run, with no intention of ever planting. It's a conspiracy."

But most of the land clearing actually does result in the establishment of working palm oil plantations. The oil palm can bear fruit for 30 years, provides employment, and produces more oil per acre than any other major oilseed crop.

Palm oil is not only one of the world's most popular edible oils but can also be mixed with diesel fuel to make one of the principal forms of biodiesel. Its asserted environmental benefits are based on the theory that the organic component of the fuel is recycled when the CO_2 emitted when it is burned is later reabsorbed by the growth of yet more oil palms. However, extensive research based on years of experience with the life cycle of the oil palm has proved that the clearing and burning of the forests where the trees are grown contributes far more CO_2 to the atmosphere than is ever reabsorbed. This life-cycle analysis is one of several factors leading to a reconsideration of the net environmental impact of biodiesel, ethanol, and other biofuels.

Indonesia has passed a law subsidizing the use of palm oil in Indonesian cars. However, most palm oil from both Indonesia and Malaysia is exported to North America and Europe to feed the demand for biofuels. Ironically, a U.S. tax incentive intended for the promotion of biofuels was a significant contributing factor in the clearing of virgin forests for the expansion of palm oil plantations. This tax loophole allowed importers of palm oil into the United States to receive a $1 per gallon subsidy if they added some biodiesel to the palm oil and then reexported the blend to European markets, where they received additional governmental subsidies intended to encourage the use of biofuel.

The net effect is that U.S. and European taxpayers were actively subsidizing the destruction of virgin tropical forests in the name of what was originally believed to be an environmental benefit. Lawmakers in the United States were successful in 2009 in closing the U.S. tax loophole.

LARGE SWATHS OF THE LAST REMAINING PEATLAND FOREST ON SUMATRA ISLAND IN INDONESIA ARE BEING LOGGED, BURNED, AND DRAINED TO CULTIVATE PALM OIL PLANTATIONS.

TIMBER FROM CLEARED SUMATRAN PEATLAND
FORESTS IS TRANSPORTED BY BARGE TO MARKET.

"What they're really doing is stealing the timber, because they get to clear it before they plant. But the timber's all they want; hit and run. . . ."

WILLIE SMITS

WILLIE SMITS: ECOSYSTEM RESTORATION AT ITS BEST

In 1989 Dr. Willie Smits, a Dutch forestry scientist, found a dying baby orangutan in a street market garbage heap in Balikpapan, Indonesia, rescued her, and nursed her back to health. Two years later, Smits founded the Borneo Orangutan Survival Foundation, now the world's largest orangutan anti-extinction project. The foundation has helped rehabilitate 1,000 young orangutans.

Smits had recently arrived from the Netherlands to do tropical forestry research when the baby orangutan came into his life. He holds deforestation, especially for the harvest of palm oil, accountable for the threat to orangutans. Deforestation is also a primary reason that Indonesia is one the world's largest greenhouse gas polluters. Rain forests have self-perpetuating water cycles: water transpires from trees and vegetation, condenses into clouds, and then falls again as rain downwind. When this cycle is broken by deforestation, the result is often increased temperatures and decreased or irregular rainfall. Smits's own data, after seven years of reforesting Indonesia's tropics, shows that the inverse is also true: reforestation helps restore the natural rain cycle.

Since then Smits's foundation has been as much about protecting and restoring the Indonesian forest as it has been about orangutans. He knew that if he wanted to save orangutans, he had to find them a habitat thick and wide enough to keep populations healthy, safe, and far from poachers. To protect the forest, Smits made local people the defenders of it, by building a conservation zone within an economic system dependent on the forest's health.

In 2002, with the Masarang Foundation, Smits founded Samboja Lestari ("Everlasting Forest"), a 5,000-acre reserve 22 miles (35 kilometers) northeast of Balikpapan, in one of the region's poorest areas. He combined intensive reforestation efforts with agriculture, planting pineapples, papayas, and beans between acacia trees. Growing crops between trees reduces competition among trees and helps the ecosystem regenerate more quickly. Deep inside this reforested area is the orangutan rehabilitation center, far from human populations.

Around the reforested area is a 328-foot (100-meter) ring of flood- and fire-resistant sugar palms.

Dr. Willie Smits with some of the orangutans his Borneo Orangutan Survival Foundation has helped.

These trees serve as an ecological buffer and also as a cash crop. The palms are tapped twice daily for sugar water that is processed at Masarang's palm sugar factory. Jobs for 3,000 people have been created here.

Today, Samboja Lestari is home to more than 200 healthy orangutans. And reforestation seems to have at least temporarily reversed some climate trends. Locally, the average air temperature has fallen 3 to 5°C (5.4 to 9° F), cloud cover has increased 11 percent, and rainfall is up 20 percent. The land, which had been reduced to a near desert, is now home to 1,800 species of trees, 137 types of birds, and 30 different kinds of reptiles.

Smits's goal is to put a value on the ecosystem that is high enough to keep it intact. To help the orangutans, he is making sure that the forests and the local people benefit.

Every one of the solutions to global warming is difficult because every cause of the crisis is deeply ingrained in patterns of behavior, commerce, and culture that have built up over long periods of time. Every one of the solutions is made more complex by the political and geopolitical complications that have long frustrated constructive action. In the case of deforestation, one of the principal difficulties has been the deep division in the modern world between wealthy, industrialized countries—mostly in the Northern Hemisphere—and poorer, less-developed countries—mostly in the tropics and subtropics. The reasons for this disparity in

in our atmosphere has come from deforestation in past centuries. According to some calculations, it was not until the 1970s that fossil fuel use overtook deforestation as the leading cause of global warming.

The total forested area on the surface of the earth is roughly nine billion acres—covering one third of the landmass. Deforestation has occurred for many thousands of years—though at much lower rates than at present. According to the World Resources Institute study, we now have only half of the forest cover we had 300 years ago. The largest areas of forest are found in

Scientists estimate that more than 40 percent of the excess CO_2 that has accumulated in our atmosphere has come from deforestation in past centuries.

wealth are, of course, rooted deeply in history and geography—and are compounded by the bitter legacy of colonialism.

The less-developed countries often point to the deforestation of North America and Europe during earlier centuries as evidence of hypocrisy on the part of wealthy countries that are condemning the ongoing forest destruction in poor countries. And of course, they have a point. Before the dramatic expansion of oil and coal use in the second half of the 20th century, deforestation was the largest source of CO_2 emissions on the planet. Even now, scientists estimate that more than 40 percent of the excess CO_2 that has accumulated

Russia, Brazil, Canada, the United States, China, Australia, the Democratic Republic of the Congo, Indonesia, Peru, and India. Among them, these nations account for two thirds of the earth's total forest area.

One third of all the remaining forests are still "primary forests" where human interventions have not yet had an impact. Though the total forest area continues to decrease, the rate of net loss is beginning to slow down.

According to the U.N. Food and Agriculture Organization, "Eighty-four percent of the world's forests are publicly owned, but private ownership is on the rise." One third of the world's forests are

Ironically, a U.S. tax incentive intended for the promotion of biofuels was a significant factor in the clearing of virgin forests for palm oil plantations.

FORMER PEAT FOREST LAND IN BORNEO, INDONESIA, IS NOW USED FOR A MASSIVE PALM OIL PLANTATION.

primarily used for the production of wood and non-wood forest products.

But there are vast differences between the temperate-zone forests of the Northern Hemisphere and the tropical forests on either side of the equator, which have a much higher density of carbon than any other ecosystem on the planet—an estimated 49 tons of carbon per acre, compared with 26 tons of carbon per acre in temperate forests. Rain forests represent a special case; though they cover only 7 percent of the earth's landmass, they have almost half of all the trees.

Moreover, the tropical soils underneath these rain forests are often surprisingly thin and nutrient poor. Although volcanic soils and floodplain soils are generally richer, virtually all of the nutrient

An eroded riverbank in the Amazon reveals how thin some tropical soils are. The soil is also usually nutrient poor. Most of the rain forest's nutrients are concentrated in the living biomass and in a layer of decomposing organic matter.

content in many rain forests is contained not in the soil but in the green cathedrals of living trees and plants above it and in the decomposing litter lying on top of the ground.

And the consequences for biodiversity loss are far more severe in tropical rain forests because so much of the biodiversity on Earth is contained there. An estimated 50 to 90 percent of all species on Earth are found in forests, and the upper end of that range is based on the widespread belief by biologists that a very large percentage of species are still unknown to science and are clustered in tropical forests. These unusually rich reserves of biodiversity are being destroyed as animal habitats shrink. Among the best-known species now at risk of disappearing are the orangutan in Borneo, the Sumatran tiger, the Asian elephant, some of the large primates of Africa (our closest relatives), countless potential sources of new medicines—and wild relatives of food crops, which depend for their survival on occasional replenishments of their gene pool (from their distant cousins in the wild) to make them resistant to blights and pests.

The cumulative impact of habitat destruction across the surface of the planet is leading to what some biologists are now calling the Sixth Great Extinction. (See "The Sixth Great Extinction," page 186.)

Norman Myers said in a speech at an Asia-Pacific forestry conference in Vietnam last year: "I'm going to give you my bottom-line message right now, up front: This is a super crisis that we are facing, it's an appalling crisis, it's one of the worst crises since we came out of our caves 10,000 years ago. I'm referring, of course, to elimination of tropical forests and of their millions of species."

Three of the previous five extinctions occurred 65 million years ago, when the dinosaurs disappeared; 200 million years ago, when 76 percent

THE ROLE OF FORESTS IN THE CARBON CYCLE

Forests play a two-part role in the movement of carbon through the ecosystem: by absorbing it from the atmosphere, and then by storing it in trees and in the soil. Through the natural process of photosynthesis, atmospheric CO_2 is taken in through tiny openings in leaves and incorporated into the tree or plant. This "fixed" carbon stays intact until plants or soils are disturbed, as when trees are burned or the ground is tilled. The process by which forests "inhale" CO_2 and "exhale" oxygen—why forests are often called the lungs of the earth—occurs at a microscopic level. Photosynthesis takes place inside chloroplasts, subcells that can number as many as 50 in every plant cell. The chloroplasts contain structures called grana, surrounded by an aqueous fluid called stroma. The grana is the home to photolysis, a process that splits water into hydrogen and oxygen. The oxygen is released by the plant, while the hydrogen moves into a second process, known as the Calvin cycle, which uses the energy produced by photolysis to combine the hydrogen atoms with CO_2 to create sugars. Those sugars make up the building blocks of more complex plant cells where carbon is stored for the long term.

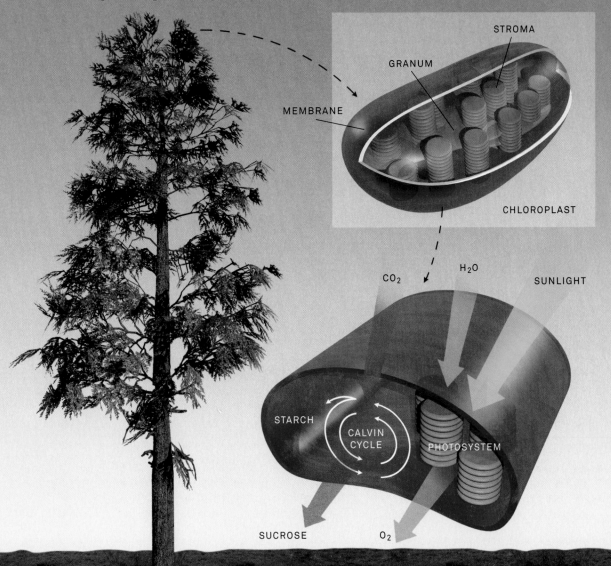

of all species went extinct for reasons that scientists do not yet fully understand; and 250 million years ago, when 96 percent of ocean species went extinct and two thirds of reptile and amphibian families on land became extinct. That largest of all extinction events coincided with the convergence of all the continents into the single land mass known as Pangaea. The two other mass-extinction events—neither of which is well understood by scientists—occurred 364 million years ago and 440 to 450 million years ago.

The undervaluation of biodiversity and forest cover, compared with the economic value of timber and subsistence agriculture, has led to serious miscalculations of the net economic impact of deforestation. Indeed, the widespread failure of market economics to account for environmental factors is particularly acute where the valuation of trees and forests is concerned. So long as there is no price on carbon, the market system will continue to encourage the massive dumping of CO_2 into the atmosphere and the wanton destruction of the forests of the earth. Once carbon is accounted for, the value of trees in absorbing large quantities of CO_2 will, in

THE SIXTH GREAT EXTINCTION

Most biologists believe that we are now living in the sixth great mass extinction in the history of the planet. Species are going extinct at many times the natural rate—in large part because of the rapid destruction of tropical rain forests and unique ecosystems that serve as home to as many as 90 percent of the known species on Earth. The distinguished biologist E.O. Wilson said in 1986, "Virtually all students of the extinction process agree that biological diversity is in the midst of its sixth great crisis, this time precipitated entirely by man."

Tom Lovejoy, one of the leading experts on biodiversity, has said, "Few dispute the proportion of species destined to disappear if current trends continue—that is, something close to half." At current rates, biologists predict this to occur within the present century unless the world finds a way to halt the destruction of forests and other important ecosystems.

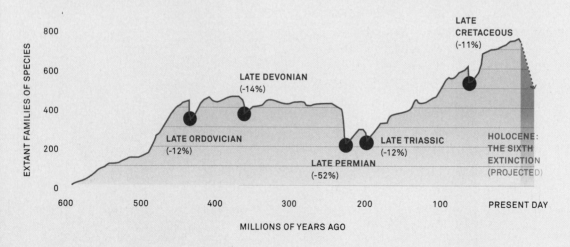

SOURCE: David Raup and John Seposki, *Science*, March 19, 1982

many cases, far outweigh their value as timber.

Larry Linden, a well-known expert on forest economics, offers the following example, which illustrates the absurdity of ignoring the value of carbon: A hectare of trees (two and a half acres), which is cleared and sold as pastureland, may bring, on average, $300. But in destroying that land, $15,000 worth of carbon is released into the atmosphere. (This assumes a carbon price of $30 per ton and 500 tons of CO_2 embodied in the same trees.) Linden's calculations indicate that, worldwide, a $30 per ton price on CO_2 would result in an 80 percent reduction in deforestation. And at a price of $20 per ton, we could achieve a 60 percent reduction. (Five-year contracts for CO_2 are now selling for $26 per ton on the European Climate Exchange; prices are expected to rise if a global treaty is finalized in Copenhagen.)

In wealthier countries, common forest management practices also typically ignore the economic value of the role trees play in sequestering CO_2. Larry Schweiger, author of the excellent book *Last Chance: Preserving Life on Earth,* uses the illustrative example of white oaks that are harvested when they reach a diameter of 12 inches at breast height. But the growth pattern of the white oaks is not unlike that of several other deciduous trees in the temperate zones of the Northern Hemisphere: it grows in a pattern that resembles a bell curve. During the early years of the tree's life, there is relatively little carbon sequestered in the tree, but as the white oak continues to grow, the amount of carbon added to its mass each year accelerates—peaking at an age of 120 years, after which the amount of added carbon begins to decline slowly until the tree dies.

If the tree-carbon sequestration curve was recognized and valued in the marketplace, the tree could still be harvested—though at an older age when substantially more carbon had been sequestered. But so long as this extra value to society is ignored in the market prices paid for harvested trees, the opportunity to sequester far more carbon in forests will be lost.

Even as the world tries to restore the health and integrity of forests as a key strategy for solving the climate crisis, scientists are expressing increasing concern about the impact of the climate crisis itself on the future ability of forests to continue sequestering CO_2.

CARBON UPTAKE IN TREES

The carbon-uptake rate of a tree over its lifetime resembles a bell curve, with a slow start over a tree's early decades and a multiyear peak before the rate falls. Paying attention to trees' growth and CO_2-uptake rate may suggest when trees should be harvested, to maximize the amount of carbon they sequester. The curve below applies to trees generally; specific species' uptake rates vary.

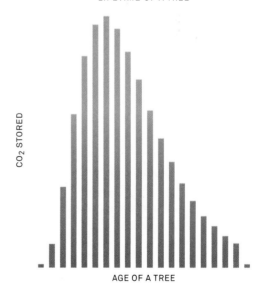

RELATIVE CO_2 FIXATION RATE OVER THE
LIFETIME OF A TREE

CO_2 STORED

AGE OF A TREE

SOURCE: Börje Kyrklund, *Unasylva* 163, 1990

One of the most significant new findings about the vulnerability of forests throughout the world to the impact of the climate crisis came in a massive study in 2009 by the International Union of Forest Research Organizations (IUFRO), which concluded that a temperature increase of 4.5°F (2.5°C) could cause many forests throughout the world to lose their role as net absorbers of CO_2. They might then instead become net contributors of CO_2 to the atmosphere.

The IUFRO found in its study: "Several projects indicate significant risks that current carbon regulating services will be entirely lost, as land ecosystems turn into a net source of carbon beyond a global warming of 2.5°C.... Moreover, since forests also release large quantities of carbon if deforested or impacted by other degrading stressors, they exacerbate climate change further."

This appears to have already begun in the boreal forests of Canada, according to a study by the Canadian Forest Service, which concluded that Canadian forests have now become net contributors of CO_2 to the atmosphere. The principal causes of this shift in the Canadian forests were rampaging mountain pine beetles in British Columbia and Alberta, which have already destroyed more than 30 million acres of mature forests. In the U.S., seven million acres have already been affected, and it is estimated that another 14 million soon will be. The beetles are no longer held back by a sufficient number of cold days to keep them in balance. Moreover, wildfires have spread at record rates—and the trees are more vulnerable to fire when warmer temperatures reduce soil moisture and expose the trees to drought as well as beetles.

Indeed, evergreen forests throughout western North America and much of Europe are experiencing a historic assault from bark beetles and pine beetles with the lengthening of summer and the sharp reduction in the number of cold snaps that kill off beetle larvae. The U.S. Geological Survey found in January that tree deaths in the old-growth forests of the American West have already more than doubled—a trend they say is likely caused by warming in the West and related drought conditions.

Independent research at the University of Arizona at the Biosphere 2 facility has confirmed that temperature increases make some trees much more likely to die during droughts. Although the study raised temperatures 4°C, the experiment showed for the first time that warmer temperatures stress these trees and make them more vulnerable to beetles and drought.

A team of researchers led by A.L. Westerling of the Scripps Institution of Oceanography published a study in *Science* magazine in 2006 documenting the fact that "large wildfire activity increased suddenly and markedly in the mid-1980s, with higher large-wildfire frequency, longer wildfire durations, and longer wildfire seasons. The greatest increases occurred in mid-elevation, Northern Rockies forests, where land-use histories have relatively little effect on fire risks and are strongly associated with increased spring and summer temperatures and an earlier spring snowmelt."

Moreover, researchers at the University of Tel Aviv found compelling evidence that future temperature increases will bring a significant increase in the average number of lightning strikes. According to their study, each additional one-degree increase in temperature is likely to bring a 10 percent increase in the number of lightning strikes—so that a 5°C (9°F) increase would bring, on average, 50 percent more lightning strikes, thus increasing the number of wildfires still more.

Some trees and plants do grow faster as CO_2 levels increase. However, while some areas may well see faster forest growth until heat and drought

WINTERS HAVE NOT BEEN COLD ENOUGH TO KILL OFF MOUNTAIN PINE BEETLE LARVAE IN COLORADO. MORE THAN 600,000 ACRES OF FOREST IN THAT STATE ARE NOW AFFECTED BY THE PEST.

exceed the CO_2 fertilization effect, many will not. Indeed, most cannot unless other inputs—like water and nitrogen—are also increased. Moreover, new research indicates that increases in average soil temperature can damage soil fertility by interfering with the availability of some volatile molecular components important for tree nutrition—like nitrogen.

And the larger point, as recent studies prove, is that temperature stress, increased beetle infestations, larger and deeper droughts, and more fires all lead to the accelerating loss of trees and forests.

Some researchers have noticed a surge in secondary-forest growth in areas once cleared in countries like Panama and Costa Rica, where the relative prosperity in cities and towns has drawn migrants from rural areas and reduced the pressure of poverty and population growth on the forests. But in most tropical countries, this phenomenon has not been replicated, and the pressure on the land continues unabated. This same demographic phenomenon is partially responsible for the natural regrowth during the 20th century of large areas of previously deforested land in North America and Europe. Europe and North America have long been restoring their net forest cover, and China is doing so now far more effectively than any other nation in the world.

The new growth usually does not fully replace the ecosystem services provided by the forests that have been destroyed. There is a wide variation depending upon the way in which the replanting is carried out. If all the trees are the same species, the lack of diversity makes them much more vulnerable to the coming increases in droughts, fires, and beetles—and makes them far less hospitable to the rich web of biodiversity that thrives in mature and diverse forests—particularly rain forests.

Moreover, it takes many decades for the new trees to reach a level in their growth curve that allows them to begin sequestering large quantities of CO_2 from the atmosphere.

On the other side of the ledger sheet, mature forests have so many trees that are past their growth peaks that the forest as a whole is often nearly CO_2 neutral. It contains an enormous amount of sequestered carbon, but the net sequestration of new CO_2 is far less than that in younger, still rapidly growing trees. As a result, when all the carbon is disgorged with the destruction of mature forests, it takes a long time for the CO_2 to be pulled back into the trees that are planted to replace the old forest.

The sequestration of CO_2 and the preservation of the earth's biodiversity are only two of the environmental benefits—sometimes called "ecosystem services"—provided to people by forests. They reduce temperature extremes, provide a source of income when managed properly, reduce soil erosion, enhance the availability of clean water, prevent desertification, protect against coastal erosion, control avalanches, provide habitat for wildlife that is needed by society, and serve as the home for up to 90 percent of the known species on the land areas of the earth. They also enhance the productivity of sustainable agriculture in areas in and around the forests.

Forests also actually bring significantly more rain than would fall if the trees did not exist—by seeding clouds with bacteria that float up from the trees and serve as "nucleators" for the formation of ice crystals that mark the first step in the formation of rain in clouds. Water vapors high in the atmosphere bind together to make crystals only at temperatures much lower than freezing. However, the bacteria from trees allow the crystallization process to occur at less frigid temperatures because

they contain a kind of scaffolding in their protein structure that allows the water vapor in the air to coalesce and bind together around the bacteria to cause rain. Brent Christner, a microbiologist at Louisiana State University, led a team that recently found that this cloud seeding by trees is far more common and more significant than had been previously understood when the phenomenon was first discovered in the 1970s.

Forests also modulate the hydrological cycle by absorbing heavy rains, enhancing the seepage of water into the ground held firm by their roots, and reducing surface runoff. In this respect, they even out the availability of water throughout the year—much as ice and snow in the mountains do. In rain forests like the Amazon, the transpiration of moisture back into the air allows the moisture to roll downwind in waves of vapor that nourish

Charles David Keeling's groundbreaking study of CO_2 levels at the Mauna Loa Observatory in Hawaii—in collaboration with Roger Revelle—revealed both the seasonal cycle of atmospheric carbon dioxide and that the amount of CO_2 in the atmosphere is increasing over time.

the entire forest. Indeed, Tom Lovejoy, one of the world's experts on the Amazon, says that it actually makes half of its own rainfall and provides moisture to other parts of Brazil lying south of the western Amazon.

For all of these reasons, the world as a whole has an incentive to place a sufficient value on carbon to encourage the preservation of the earth's ability to more quickly reabsorb the man-made CO_2 put into the atmosphere and to avoid adding yet more CO_2 through continued forest destruction.

Most proposals for a global solution to reducing CO_2 emissions include, in one form or another, a grand bargain between the North and the South, within which flows of aid from wealthier countries to less-developed countries are increased to finance

THE KEELING CURVE

The enormous role played by forests in sequestering CO_2 can be seen in the famous Keeling curve, which measures the rapid accumulation of CO_2 in the atmosphere since measurements began in 1958. The stair-step pattern reflects the annual tilting of the Northern Hemisphere toward the sun in summer and away from the sun in winter. When the deciduous trees in the Northern Hemisphere (much larger than those in the Southern Hemisphere) lose their leaves, the exhalation of CO_2 into the atmosphere causes a sharp jump in CO_2. When the same trees grow leaves anew the following spring and summer, the amount of CO_2 in the atmosphere comes back down. The fact that these concentrations continue to increase from one year to the next reflects the massive burning of coal, oil, and natural gas and the massive amount of deforestation taking place.

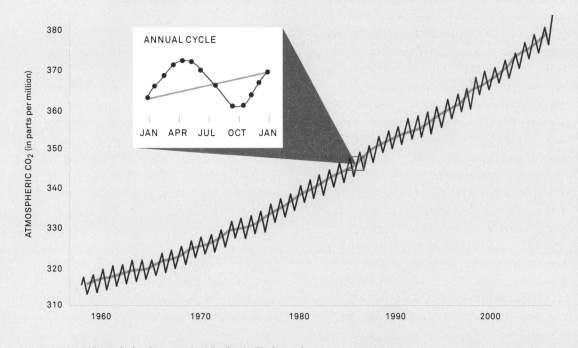

SOURCE: NOAA/Scripps Institution of Oceanography, University of California, San Diego

the changes necessary to fight poverty while stopping rampant deforestation. As a result, a great deal of attention has been focused on ways for the world community to assist tropical countries in changing the prevailing pattern of deforestation.

The Kyoto Protocol of 1997 specifically addressed the need to control deforestation and to maintain the forests of the world through afforestation and reforestation, in Article 3.3. However, the treaty did not include any mechanisms for accomplishing this goal—because of deep concerns on the part of developed countries that there was no reliable and accurate measuring system for determining which forests would have been preserved anyway in the absence of the treaty, nor for detecting "leakage" in the form of deforestation that is not reported by the countries where it has taken place.

Since then, satellite measuring techniques have improved dramatically, to the point where scientists now have confidence that they can accurately monitor exactly what is happening on the ground throughout the world—in some areas, literally tree by tree. It is now possible to establish highly accurate "national baselines" that can virtually eliminate the problem of leakage. In addition, the explosive growth of environmental NGOs (nongovernmental organizations) in every country over the past decade has dramatically improved knowledge of land use policies and intentions relating to almost every forest in the world.

As a consequence, negotiators in Bali, in December 2007, were able to couple deforestation goals with reductions in industrial and transportation emissions. Although this conceptual framework must be fleshed out and ratified during the Copenhagen negotiations in late 2009, it has served as the basis for a truly global framework for reductions of CO_2 and other greenhouse gas emissions that includes, for the first time, both land use and the burning of fossil fuels.

Several global certification systems have been organized to identify which forests are sustainably managed, so that purchasers can avoid contributing to deforestation. Coupled with regulations cracking down on illegal forest practices—and, hopefully, the placing of a price on carbon—these efforts to affect the demand for products that are sustainably grown and harvested represent a key part of the solution to deforestation.

A second initiative, the Forests Dialogue, was formed in 1999 by the World Bank, the World Business Council for Sustainable Development, and the World Resources Institute. Its mission is to foster a constructive dialogue among all the important stakeholders in the outcome of forest practices and to address the key issues that must be solved through mutual understanding and agreements among stakeholders with different priorities and incentives. They seek to provide substantial additional funding for the building of institutional capacity in forest countries to tackle the drivers of deforestation and support sustainable development while remaining accountable for good forest governance.

In many areas, success will require attacking corruption among local, regional, and national officials that results in the poor enforcement of existing laws and regulations designed to stop destructive deforestation practices and illegal logging and land clearing. In many areas, including the Amazon and the central Congo Basin, the lack of clear property rights for indigenous peoples and longtime residents who live in the forest is also a cause of this ongoing destruction.

As most experts have pointed out, there is no one-size-fits-all approach, but in preparation for a global effort that is expected to be part of a comprehensive climate treaty, many organizations

IN CHINA, THE GOVERNMENT HAS PLANTED MILLIONS OF TREES, INCLUDING BROAD BELTS OF TREES AROUND BEIJING, AS PART OF ITS EFFORT TO "GREEN" THE CITY AND REFOREST DEGRADED LAND.

are working toward the establishment of national baselines as a first step toward the creation of national accounting systems and credible country-level monitoring systems. The ultimate key to success in deforestation is the placing of a price on carbon.

Some nations are proceeding with national reforestation and afforestation programs, even in the absence of a global treaty. China leads the world in tree planting, with a highly effective afforestation and reforestation program that has planted two and a half times more trees in each of the last several years than has the rest of the world combined. In fact, China has been planting so many trees that it is now planting more by a third each year than the largest deforesting nation, Brazil, is cutting down. It should be noted, however, that while it is protecting and expanding its own forests, China has contributed greatly to the demand for wood from tropical forests, and among its recent African land purchases are 6.9 million acres in Africa's Congo Basin, which it says it intends to convert into palm oil plantations.

China stopped deforestation more than 10 years ago, and in 1981, the National People's Congress declared that all citizens of China above the age of 11 (and until age 60) have a duty to plant at least three trees each year. The planting usually takes place in March and April, during spring for most of China. The Chinese tree-planting program is driven by the central government in Beijing, with cooperation from regional leaders. The Chinese people planted 11.7 million acres of forests in 2008 alone—a 22 percent increase over 2007, according to statistics released by the Chinese National Greening Committee.

Chinese schools require each student to plant at least one tree before graduating, and most schools set aside time for a "green education"

program. The nation announced last year that it will spend almost $9 billion on its tree-planting program for the year and set a goal of covering 20 percent of the nation in forests by next year. The president of China, Hu Jintao, has personally taken part in tree planting to underscore its importance as a national priority.

Professor Wangari Maathai, Nobel Peace Prize laureate for 2004 and founder of the Green Belt Movement in Kenya, has been responsible for planting more than 30 million trees in Kenya and 11 other African countries over the past 30 years. She has also been responsible for convincing the United Nations Environment Program to launch a major initiative, Plant for the Planet: Billion Tree Campaign, which has already succeeded in planting more than three billion trees and has now set a new goal of planting seven billion trees.

In addition to the Chinese program, the next-largest tree-planting programs are in Spain, Vietnam, the United States, Italy, Chile, Cuba, Bulgaria, France, and Portugal.

Recently, Brazil proposed a program to pay small farmers to plant new trees in areas of the Amazon that have been deforested, but thus far, the program has had little apparent impact.

The World Agroforestry Center recommends that deciduous trees be planted in regions where water supplies are a problem because they require less water than evergreens and are more adaptable to periods of water shortage. If every person on Earth planted and cared for at least two tree seedlings each year, the world could make up for the loss of trees to deforestation over the previous 10 years in the next 10 years. International cooperation could provide funding for jobs in tropical countries planting trees and thereby fight poverty and the climate crisis simultaneously. Just as important, we could help to solve the extinction crisis.

VILLAGERS IN BURKINA FASO SHOW A SNAPSHOT
OF THEIR LAND TAKEN IN 1986, BEFORE THEIR
REFORESTATION EFFORTS BEGAN.

CHAPTER TEN

SOIL

SOIL IS THE LIVING SKIN OF THE EARTH. A
SCIENTIST STUDIES HISTORIC SOIL PROFILES IN
WASHINGTON STATE'S PALOUSE RIVER BASIN,
WHICH HAS SUFFERED FROM EROSION.

As a boy growing up during the summers on my family's farm in Tennessee, I learned from my father how to recognize the richest and most productive soil: In a word, it's black. It's also porous and moist. But it was not until much later in my life that I learned the reason fertile soil is black: it's the carbon.

The soils of the earth contain between three and four-and-a-half times as much carbon—just in the first few feet—as do the plants and trees, and more than twice as much as is currently in the atmosphere. With improved agricultural and land management practices, we can significantly increase the amount of CO_2 that is pulled from the atmosphere by vegetation and left sequestered in the soil, while enhancing agricultural productivity and food security—and restoring degraded lands—at the same time.

As with other climate solutions, however, the success of this promising strategy depends upon large-scale changes in long-established patterns.

When my father was a young man, the greatest threat to the productivity of the land in America was soil erosion. Farmers and landowners in his generation were enlisted by Franklin Delano Roosevelt in a nationwide struggle to stop the soil erosion that had led to the Dust Bowl of the 1930s and had left many farms cut with deep gullies, through which the best soils for agriculture were washed away. To this day, I remember the lessons he taught me. For example, when walking across the farm, keep an eye out for the first hint of erosion; stop the beginning of a gulley before it can start to deepen.

The battle against soil erosion and the degradation of soil quality on farms in the United States is a success story. And by slowing soil erosion, the United States has also begun to restore the carbon content of the soil. The crucial layer of humus in healthy soils is made up, on average, of 58 percent carbon. There are, however, new threats to the quality of soil and new opportunities to sequester much more carbon in the soil.

More important, in much of the developing world—particularly in sub-Saharan Africa—the degradation of soil quality continues to worsen and has reached levels that threaten food security for hundreds of millions of people. The loss of soil carbon in degraded African croplands has already exceeded the magnitude of losses in the United States 80 years ago, just prior to the Dust Bowl. Addressing this threat effectively can not only make African soils more fertile and productive, it can simultaneously pull enormous amounts of CO_2 from the atmosphere of the planet and sequester it in the soil as it is restored to health.

The thin layer of soil that lies atop the outer shell of the earth's crust is roughly analogous to the "skin" of the earth. It is alive, in the sense that it is filled with microbes, fungi, worms, minerals, and nutrients, all of which make possible the growth of plants and trees—which use photosynthesis to combine CO_2 from the air with water, nitrogen, organic carbon, and other nutrients and minerals

CARBON IN OUR SOIL

Soil plays an active role in the earth's carbon cycle, storing an estimated three to four-and-a-half times as much carbon as all of the earth's plant matter combined. Carbon moves into soil through plant roots and decaying organic matter, such as leaves and tree limbs. Some of that carbon is soon cycled back into the atmosphere, but much of it remains in the soil. Fungi, bacteria, and other microorganisms that aid in the decomposition of organic matter are a significant pathway through which carbon becomes part of the soil.

FUNGI

HUMUS

BACTERIA

from the soils. The complex biogeochemical process of plant growth relies heavily on symbiotic relationships between the vegetation and microbes that are responsible for shepherding the exchange of molecules back and forth between the roots of the plant and the soil.

When these processes are disrupted by exploitive land-use methods, by the cutting and burning of trees and vegetation, by plowing, and by the excessive and indiscriminate use of synthetic chemical compounds on the land, the result can be short-term increases in crop yields at the expense of long-term fertility in the soil. And the loss of soil fertility inevitably results in the disgorging of those large amounts of carbon normally stored in healthy, regenerating soils.

The Agricultural Revolution began soon after the end of the last ice age more than 10,000 years ago in the Fertile Crescent, stretching from Egypt through Mesopotamia, and emerging in the same era in India and southern China. Early in the development of agriculture, the first primitive version of a plow was used—a vertical piece of wood dragged by two people through the top layer of the soil. Two millennia later, when oxen were domesticated in Mesopotamia, plowing became more efficient. Later still, around 3500 B.C., the plowshare was developed, with the addition of an iron edge to the wooden blade of the plow to more effectively loosen the topsoil. The all-iron Roman plow, dated to the year A.D. 1, was an improvement in this technology that was used for a thousand years until it evolved into the basic moldboard design that turns over the soil as it is broken.

In 1837, a blacksmith in the American Midwest, John Deere, developed and marketed an improved cast-iron moldboard plow that American settlers began using to break topsoil as they moved westward. When the first tractors were hitched to John

SOME OF THE WORLD'S WORST SOIL EROSION
IS SEEN IN THE LOESS PLATEAU IN SHAANXI
PROVINCE, CHINA. ONLY SMALL FIELDS FOR CROPS
REMAIN AS THE LAND IS STRIPPED AWAY.

CONVENTIONAL PLOWING TECHNIQUES CAN
ACCELERATE THE LOSS OF TOPSOIL AND LESSEN
THE AMOUNT OF CARBON STORED IN SOIL.

Deere's plow at the beginning of the first decade of the 20th century, the prairies were broken open, releasing enormous amounts of carbon from Midwestern soils, loosening them and beginning the soil erosion calamity that became the Dust Bowl of the 1930s. That's what led to the emphasis on soil conservation and the lessons my father later taught me when I was a child.

From the late 1800s through the beginning of World War II, the carbon content of U.S. soil decreased by more than 50 percent. Agricultural reformers began to advocate the abandonment of plowing after the publication in 1943 of *Plowman's Folly* by Edward Faulkner. But the "no-till" and "conservation-till" movement did not pick up speed until the introduction of herbicides, after World War II, when chemical-weapon stocks were converted en masse into herbicides.

The principal purposes of plowing are to make it easier to put seeds into the soil and to control weeds. The loosening of the soil also makes it more porous, but the experience of the 20th century proves decisively that any benefits that might result from quicker absorption of water and fertilizer are all too often far outweighed by making the soil more vulnerable to erosion.

The combination of herbicides and the mechanical planting drill now make plowing largely unnecessary. Herbicides, however, bring problems of their own: numerous health risks are associated with the more potent herbicides, and each pound causes six pounds of carbon (22 pounds of CO_2) to be released during its manufacture. Moreover, the planting drill requires larger tractors with more horsepower (and higher fuel consumption) than is commonly found in less-developed countries.

The management of plant residue is also an important factor in the conservation of soil and soil carbon. If these residues are removed from the land for livestock feed or biofuels, so is one of its most valuable protections against soil erosion by water and wind. These residues are also important sources of soil fertility regeneration and are food for soil organisms. Indeed, this is one of the arguments in favor of no-till methods of cultivation, which leave the soil less disrupted, less depleted, and less vulnerable to erosion.

However, of the 3.75 billion acres under cultivation in the world today, less than 250 million acres now utilize the no-till technique, mostly in the United States, Brazil, Argentina, Canada, and Australia. That is one of the reasons why soil carbon and soil fertility are still being degraded at much higher rates in developing countries. Moreover, many African farmers are granted land tenure only one year at a time—which encourages them to remove the plant residue for an additional source of income, rather than leaving it on the ground to protect and regenerate the soil for the following year.

Although the beginnings of the climate crisis are usually dated to the beginning of the Second Industrial Revolution a century and a half ago, the first large additions to the atmosphere of man-made CO_2 actually came from the vast changes in land use across the surface of the planet when forests were cleared as the Agricultural Revolution gained momentum. Later, the mass use of plowing disgorged large additional amounts of CO_2 from vegetation and soils.

Indeed, some scientists calculate that it was not until the 1970s that fossil fuel burning became a larger source of global warming pollution than the combination of agriculture and deforestation. Rattan Lal, a land-use expert at Ohio State University, estimates that during the last 10,000 years, roughly 470 gigatonnes of carbon (four gigatonnes equals one part per million of CO_2 in the atmosphere) have come from cutting and burning trees and degrading soils, while about 300 giga-tonnes have come from the burning of fossil fuels.

That ratio has changed dramatically in the past half-century, of course. CO_2 emissions from fossil fuel burning have accelerated considerably in recent decades. Lal's calculations indicate that more than four times as much global warming pollution now comes from fossil fuel burning compared with changes in land use.

Furthermore, much of the CO_2 released during the early stages of the Agricultural Revolution has long since been recycled back into the land and its vegetation. Charlotte Streck, one of the founders of Climate Focus, said recently, "Flux of soil organic carbon between the earth and the atmosphere is among the largest global carbon flows on the planet." Dr. William H. Schlesinger, president of the Cary Institute of Ecosystem Studies, has calculated that approximately 10 percent of atmospheric CO_2 passes through the soil each year, though Lal puts the figure at 7.5 percent.

Nevertheless, the current pattern of agriculture and soil degradation continues to be responsible for an enormous amount of global warming pollution. In spite of America's success in dealing with soil erosion, the mechanization of agriculture during the 20th century, the quadrupling of human population, changes in diet, the availability of plentiful sources of oil-based diesel and gasoline, and the use of synthetic nitrogen fertilizers have all combined to make modern agriculture one of the largest sources of global warming pollution.

The Intergovernmental Panel on Climate Change (IPCC) indicates that agricultural land use contributes 12 percent of global greenhouse gas emissions, just from methane and nitrous oxide, both of which are far more powerful—molecule for molecule—than CO_2 in trapping heat. In the United States, government research shows that,

with chemical fertilizers, herbicides, and heavy fossil fuel use, U.S. agriculture contributes nearly 20 percent of U.S. CO_2 emissions.

Moreover, the continuing degradation of soils and widespread slash-and-burn land-use strategies in less-developed countries contribute greatly to the destructive impact our current global agriculture system has on the climate of our planet.

By changing this pattern, we can not only reduce emissions of CO_2, methane, and nitrous oxide, we can also pull a significant percentage of the CO_2 already accumulated in the atmosphere into the soils, where much of it can be sequestered for hundreds or even thousands of years. And, as is the case with most of the solutions to the climate crisis, the co-benefits are also of great value to human civilization.

Ironically, the greatest opportunities for sequestering CO_2 in soil are on already degraded lands. For example, the restoration of prairie grasslands throughout the world represents an unparalleled opportunity to pull CO_2 out of the atmosphere and into the soil. Grasslands are particularly effective in sequestering carbon because of the high carbon input from the roots of tall grasses.

Africa, in particular, is believed by many experts to be facing serious impending food shortages because of degraded soils and rapidly growing populations. Hans van Ginkel, former under secretary-general of the United Nations, said: "The low fertility of African soils is the single most critical impediment to the region's economic development. We cannot begin to make real progress in the battle against poverty and malnutrition in Africa until the problem of degraded soil is addressed."

According to Lal, throughout sub-Saharan Africa, "most agricultural soils have lost 50 to 70 percent of their original soil organic carbon pool, and the depletion is exacerbated by further

Modern agriculture is one of the largest sources of global warming pollution.

PESTICIDES ARE SPRAYED FROM THE AIR OVER
THIS FIELD NEAR MEMPHIS, TENNESSEE.

soil degradation and desertification.... In West Africa, extractive farming practices, overgrazing, and demand for fuelwood has led to serious land degradation."

Streck pointed out recently, "More than 80 percent of farmland in sub-Saharan Africa is plagued by severe degradation due to population growth, inability to afford fertilizer, deforestation, and use of marginal lands."

The largest amount of carbon presently sequestered in soils is in wetlands—including peatlands. Many experts contend that the single biggest

the soil because the low temperatures inhibit the microbial activity that releases carbon.

If global warming continues to thaw these frozen soils of the Far North, large quantities of CO_2 and methane could be released into the atmosphere relatively quickly. The only way to prevent this catastrophe is to slow and then reverse the accumulation of global warming pollution that is driving up global temperatures.

As scientists focus on ways that improved agricultural techniques can sequester more carbon in the soil, the productivity of agriculture worldwide

"We cannot begin to make real progress in the battle against poverty and malnutrition in Africa until the problem of degraded soil is addressed."

HANS VAN GINKEL

change needed in land-use patterns to reduce global warming pollution is to ban the clearing, drying, and burning of peatlands. Indeed, Thomas Lovejoy, biodiversity chair of the Heinz Center, has proposed that exposed peatlands, where the tree cover has been removed, should be "rewetted" to protect them from drying and releasing their CO_2.

The second-largest amount of carbon contained in soils is found in the cold and frozen ground of the Far North—in the tundra and the boreal soils of the Arctic and sub-Arctic. Even though they have less vegetation aboveground, they contain enormous quantities of carbon that has accumulated in

is also threatened by the climate crisis. Although the impacts are different from one region to another, the world as a whole would experience a serious decline in agricultural productivity with continued global warming. Less-developed countries in the tropics and subtropics are at increased risk for significant declines in agricultural productivity because temperatures there are already at or above crop tolerance levels.

More water is needed by plants to keep cool in the face of higher temperatures, and some plants have an upper limit to the temperatures they can tolerate without dying. This will be a particularly

IN KEITA, NIGER, VILLAGERS JOURNEY SIX MILES TO COLLECT FIREWOOD. THE LOCAL WOODLANDS WERE DESTROYED YEARS AGO, SPEEDING THE DEGRADATION OF THE SOIL.

RECORD-BREAKING DROUGHT AND HEAT IN AUSTRALIA HAS BEEN DISASTROUS FOR THE COUNTRY'S AGRICULTURE. IN VICTORIA STATE, APPLES HAVE BEEN SUNBURNED ON THEIR TREES.

serious threat if temperatures increase by as much as is now predicted under the business-as-usual scenario: 11°F in this century.

One of the countries predicted to experience the harshest agricultural impact is India, which scientists say could suffer a decline in agricultural productivity of 30 to 40 percent during this century under a business-as-usual scenario—even as it overtakes China as the most populous nation in the world. Even worse, Sudan faces a decline of as much as 50 percent in agricultural production, and Senegal is projected to experience a 52 percent decline. Mexico's agricultural productivity is projected to decline by a third.

The erratic nature of changes in long-established climate patterns also complicates the ability of farmers to predict the right times for planting. Dr. Jerry L. Hatfield, director of the National Soil Tilth Laboratory, testified this year that "extreme events, like heat waves and regional droughts, have become more frequent and intense in the past 50 years and affect agricultural operations and decision-making."

Arthur Yap, the secretary of agriculture in the Philippines, told me last year that, throughout his life, farmers have been able to confidently predict the arrival of seasonal rains during the first two weeks of June, but now, that long-stable pattern has been disrupted and he cannot advise his nation's farmers what to expect and when to plant. The scrambling of seasons poses a particular problem for plants that emerge earlier with warmer and earlier springs but then become vulnerable to frost when cold snaps nip them in the bud.

The IPCC has warned that changes in the availability of water—in the right amounts and at the right times—will have a particularly disruptive impact in many growing areas. The number of extreme downpours, with associated flooding and soil erosion, is predicted to increase significantly. At the same time, the number and severity of droughts in most mid-continent regions will also increase. The combination of heavier downpours and deeper droughts is expected to markedly reduce crop yields, on average, around the world. Since approximately 95 percent of U.S. agriculture depends solely on rain, and not irrigation, the increasingly erratic pattern of alternating heavy downpours and extended droughts would have a harsh impact. According to the Federal Emergency Management Agency, the United States already suffers losses from droughts averaging $6 billion to $8 billion per year, much of which is in crop loss.

For those growing areas in the West that depend on seasonal melting from the snowpack in the mountains, the rapid loss of snowpack, along with earlier spring melting and a pronounced shift from snowfall to rainfall, is already posing serious difficulties that are expected to worsen with continued global warming.

The worst impacts in the United States are projected to occur in the Southeast and in the southwestern plains, with predicted declines of 25 to 35 percent in agricultural productivity. And these projections do not include the impact of additional insects, more drought, and less water for irrigation. Some states in the upper Midwest could see increased agricultural productivity, depending upon what happens to rainfall in the region.

One consequence of modern agriculture's increased reliance on a few varieties of hybrid crops is their adaptation to the narrow range of ecological conditions, including temperature and rainfall, in the areas in which they are grown. This specialization can also make them more vulnerable to the predicted increase in extreme-weather events—particularly periods of very high temperatures that last for more than a few days. Some plants are especially vulnerable to prolonged periods of higher nighttime temperatures, which speed the rate of development and hasten the reproductive stage, effectively shortening the growing period and productivity of the plant.

Another consequence of increased heavy downpours, coupled with fewer precipitation events, is delayed spring planting—which has its harshest impact on those farmers who depend upon the premium paid for early production of high-value crops. Similarly, flooding at harvesttime causes particularly heavy losses.

Higher temperatures also have a pronounced impact on soil moisture evaporation rates, compounding the impact of intermittent droughts. Higher average temperatures—especially during summer months—will have the harshest impact in areas where high temperatures already impose a limit on agricultural production. In some drier areas, increased loss of soil moisture can also lead to dust storms. In 2009, the American West suffered an unusual increase in dust storms. Longer and hotter heat waves also cause stress in animals and have been responsible already for livestock deaths.

Rattan Lal includes in his "Laws of Sustainable Soil Management" law number four: "The rate and susceptibility of soil to degradation increase with increase in mean annual temperature and decrease in mean annual precipitation."

The threat to agriculture from insect infestations is also projected to increase in a warmer world. Longer growing seasons are accompanied by more generations and larger populations of insects.

As noted in Chapter 9, higher levels of CO_2 can stimulate plant growth, but only where water and other nutrients (especially nitrogen) do not impose limits on the extra growth that would otherwise occur. In addition, those plants benefiting from higher carbon dioxide levels often grow larger but become less nutritious because of a reduction in nitrogen and protein content. This appears to be especially true for grasses, causing animals raised in pastureland to graze more in order to get the same protein.

Moreover, higher CO_2 levels stimulate the growth of weeds much more than of food crops. And these CO_2-enhanced weeds become more resistant to herbicides.

One of the unwanted weeds that thrive in higher CO_2 environments, by the way, is poison ivy—which not only grows larger with more CO_2 but also produces a much stronger form of urushiol, a poison that 80 percent of people are vulnerable to. According to the United States Global Change Research Program, "Given continued increases in carbon dioxide emissions, poison ivy is expected to become more abundant and more toxic in the future."

There is also projected to be a northward expansion of the range of some weeds that are now confined to southern latitudes.

Increased wildfires and strong wind events are also predicted to take a heavy toll on agriculture. Increased low-level ozone, which is toxic to plants and inhibits growth, will also, in some areas, offset any increased growth from CO_2 enrichment.

One of the other lessons my father taught me on our family farm was that the corn we grew each summer to supplement winter feeding had

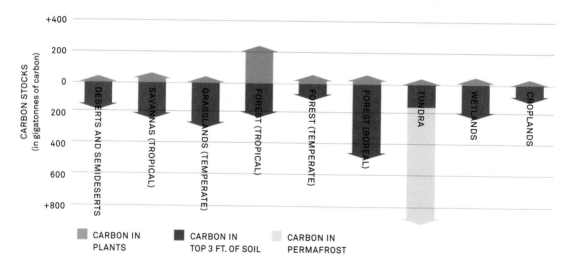

CARBON IN THE ECOSYSTEM

Vegetation and soil both serve as enormous sinks of carbon. There is even more carbon contained in soils than in trees and other plants. Scientists have discovered that the most important soil sinks are wetlands, grasslands, and peatlands. Some methods of land management can increase the amount of carbon in the soil; many farming methods, however, reduce it.

SOURCE: IPCC

ORGANIC FARMING, OFTEN PRACTICED ON A
SMALL SCALE, HAS BEEN PROVED TO INCREASE
THE CARBON CONTENT OF SOIL. FARMER AND
SON WORK IN WHATCOM COUNTY, WASHINGTON.

to be rotated each year from one field to another
so that the soil in which the previous year's
crop was grown could be replenished with nutrients, by spreading cow manure or planting clover,
alfalfa, or other plants that put nitrogen back into
the soil.

Crop rotation and the spreading of animal
manure used to be the primary ways of restoring
nitrogen in soils depleted by corn and other crops
that pull large amounts of nitrogen from the soil as
they grow. But modern industrial agriculture has
disrupted the age-old ecological balance between
animals and plants on farms. Most of the animals
have been removed from farms and concentrated
instead in large feedlot operations, where the
manure is no longer valued as a beneficial fertilizer but must be dealt with as a large source of

pollution. Indeed, the unnatural force-feeding
of corn to cattle, whose digestive systems evolved
to eat grass, makes their stomachs—and their
manure—more acidic.

The net result is that the manure from factory
farms is unfit for use as fertilizer because its newly
engineered toxicity stunts the growth of plants and
seriously damages the fertility of soil. So farmers,
wisely, won't use it.

Before the discovery in Germany in 1909 of a
workable method to create synthetic ammonia,
farmers relied on the spreading of manure on fields
and the use of crop rotation techniques to periodically replenish nitrogen in their soil in order to
maintain its productivity. After World War II, however, the reliance on synthetic nitrogen fertilizers
grew quickly when the German technology for

synthesizing ammonia was applied to the large stockpiles of ammonium nitrate left over from the manufacture of munitions during the war. Cheap oil from the Middle East, and the later development of large reserves of natural gas, provided the enormous inputs of energy and hydrogen that made possible the manufacture of vast quantities of synthetic nitrogen fertilizer.

As Michael Pollan skillfully describes in *The Omnivore's Dilemma,* the surge of new supplies of affordable synthetic nitrogen fertilizer coincided with the introduction of new varieties of hybrid corn and changes in agricultural policy to subsidize the overproduction of grain. Indeed, U.S. agricultural policy since the early 1970s has heavily subsidized the production of as much food as possible, with the political objective of keeping food prices as low as possible and stimulating farm exports to the rest of the world. Most large-scale farming operations today would lose enormous sums of money if not for the taxpayer subsidies that encourage the continuation of wasteful and harmful practices. Industrial agriculture now uses 10 calories of energy from fossil fuels to produce one calorie in food.

What was not understood at the time this new pattern of production was adopted is that the application of nitrogen fertilizer stimulates not only the growth of plants but also the growth of carbon-hungry bacteria in the soil. Every ton of synthetic nitrogen fertilizer requires the burning of enough natural gas to release 1.25 tons of carbon (or 4.6 tons of CO_2) into the atmosphere.

When the fertilizer is spread on top of the soil, every ton consumed by bacteria in the soil causes them to also consume 30 tons of carbon. When the nitrogen fertilizer is applied quickly, the bacteria multiply quickly and deplete carbon from the soil. Like all such processes, this one is resistant to

oversimplification; the net result includes the removal of soil carbon in large quantities, yet nitrogen and other elements are necessary for conversion of carbon in the biomass into humus.

This heavy use of nitrogen fertilizers causes an overload of nutrients in the waterways into which they drain, which stimulates the growth of algae in the areas of the ocean—like the Gulf of Mexico—into which these streams eventually run. When the stimulated algae blooms die, their decomposition starves the water of oxygen, kills fish, and creates what scientists call dead zones. The number and extent of these dead zones in the ocean are now increasing rapidly.

Many have described our current reliance on synthetic nitrogen fertilizers as a Faustian bargain, in that it ensures a plentiful bounty in the short term, but at the expense of the depletion of the necessary carbon from the soil in the long term. The use of synthetic nitrogen fertilizers is a little like the use of steroids by athletes. Their muscles grow at an unnatural rate, but the health and integrity of their bodies is degraded in ways that are hard to detect at first but can become quite serious over time.

In recent years, natural gas has been the primary source of energy and hydrogen for the making of synthetic ammonia—the source of synthetic nitrogen fertilizer. According to Ford B. West, president of the Fertilizer Institute in the United States: "As much as 90 percent of the cost of producing a ton of ammonia, the building block for all other nitrogen fertilizers, can be tied directly to the price of natural gas. This makes nitrogen production one of the most energy-intensive manufacturing processes that exist."

Domestic manufacturers supplied 85 percent of nitrogen used by U.S. farmers until the year 2000. Since then, 26 facilities have closed, according to West, "due primarily to the high cost

Industrial agriculture now uses 10 calories of energy from fossil fuels to produce one calorie in food.

DIESEL TRUCKS AND FORKLIFTS ARE USED TO MOVE LARGE BLOCKS OF ALFALFA DESTINED TO FEED DAIRY CATTLE IN NORTHERN CALIFORNIA.

It takes more than seven pounds of plant protein to produce one pound of beef—and more than 6,000 gallons of water!

AT INDUSTRIAL-SCALE LIVESTOCK OPERATIONS, ANIMAL MANURE MUST BE TREATED AS WASTE AND CANNOT BE USED FOR FERTILIZER BECAUSE IT IS TOO ACIDIC.

of natural gas." More than half of the synthetic nitrogen fertilizer used in the U.S. is now imported.

It should be noted, however, that some areas of the world with severely degraded lands are desperately in need of more fertilizer, not less. The impact of nitrogen fertilizer is not the same in all soils, or in all countries. For example, farmers who put straw on their land should add nitrogen fertilizer to enhance the benefits of the straw. Other fertilizers, including phosphorus and sulfur, are also beneficial.

In sub-Saharan Africa, the carbon content of degraded cropland soils is, on average, less than 10 to 20 percent of the minimum threshold necessary for adequate yields. By adding only eight pounds of fertilizer per acre—less than 2 percent of what is commonly used in the American Midwest, and less than 0.5 percent of what is used in northern China—these countries are averaging yields of only one ton of cereal grain per hectare (two and a half acres), compared with an average of 10 tons per hectare in much of the American Midwest and five to six tons in Asia.

Changes in diet have also enhanced the impact of modern agriculture on the climate crisis. Most people eat much more meat than their parents or grandparents did. On average, it takes more than seven pounds of plant protein to produce one pound of beef—and more than 6,000 gallons of water! The explosion in the numbers of cattle, pigs, and chickens has also dramatically increased methane emissions from agriculture (*Science Daily* recently reported a study showing that 72 percent of all methane emissions in Canada are from cattle). Some researchers are now discovering ways to change the diet of the animals to reduce the amount of methane they burp without reducing the quality or hurting the taste of the meat.

The per capita consumption of meat in developed countries has increased 50 percent over the past half-century, but 200 percent in developing countries. The health consequences of a more meat-intensive diet (higher rates of heart disease, hypertension, cancer, and obesity), which first appeared in wealthier countries, are now spreading rapidly to poorer countries.

The dependence on large-scale industrial agriculture—as opposed to the many small farms that used to be located near every town and city—increases the amount of fuel required to transport the food from the field or the feedlot to the table. Moreover, the huge amounts of fossil fuel energy required to drive the tractors and trucks, manufacture the fertilizers and herbicides, and deliver the food—often from one continent to another—all result in significant additions of greenhouse gases to the atmosphere.

The good news is that all of these patterns can be changed in ways that enable the use of agriculture and land management as an important solution to the climate crisis. Rattan Lal has said that better soil management has the potential to sequester up to 15 percent of the world's annual fossil fuel emissions in soils. Recarbonizing the planet by sequestering carbon in soils and trees has the capacity to remove as much as 50 parts per million of atmospheric CO_2 over the next 50 years.

Dr. Timothy J. LaSalle, CEO of the Rodale Institute, is a leader among those who have proposed a dramatic shift in agricultural policy and practices to focus on building soil organic matter—including the restoration of carbon to the soil—by organically managing and enriching soils. LaSalle points out that the application of soluble nitrogen fertilizers stimulates "more rapid and complete decay of organic matter, sending carbon into the atmosphere instead of retaining it in the soil as the organic systems do." By using no-till techniques and by enriching soils with natural sources of nutrients, farmers can cut costs and improve both productivity and profitability.

"Successful implementation of organic regenerative farming practices on a national basis will depend on two factors," LaSalle said. "A strong bottom-up demand for change and a top-down shift in state and national policy to support farmers in this transition…. Farmers should be paid on the basis of how much carbon they can put into and keep in their soil, not only how many bushels of grain they can produce. Incentives will encourage resource conservation and other carbon-enhancing means of producing crops for food, feed, and fiber."

LaSalle is more optimistic than almost any other analyst about the potential scale of what could be accomplished by such a dramatic shift in agriculture. He believes that regenerative agriculture, if practiced on all of the planet's tillable acres, "could sequester nearly 40 percent of current CO_2 emissions." Other experts believe LaSalle and Lal are too optimistic. Another leading expert on the carbon content of soils, William Schlesinger, believes that "it would require the planting of lands equivalent to the size of Texas to accumulate even 10 percent of the nation's annual CO_2 emissions from fossil fuel combustion in trees and soils." Nevertheless, as Schlesinger acknowledges, "if we could recapture even a small percentage of the historical losses of soil carbon, a large amount of atmospheric carbon dioxide might be sequestered in soils."

The difference in projections of what is possible between optimists and pessimists in the field of soil science is partly attributable to their differing views on the potential of recently discovered techniques for dramatically increasing soil carbon content fairly quickly. One of the most exciting new strategies for restoring carbon to depleted soils, and sequestering significant amounts of CO_2 for 1,000 years and more, is the use of biochar.

Biochar is a form of fine-grained, porous charcoal that is highly resilient to decomposition in most soil environments. It occurs naturally but can be manufactured cheaply in large quantities by burning wood, switchgrass, manure, or other forms of biomass in an oxygen-free or low-oxygen environment that transforms the biomass into what is more than 80 percent pure carbon.

The process by which biochar is made can also be designed to produce gas or liquid fuel that can be used to make electricity and can serve as an energy source for the making of more biochar. Moreover, a new design for cookstoves for those in less-developed countries who routinely burn wood or manure allows the burning of only the oils and gases in the wood to produce a cleaner, less air-polluting stove and produce biochar at the same time.

Burying biochar in soil replenishes the carbon content, protects important soil microbes, and helps the soil retain nutrients and water. It also reduces the accumulation of greenhouse gas pollution by avoiding the releases that would occur from the rotting of the biomass on the surface, by sequestering the CO_2 contained in the biochar, and by assisting the process by which plants growing in the soil pull CO_2 out of the air with photosynthesis.

It also increases the organic health of the soil by stimulating the growth of rhizobium bacteria and mycorrhizal fungi, both of which further improve the overall quality of the soil. David Shearer, a biochar entrepreneur who has studied the science extensively, says: "If you put biochar in the soils, it literally has a residence time on the order of centuries to millennia. It's a carbon lattice; it creates a habitat for fungi and bacteria—a habitat that creates a great deal of conductivity for exchange capacity."

FOR THE PAST 30 YEARS, THE RODALE INSTITUTE IN PENNSYLVANIA HAS EXPLORED ORGANIC FARMING METHODS TO REGENERATE THE PRODUCTIVITY OF THE SOIL.

IN WEST VIRGINIA, A POULTRY FARMER PRODUCES
BIOCHAR FROM CHICKEN WASTE AND WOOD CHIPS,
TURNING IT INTO A VALUABLE RESOURCE FOR
RECARBONIZING SOILS.

In recent decades, soil scientists discovered that Amazonian Indians were using biochar at least a thousand years ago to create fertile black soils that are still far more productive than the soils around them, even though the biochar was buried a millennium ago. These soils, called *terra preta* (Portuguese for "black earth"), provide a unique way of assessing the longevity of the benefits conferred on soils by using biochar. In addition, it appears that these rich soils gain the ability to regenerate themselves.

Climate experts—including Tim Flannery in Australia and James Lovelock in the U.K.—are enthusiastic about the potential for a global

burying it in the soil. Then you can start shifting really hefty quantities of carbon out of the system and pull the CO_2 down quite fast.... This is the one thing we can do that will make a difference, but I bet they won't do it."

In an open letter in 2008, Flannery wrote: "Biochar may represent the single most important initiative for humanity's environmental future. The biochar approach provides a uniquely powerful solution, for it allows us to address food security, the fuel crisis, and the climate problem, and all in an immensely practical manner. Biochar is both an extremely ancient concept and one very new to

"Biochar may represent the single most important initiative for humanity's environmental future."

TIM FLANNERY

biochar agricultural strategy. Johannes Lehmann, a soil scientist at Cornell University, said, "Any organic matter that is taken out of the rapid cycle of photosynthesis . . . and put instead into a much slower biochar cycle is an effective withdrawal of the carbon dioxide from the atmosphere."

Lovelock, who has consistently been the most pessimistic expert on the future course of the climate crisis, said in 2009: "There is one way we could save ourselves, and that is through the massive burial of charcoal. It would mean farmers turning all their agricultural waste—which contains carbon that the plants have spent the summer sequestering—into nonbiodegradable charcoal and

our thinking." He described the biochar strategy as "the most potent engine of atmospheric cleansing we possess."

In response to the unbridled enthusiasm for biochar demonstrated by many climate and soil scientists, a few environmental activists have expressed concern that a global biochar strategy—if poorly designed—could mimic the palm oil boom that has driven the destruction of tropical forests for the planting of palm oil plantations in Southeast Asia. Their nightmare vision of "biochar plantations" assumes that virgin forests would be cleared to make biochar and then replanted with trees that are optimized for the process.

It is all too easy to understand the basis for their concern. The subsidies for the production of ethanol—which I supported when I was in elective office—have been implemented so far in ways that do far more harm than good. Life-cycle analyses of the ethanol production process confirm that in many cases more greenhouse gases are added to the environment than are removed. Moreover, the impending use of cellulosic ethanol processes sometimes presumes the use of corn stover, sugarcane bagasse, and other plant residues as inputs for the manufacture of alcohol fuels. But crop waste, properly understood, is not really waste. It is important for the protection and regeneration of soils.

However, advocates of the biochar strategy point out that biomass sources like manure, switchgrass, algae, and rice husks would be far more cost-effective inputs for the making of biochar. Any trees and tall grasses planted for biochar could be planted on already degraded land and then harvested and managed in ways that both restore the land and produce large quantities of biochar for the replenishment of soil nutrients elsewhere.

The principal barrier to the use of this strategy is the lack of a price on carbon that would drive the economy toward the most effective ways to sequester it. There is presently no formalized network of biochar distribution channels or commercial-scale production facilities. But a stable price on carbon would cause them to quickly emerge—because biochar holds such promise as an inexpensive and highly effective way to sequester carbon in soil. Already, at least one company, Mantria Industries in Dunlap, Tennessee, has built a commercial-scale biochar plant that opened in August 2009 and sells biochar under the brand name EternaGreen. The company describes itself as the "world's first commercial-scale biochar facility."

There is also growing excitement and enthusiasm about the role that can be played by adding rhizobium bacteria and mycorrhizal fungi to the soil during the planting of seeds. Rhizobium is a highly specialized bacteria that pulls carbon from the roots of leguminous plants and sequesters it in the soil, making it more fertile in the process, while "fixing" nitrogen atoms (breaking apart the two atoms in organic nitrogen and attaching them to hydrogen in order to feed nitrogen to the plants). As agronomists have long taught, carbon provides the bulk of plant matter, such as the walls of plant cells, while nitrogen is used as the building block for amino acids and enzymes, which form the protein, chlorophyll, and the other elements of the plant that determine its quality and value. These bacteria are the real reason why crop rotation was and is a successful strategy for restoring the health of soils.

In recent decades, scientists have also been surprised to learn the large role played by mycorrhizal fungi in stimulating plant growth naturally while enhancing the sequestration of carbon in the soil. Some fungi specialists, including Paul Stamets of Fungi Perfecti, have proposed large-scale projects to use mycorrhizal fungi to enhance productivity.

Fungi, which were among the first forms of life to populate the land, are actually somewhat closer to the animal kingdom than to the plant kingdom. For example, they take in oxygen and emit CO_2. But while some of their carbon output goes into plant growth, much of it remains in the soil. Mycorrhizal fungi have developed an intricate symbiotic relationship with plants. They extend microscopic networks of mycelium—ultrathin strands visible only through a microscope—throughout the soil and produce a gluelike substance called glomalin that helps hold the soil together, makes it resistant to erosion, and enhances its absorption of water. These fungi decompose material from the roots of the plants to produce more nutrients and sequester more carbon. They feed nutrients to the plants and, in return, the plants feed them tiny drops of sugar. This mutually beneficial symbiosis enhances productivity and increases nutrients and carbon in the soil simultaneously. Among the other harmful consequences of heavy plowing is the disruption of these delicate mycorrhizal networks.

If we made urgent the rescue of the global environment—including, most prominently, solving the climate crisis—we would change the basis for farm subsidies from one that rewards overproduction to one that rewards the accumula-

contour hedges or buffer strips to protect against erosion and to replenish nitrogen in the soil.

▶   Return animals to farms and use the manure as natural fertilizer.

▶   Eat less meat.

▶   Eat from local agricultural sources as much as possible.

▶   Support farmers' markets.

▶   Leave the crop residue on the land.

▶   Use biochar in a carefully managed, publicly subsidized global program (but take care to use the right sources for the biochar; don't use corn stover

Change the basis for farm subsidies from one that rewards overproduction to one that rewards the accumulation of carbon in soils and restores the productivity of the land.

tion of carbon in soils and restores the productivity of the land.

A global plan for sequestering more carbon in the soil would involve several elements:

▶   Restore wetlands and prohibit the draining and cultivation of peatlands.

▶   Sharply reduce plowing and convert as much land as possible to no-till cultivation with the use of mulch.

▶   Plant cover crops every other year in a complex crop rotation cycle.

▶   Plant leguminous trees every 30 feet or so as

or other crop waste that should be returned to the land as protective and regenerating cover).

▶   Add rhizobium bacteria and mycorrhizal fungi to soils as a means of speeding the recovery of soil fertility and the sequestration of soil carbon.

▶   Conserve, harvest, and recycle water within a watershed or farm.

An effective plan would require the creation of positive carbon and nutrient budgets for agroecosystems, budgets in which carbon inputs consistently exceeded carbon outputs. Within these budgets, integrated nutrient management, coupled with careful land-use strategies, could—in the opinion of many experts—have a dramatic impact in reducing the amount of CO_2 in the atmosphere. And, as previously noted, the co-benefits of this strategy would include increases in the productivity of soils and progress in fighting poverty, malnutrition, and hunger.

The key to implementing such a strategy is to put a price on carbon and include soil carbon inflows and outflows as part of a global treaty. It took many years to work out the basis for integrating forests into the framework for an international treaty reducing carbon emissions, and the work is due to be completed in the text of the Copenhagen treaty.

Unfortunately, negotiators from developed countries have claimed that difficulties in measuring the amounts of soil carbon for the establishment of national baselines and the monitoring of losses and additions on a regular basis are still too great to warrant the inclusion of soil carbon in a global treaty. At present, the largest emissions trading system, the European Union Greenhouse Gas Emission Trading System (EU ETS), bans both forests and soil carbon from participation. Only the Chicago Climate Exchange (CCX) recognizes certified emissions reductions from soil carbon. The world must adopt standards as a prerequisite for creating credible emissions reductions from soil carbon.

However, the enormous opportunity for sequestering carbon in soils should drive negotiators toward solutions for the problems they have cited as obstacles for this improvement in the treaty. Moreover, sub-Saharan Africa would be the biggest loser in any agreement that failed to recognize soil carbon sequestration, because the restoration of soil carbon in Africa represents a huge opportunity not only for removing carbon from the atmosphere but also for restoring the fertility of soils in Africa.

In 2009, with seed money from the Bill and Melinda Gates Foundation, a group of soil scientists launched an ambitious project to create a global digital map of soils everywhere in the world. This effort, named GlobalSoilMap.net and led by Alfred Hartemink of the International Soil Reference and Information Centre, is prioritizing the mapping of African soils. "People are realizing that food comes from the land, and if you want to end hunger, you need to know your soil and the soil needs to be in good condition," Hartemink said.

As for the difficulty of measuring soil carbon, many of the leading experts on soils strongly disagree and argue that this view by negotiators from developed countries has been made obsolete by science. A great deal of work has improved the confidence of experts that this task is eminently manageable.

The Brookhaven National Laboratory has innovated the use of a technology, called inelastic neutron scattering, for accurately reading carbon content in the soil to a depth of two feet when a tractor passes over it. Satellite-based, near-infrared spectroscopy can also be used in monitoring soil carbon content. The Los Alamos National Laboratory has developed laser-induced breakdown spectroscopy, which can analyze representative samples for their carbon content.

When coupled with sophisticated modeling of soil types in different areas, these and other emerging techniques can produce highly accurate measurements of soil carbon in every country and facilitate accurate monitoring of how much soil carbon is being lost or sequestered.

The benefits both for the global environment and for cementing a bargain between wealthier countries and poorer countries make meeting this challenge an extremely important one.

PALOUSE RIVER BASIN, WASHINGTON.
GRASSLANDS ARE PARTICULARLY EFFECTIVE
IN SEQUESTERING CARBON.

CHAPTER ELEVEN

POPULATION

OSHODI MARKET, LAGOS, NIGERIA. MORE THAN
90 PERCENT OF URBAN POPULATION GROWTH IS
EXPECTED TO OCCUR IN DEVELOPING COUNTRIES.

The spectacular growth in world population since the 18th century—and particularly during the 20th century, when it almost quadrupled—is obviously one of the principal causes of a radical change in the relationship between human civilization and the ecological system of the earth. The impact of larger numbers of human beings would be far less, of course, if the average consumption of natural resources were less and if the technologies we use at present to exploit the earth's bounty were replaced by better and far more efficient technologies that minimize the environmental damage we cause.

As the population increased from 1.6 billion in 1900 to almost 6.8 billion today, powerful new technologies spread throughout the world even more rapidly—especially during the second half of the century. These new technologies fueled a dramatic expansion of economic activity that accelerated during the post–World War II boom. The globalization of industry and commerce led to huge increases in the amount of coal and oil used—and thus ever more rapid production of CO_2—and comparable increases in the production of methane and other greenhouse gases. Moreover, we will add the equivalent of another China's worth of people by 2025, mostly in low-income, developing countries. In all of these countries, there is growing access to more powerful technologies that often magnify harmful impacts on the environment.

For all of these reasons, any plan to solve the climate crisis must deal with the challenge of stabilizing global population as quickly as is feasible. Yet most discussions about solving the climate crisis rarely touch upon population, leading some observers to wonder if it has somehow become a taboo subject. Indeed, people often ask me why population is not more prominently mentioned in debates over how to save the earth's climate balance.

The answer is that the world's effort to stabilize the growth of human numbers is actually an historic success story, albeit in slow motion. Even so, the question of how to stabilize population is, by and large, one of the rare questions around which there is a global consensus and a proven track record of emerging success.

In the last quarter-century, demographers and other social scientists have made dramatic advances in their detailed understanding of the complexities involved in population growth and stabilization. And the application of their new knowledge in countries throughout the world has already led to a dramatic slowdown in the rate of

A.D. 1	50	100	150	200	250	300	350	400	450	500	550	600	650	700	750	800	850	900	950	1000

SOURCE: U.S. Census; United Nations, *World Population to 2300*, 2004; Carbon Dioxide Information Analysis Center; *AAAS Atlas of Population and Environment*

GLOBAL POPULATION GROWTH AND CARBON EMISSIONS

Global population is still growing, but it is expected to plateau at slightly more than nine billion people halfway through the 21st century. However, even as human population stabilizes, greenhouse gas emissions rates are increasing. Annual carbon emissions have quadrupled since 1950, and their rate of growth sharply increased between 2000 and 2008. Many scientists say that CO_2 concentrations must be stabilized at 350 parts per million in the atmosphere, which would require a real reduction from the present concentrations.

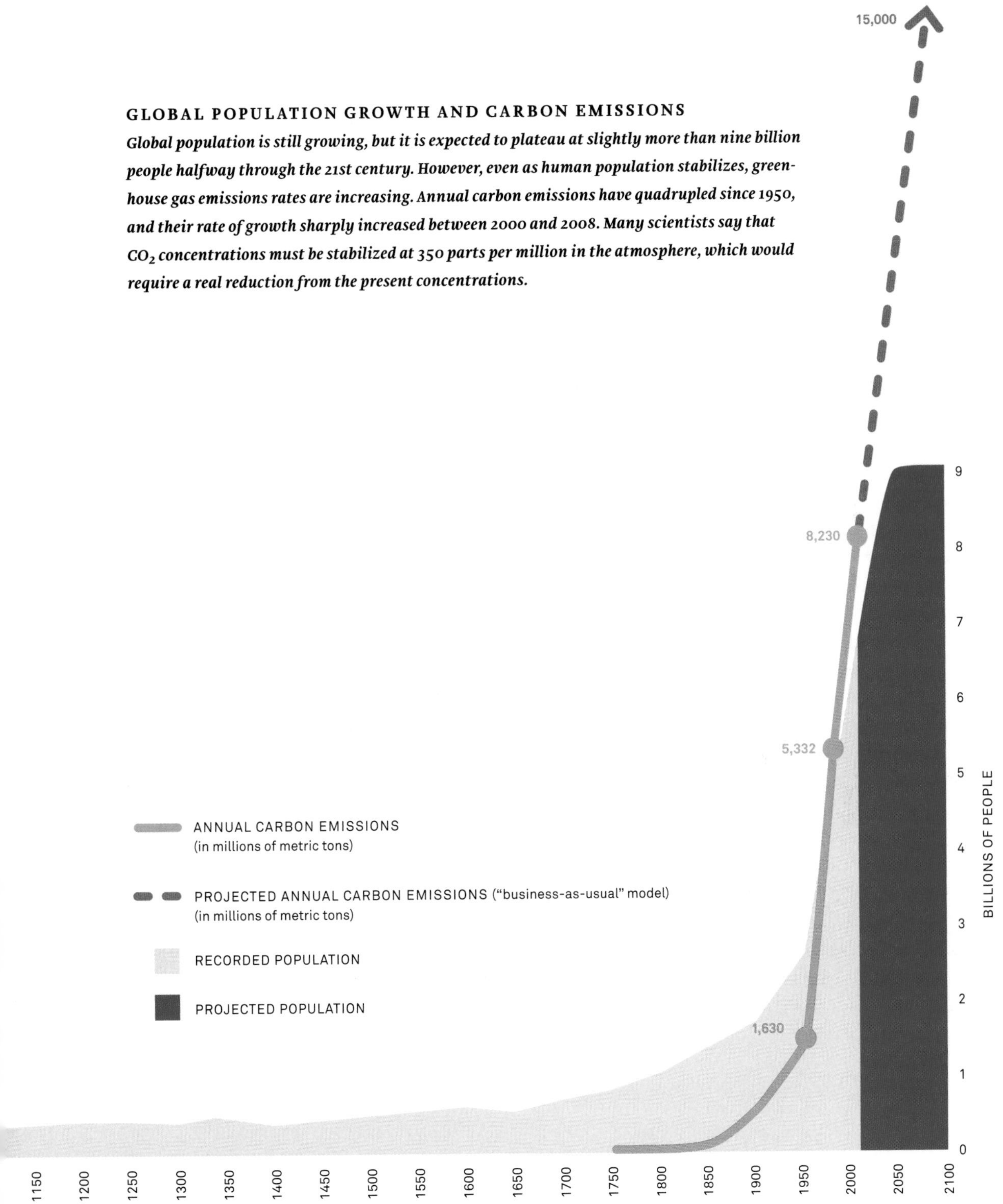

15,000

8,230

5,332

— ANNUAL CARBON EMISSIONS
(in millions of metric tons)

– – PROJECTED ANNUAL CARBON EMISSIONS ("business-as-usual" model)
(in millions of metric tons)

RECORDED POPULATION

PROJECTED POPULATION

1,630

BILLIONS OF PEOPLE

9
8
7
6
5
4
3
2
1
0

1150 1200 1250 1300 1350 1400 1450 1500 1550 1600 1650 1700 1750 1800 1850 1900 1950 2000 2050 2100

population growth. With continued persistence, the world can now look forward to a stabilized global population of slightly more than nine billion people halfway through the 21st century.

There is great uncertainty surrounding this projection, however, and doubt among many experts that the world's political leaders will continue to make the advances necessary to accomplish this goal. And while the subject of population is certainly not taboo in discussions of climate, some related subjects—such as contraception and equality in the status of women—are tricky to deal with in numerous countries worldwide.

Decades ago, demographers noticed that nations with higher incomes had slower population growth, and that those same countries had higher levels of industrialization. Based on this correlation, they concluded that the best way to slow population growth was to speed up industrialization of the low-income countries with rapidly expanding populations. Years later, however, they realized that higher incomes were correlated with slower population growth primarily because higher-income countries were doing other things that were actually responsible for the shift toward smaller families.

The new consensus understanding, which was embodied in the visionary agreement reached at the 1994 United Nations International Conference on Population and Development in Cairo, is that the dynamic at work in any national population is actually a complex system that shifts over time from one pattern—characterized by high death rates, high birth rates, and large families—to a new equilibrium pattern, characterized by low death rates, low birth rates, and small families.

Furthermore, the experts in demography

THE FOUR FACTORS TO STABILIZE POPULATION

PRIMARY SCHOOL IN BALOCHISTAN, PAKISTAN

VOTING IN TEHRAN, IRAN

1. The widespread education of girls.

2. The social and political empowerment of women to participate in the decisions of their families, communities, and nations.

have isolated the four factors that bring about a shift from the first pattern to the second. These factors are:

▸ The widespread education of girls.

▸ The social and political empowerment of women to participate in the decisions of their families, communities, and nations.

▸ High child-survival rates, leading parents to feel confident that most or all of their children will survive into adulthood.

▸ The ability of women to determine the number and spacing of their children.

All four of these factors are linked, and all four must be present for the change in population dynamics to occur. But the good news is that experience now proves conclusively that when all four factors are present, the shift to smaller families and slower population growth is inexorable. Indeed,

this transition—referred to as the Demographic Transition—is under way in almost every nation. This shift has taken place so quickly and so powerfully in the wealthier countries that 44 of the 45 developed nations with populations over 100,000 now have fertility rates lower than required to maintain their populations at the present size. New demographic analyses completed in 2009 show the beginning of a surprising new trend: many advanced countries, when they reach even higher levels of development, apparently experience a slight upturn in fertility. While experts are still trying to understand the meaning and possible persistence of this new trend, it is not expected to alter overall global population growth because few countries have the level of development at which this phenomenon appears. It is welcome news for wealthy countries with aging populations

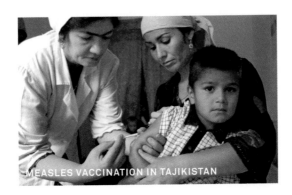

MEASLES VACCINATION IN TAJIKISTAN

FAMILY PLANNING CLINIC IN IVORY COAST

3. High child-survival rates, leading parents to feel confident that most or all of their children will survive into adulthood.

4. The ability of women to determine the number and spacing of their children.

A SETTLER PLANTS PASTURE IN RECENTLY BURNED FOREST NEAR VILA CANOPUS, BRAZIL.

and growing concerns about the workforce-to-retirement population ratio. For the time being, however, without immigration, the developed countries taken as a group would actually be declining in population. (Incidentally, the growth in U.S. population, compared with Europe, is a significant reason why CO_2 emissions have grown more rapidly in the United States than in Europe.)

In fact, even though population measurements are usually reported as national totals, the actual statistics within each nation break down quite differently for groups whose circumstances differ from those of the majority. Such differences can be especially pronounced in minority communities with less income, empowerment, and access to good child and maternal health care. One reason the United States is still growing faster in population than other developed countries is that the U.S. has the greatest disparity between rich and poor of any advanced nation, and poor families with less access to health care and high-quality education have higher fertility rates. By and large, however, the differences between rich nations and poor nations are still the most relevant factor, particularly because national policies are most relevant to the global effort to stabilize population growth.

In the less-developed, poorer nations, the Demographic Transition is also beginning to occur, even though many of these countries still have high population growth rates and will continue growing for a few decades at least. But the rate of growth is slowing even in these countries.

Still, the continuing rapid increase of population in poor countries is putting unprecedented pressure on both the environment and the social fabric of families and communities. In many less-developed countries, impoverished subsistence farmers push into areas that used to be covered by forests, cutting and burning them in order to eke

out a living. In the process, more of the world's forests—invaluable for their ability to absorb CO_2 and to provide for the rich web of biodiversity—are lost, especially in the tropics and subtropics. And as noted in Chapter 9, the conversion of forests to cropland often diminishes rainfall.

Freshwater shortages are threatening many of these same areas as larger numbers of people overwhelm supplies of drinking water and the infrastructure for delivering potable water to urban residents in rapidly growing cities. In several countries, growing populations have depleted reservoirs and aquifers. In Mexico City, for example, the largest city in the Western Hemisphere, low levels of rainfall in 2009, coupled with inadequate infrastructure and a rapidly growing population, led the city to cut off water supply to hundreds of thousands of people on five occasions. In addition, untreated waste has a harsh effect on water quality and worsens the threat from waterborne diseases like cholera.

The social fabric is also stressed—not only by the increase in absolute numbers but also by a related demographic trend: urbanization. For most of human history, no more than 15 percent of the population lived in cities. In the last century, however, that pattern has changed radically. By midway through 2008, for the first time, more than half of the world's people were living in cities. During the 20th century, while global population quadrupled, global *urban* population increased more than tenfold. On a worldwide basis, all future population growth for the foreseeable future will be in urban areas, and rural population is predicted to actually decrease. Especially in the developing world, larger numbers of people are moving from rural areas into cities. More than 90 percent of the world's continuing urban population growth is expected to take place in developing countries. In China, for example, the urban population has increased in

the last 30 years from under 200 million people to more than 600 million people, and demographers project that during the next 15 years the urban population of China will increase by another 350 million people, more than the entire population of the United States. When Chairman Mao took power in 1949, only 11 percent of China's population lived in cities. Within 20 years from now, 70 percent of China's population is expected to live in cities.

It is also worth noting that throughout the world, a significant percentage of the new urban population has been moving into low-lying areas of coastal cities that are vulnerable to sea-level rise from accelerated ice melting in Greenland and Antarctica this century due to global warming.

The rate of urban population growth worldwide has actually slowed. But the *rate* of urban growth, according to the United Nations Population Fund, is now less significant than the absolute *size* of the increments, especially in Asia and Africa. And even a slower rate of growth in large, newly urbanized areas poses great challenges. In both Africa and Asia, overall urban population is predicted to increase 100 percent during the next 20 years. In Nigeria, Lagos, the largest city, has grown from 1.9 million people in 1975 to 9.5 million in 2007, and is projected to reach 15.8 million in 2025. Lagos; Kinshasa, Congo; and Dhaka, Bangladesh, are the three fastest-growing megacities in the world.

The rapid rate of urbanization offers many opportunities for more efficient buildings and transportation systems as well as sharp increases in energy efficiency and conservation. Some of the most important solutions to the climate crisis involve new policies that seize and develop the rich potential for the world's new urban growth to become much more energy efficient. However, urbanization has been accompanied by a change in the lifestyles of former rural residents and has thus

RESIDENTS SCRAMBLE FOR WATER IN A DELHI SLUM. HOME TO 4,000 WORKERS, IT HAS NO CLEAN WATER SUPPLY.

URBANIZATION AND THE GROWTH OF MEGACITIES

For the first time in human history, more than half the world's population lives in cities. As global population quadrupled over the past 100 years, the earth's urban population increased tenfold. By 2025, the world may have as many as 27 megacities—urban areas with more than 10 million inhabitants. The rapid growth of megacities presents an opportunity to reduce greenhouse gas emissions globally. Densely populated urban areas with energy-efficient infrastructure—including compact neighborhoods, shorter travel distances, mass transit, and vertical housing— create fewer greenhouse gas emissions per capita than suburban and rural areas. New Yorkers have carbon footprints less than one third the size of the average American; residents of São Paulo have 18 percent the emissions of the average Brazilian. Ninety percent of urban growth is expected to occur in developing nations.

PARIS

LOS ANGELES

NEW YORK

MEXICO CITY

LAGOS

KINSHASA

RIO DE JANEIRO

SÃO PAULO

BUENOS AIRES

SOURCE: United Nations Population Division; CIA World Factbook

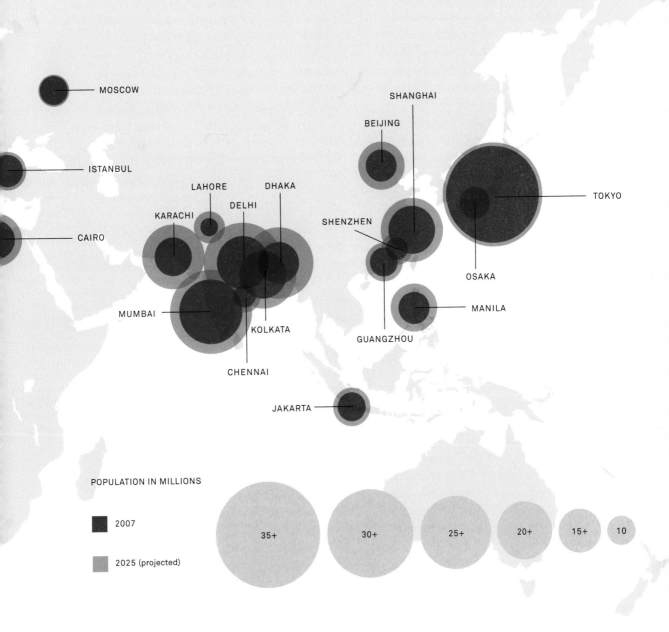

MOSCOW

SHANGHAI

BEIJING

ISTANBUL

LAHORE

DELHI

DHAKA

TOKYO

KARACHI

SHENZHEN

CAIRO

OSAKA

MUMBAI

MANILA

KOLKATA

GUANGZHOU

CHENNAI

JAKARTA

POPULATION IN MILLIONS

2007

2025 (projected)

35+   30+   25+   20+   15+   10

far led in most developing countries to significant increases in energy use by urban residents compared with rural residents.

Moreover, the current pattern of urbanization puts an outsize emphasis on accommodating cars and trucks rather than modern mass transit. This leads to massive increases in asphalt and concrete, and to traffic jams, air pollution, higher energy use, and greater CO_2 emissions. One expert on urban planning quipped that the overriding principle that seems to be driving the design of most cities today is "make sure that all the cars are happy."

Even though global population increases are now entirely urban, demographic experts point out that migration from rural areas is no longer the largest factor in most urban population growth. Rather, it is natural population increase. As a result, the effective strategy for stabilizing urban population growth is actually the same strategy that has already begun to work in stabilizing population growth generally. It all comes back to the same four factors that experience proves will lead inexorably to smaller families, slower population growth rates, and a stable world population that stops increasing halfway through the century—probably at or around 9.1 billion people.

The most powerful among the four factors that jointly bring about the Demographic Transition appears to be the first: the education of girls. When girls are well-educated in large numbers, they find ways, when they grow up to be women, to accelerate their own empowerment. They often delay the age at which they marry and/or begin having children. They add their voices to those in their respective nations who are working for better child and maternal health care, thus raising child-survival rates. Educated and empowered women also seek out the means for managing their own fertility so that they can determine the number of children

they desire and space them as they wish.

It is important to emphasize, however, that all four of these factors are necessary, and that none of the four can be left out. For example, in nation after nation, declines in the death rate precede declines in the birth rate by half a generation or more. This stark fact puts emphasis on the central importance of raising child-survival rates.

This equation masquerades as a contradiction, and at first glance many casual observers are puzzled that lower death rates lead to smaller populations. The explanation for this seeming paradox was best expressed more than 60 years ago by an African head of state, Julius K. Nyerere, who said that "the most powerful contraceptive is the confidence by parents that their children will survive."

The connection between lower death rates and smaller populations is not the only seeming paradox uncovered by demographers. Many nations that outlaw abortion have much higher abortion rates than the United States, where abortion is mostly legal during the first two trimesters, with guidelines that become more restrictive during the second trimester and nearly prohibitive during the final trimester. Data from countries throughout the world appear to show that when women have access to a full range of reproductive health services, there are many fewer unwanted pregnancies that lead to abortion.

The global communications revolution—particularly satellite television, the Internet, and cell phone technology—seems to have sped up the movement toward more education for girls and more empowerment of women. For example, in Saudi Arabia, which used to have one of the highest population growth rates in the world, 55 percent of college graduates are now women. And sure enough, the population growth rate in Saudi Arabia has been slowing significantly. From 1975

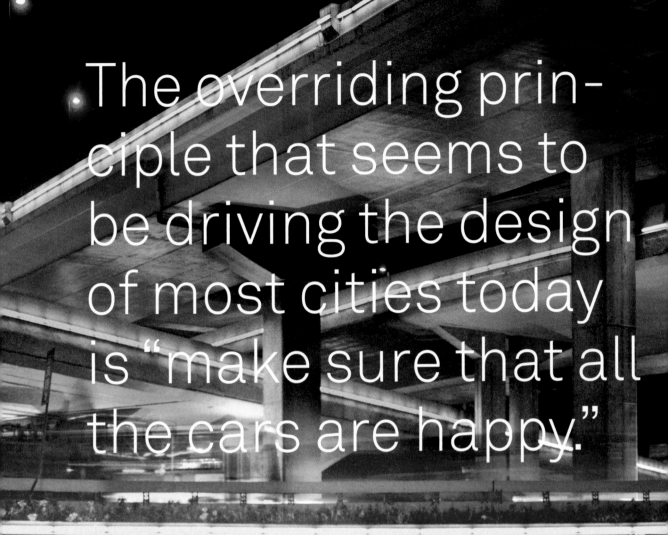

The overriding principle that seems to be driving the design of most cities today is "make sure that all the cars are happy."

HIGHWAYS IN SHANGHAI, CHINA

"The most powerful contraceptive is the confidence by parents that their children will survive."

JULIUS K. NYERERE

A MOTHER IN DUBAI LOOKS AT HER BABY IN AN INCUBATOR.

to 1980, the average number of children in a family was 7.3; now it's 3.2.

On a recent trip to one of the Persian Gulf nations, I met a woman who worked as the pilot of a 747. She wore her captain's hat over her headscarf. I didn't ask her how many children she had, but I'd be willing to bet it was a smaller number than her mother or grandmother had.

Conversely, the poorest countries with the highest death rates often have the fastest rates of population growth. They also typically have the youngest average age. Niger, with one of the highest death rates in the world, has the youngest population of any nation: median age, 15.

The speed with which population increased in the 20th century, coupled with the speed of the Demographic Transition during the last few decades, has led to historic and challenging disruptions in the generational makeup of many national populations. In Japan, for example, the average number of children per family has fallen so quickly (and immigration is so tightly controlled) that overall population started declining four years ago. Predictably, the number of old people as a percentage of the population is at an unprecedented high. The median age in Japan is not 15, as in Niger, but 43, almost three times as high.

In a number of countries, there are simultaneously high percentages of old and young people, producing what demographers call a "high dependency ratio." In 1960, there were 5.1 people in the U.S. workforce paying into the Social Security system for every retired person drawing a Social Security check. But today, that ratio has fallen to three working people for each retired person, and is on the way to a two-for-one ratio. Obviously, the anticipated financial burden for each working person is much higher when there are fewer in the workforce and more who are past retirement age.

In most developing countries, there is no national safety net for the elderly comparable to the U.S. Social Security system, and thus, the normal desire for children is reinforced by the need to rely on adult children for care in later life. In many cultures, this natural desire is also enhanced by religious and cultural practices that emphasize the benefits of carrying on a family name and tradition—and by the lingering habits of formerly agricultural societies that valued larger numbers of children who could help bring in the family's crops.

Experience has proved, however, that when children survive into adulthood at very high rates, and when the other three factors are also present, the natural desire on the part of most people for fewer children seems to override all other factors spurring the drive toward larger families.

The speed of population growth has itself been a major social and political challenge for the high-growth countries. All of the support systems—for health care, education, social security, and all the rest—are stretched to their capacity. And politicians in these countries typically have a difficult time expanding these services quickly enough to meet the needs of their growing populations.

Moreover, when the number of new jobs lags far behind the rapidly increasing numbers of people desiring to enter the workforce, social unrest—particularly among young men—can lead to political instability and worse. Civil unrest, lack of work, and environmental degradation also trigger large migrations of illegal immigrants across national borders in search of prosperous, peaceful, and productive lives. And of course, the impact of the climate crisis—particularly in dry nations suffering the harsh effects of higher temperatures and soil moisture evaporation—magnifies the pressure to migrate.

When populations of impoverished migrants

move into areas that are already home to peoples with different cultures, traditions, belief systems, and languages, the potential for conflict and violence increases. Secretary-General of the United Nations Ban Ki-moon has cited the impact of climate change in eastern Chad as one of the principal causes of the horrific violence in the neighboring Darfur region of western Sudan. When Lake Chad was dramatically reduced by drought, large numbers of Chadians moved across the border into Sudan. And although the principal cause of the genocidal violence there lies with political decisions made by the government of Sudan and the failure of the international community to intervene, the stage was set by the interaction of the climate crisis with demography.

It's ironic that, during the Cold War, many conservatives in the United States strongly supported fertility management in developing countries as a strategy for avoiding instability that could be exploited by the Soviet bloc in its effort to inspire Communist revolutions. At an early stage in his career, George H.W. Bush was a strong advocate of international family planning, and President Dwight Eisenhower was co-chairman of Planned Parenthood. In more recent years, the Republican party in the United States has made opposition to abortion one of the most prominent elements in its political agenda. Republicans have tended to conflate opposition to abortion with opposition to all fertility-management assistance in developing countries because of the asserted inability to support the latter without giving funds to organizations that also make the former available.

This new political dynamic in the United States has had a profound effect on the international efforts to expand the availability of fertility-management assistance for developing countries. Largely as a result, among the four essential factors that bring about slower population growth, the one most hampered by political interference is the wider availability of contraception. Even if all three of the other slow-growth factors are present, women must have the ability to determine the number and spacing of their children.

The most successful programs addressing the inability of poor women to manage their own fertility are those that provide "one-stop shopping" for health care directed at both mothers and children. Pregnant mothers can receive prenatal care at such clinics that is invaluable for the healthy development of their babies, receive attention for diseases and conditions afflicting themselves and their children, and get whatever advice and help they desire to manage the spacing of their next child—including advice and help if they choose not to conceive.

There appears to be a correlation between women who seek further education and further empowerment and those who choose to have fewer children. The opposition to enhancing the ability of women to manage their own fertility has come, of course, primarily from groups steeped in cultural and religious traditions that have long resisted the empowerment of women. In some areas, fundamentalist orthodoxy has led to an effort to prevent even the education of girls and women. In areas of Afghanistan and Pakistan controlled by the Taliban, for example, there has been a renewed push to close schools for girls and roll back any and all gains in the empowerment of women.

In most cases, however, opposition to the empowerment of women takes far more benign (but still onerous) forms. Often this opposition reflects broader societal reactions to rapid and disruptive upheaval brought about by globalization, access to modern communications, and other cultural changes that threaten the stability and power of long-established sources of patriarchal authority.

FEMALE STUDENTS LEARN BIOLOGY AT
A BOARDING SCHOOL IN INDONESIA.

Slowly the resistance to making available to women the means to manage their own fertility is giving way to the rising desire of people everywhere to have smaller families.

In 1994, I led the U.S. delegation to the U.N. meeting on population and development in Egypt and both enjoyed and learned from extensive negotiations with representatives of nations, religious groups, NGOs, and others from around the world.

Fifteen years after the development of the Cairo consensus, it is clear that the policies adopted there are working. But many wealthy, developed countries have failed to make good on the pledges of assistance that were made in Cairo and since. One of the principal reasons for this shortfall was the refusal of the Bush-Cheney administration to support international fertility-management programs if there was even an indirect connection between contraception assistance and organizations

providing access to legal abortions. Partly as a consequence of the Bush policy, international organizations were forced to shift their emphasis away from fertility-management assistance.

There is a huge unmet need for access to fertility management, and any comprehensive plan to address the climate crisis must include measures to meet it. Continued progress in stabilizing global population at or below 9.1 billion depends upon the willingness of the wealthier countries to keep their word and provide the support they have promised.

The U.S. policies have been reversed by President Obama. Additional resources are needed to accelerate progress in the education of girls, the empowerment of women, child and maternal health care, and particularly to provide enhanced access, on the part of women in developing countries, to the means for self-determination regarding the number and spacing of their children.

HOW WE USE ENERGY

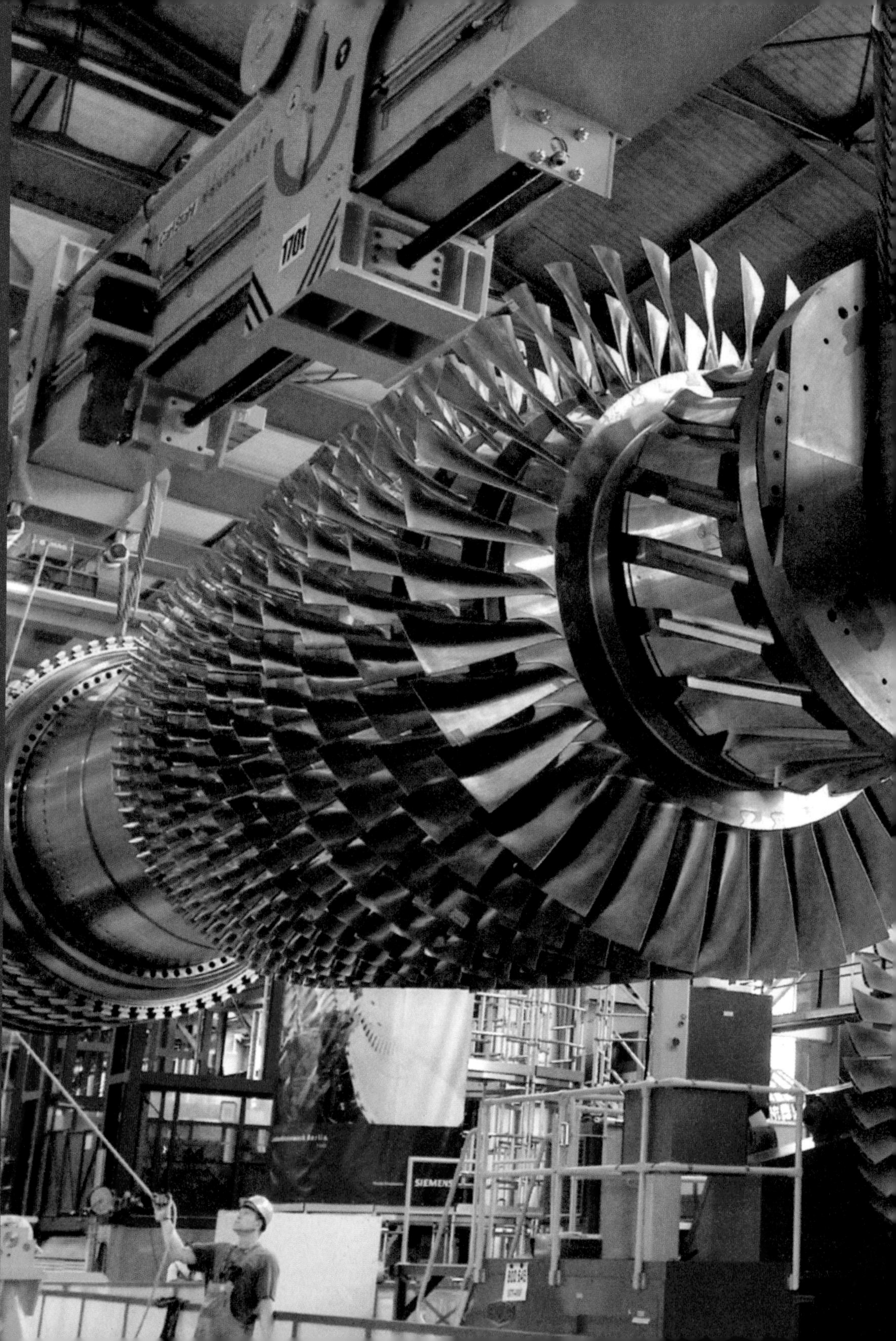

CHAPTER TWELVE

LESS IS MORE

BECAUSE TURBINES ARE AT THE CENTER
OF ELECTRICITY GENERATION, ANY GAINS IN
TURBINE EFFICIENCY HAVE HUGE IMPACTS.
NEW, PRECISELY MADE TURBINES ARE NOW
ALMOST TWICE AS EFFICIENT AS OLDER MODELS.

Improvements in the efficiency with which we use energy offer the biggest opportunities to reduce energy consumption—and CO_2 emissions—while saving money and increasing productivity at the same time. Efficiency improvements are also far and away the most cost-effective among the solutions to the climate crisis and can be implemented faster than any of the others.

Moreover, the experience of those who have focused on this resource demonstrates that it is virtually inexhaustible, because innovation in efficiency is, in a very real sense, inherently renewable. Gains from efficiency continue for the life of the process or building. In nations and organizations that choose this path, success begets success. When the culture of efficiency is infused throughout an organization, the best ideas for further gains often come from employees at every level in the organization.

The opportunities for enormous gains can be found in every sector of the economy—particularly in industrial processing; residential and commercial buildings; the electricity generating sector; the transportation sector; and even in the design of cities themselves.

The huge cost savings and CO_2 reductions from improvements in efficiency are not speculative; they are real. Overall U.S. electrical generation converts only 33 percent of fuel to electricity, but combined heat and power (CHP) plants extract more than twice as much useful energy by using energy twice. A recent study by the World Alliance for Decentralized Energy found the United States could cut 20 percent of its total CO_2 emissions by installing CHP plants and save $80 billion per year. And there are more savings possible in the use of electricity. For example, the International Energy Agency (IEA) found that, "On average, an additional $1 invested in more efficient electrical equipment and appliances avoids more than $2 in investment in power generation, transmission, and distribution infrastructure."

Virtually every study reaches these same conclusions, and yet we have had difficulty choosing to implement efficiency measures on a large scale.

According to efficiency expert Robert Ayres, in 1900 the U.S. energy system converted 3 percent of the potential into useful work. After more than a century of technical progress, the United States converts only 13 percent of the potential work in the fuel we burn into useful work, thus still wasting 87 percent of that potential. One of the reasons for our failure to do so is that the opportunities are spread out over so many different technologies in so many different settings.

But energy-efficiency experts have identified several opportunities where some of the greatest energy savings can be found:

▶ The capture and recycling of wasted heat energy from electricity generation and from heat-intensive industrial processes.

One dollar invested in more efficient electrical equipment . . . avoids more than two dollars in investment in electricity supply.

INTERNATIONAL ENERGY AGENCY

THE U.S. LEED PROGRAM CERTIFIES NEW, HIGHLY EFFICIENT ARCHITECTURE. THE ALDO LEOPOLD LEGACY CENTER IN WISCONSIN HAS EARNED THE HIGHEST LEED HONORS.

THE (OFTEN NEGATIVE) COSTS OF REDUCING GREENHOUSE GASES

Starting in 2006, McKinsey & Company launched a now-famous study of different ways to reduce emissions of greenhouse gases (GHGs). The chart below, known as the global GHG abatement cost curve, displays the abatement potential and costs associated with about 200 of the most important GHG abatement opportunities.

One significant finding is that nearly 40 percent of the potential reduction in emissions worldwide can actually save money in the near term!

The options in the green portion of the chart have a negative cost, which indicates a net benefit or savings over the lifetime of each option. The study's conclusion is that the world can reduce CO_2e emissions to stabilize concentrations in the atmosphere at 450 parts per million of CO_2e with investments of as little as 0.6 percent of the GDP—largely because of the savings made possible by gains in efficiency.

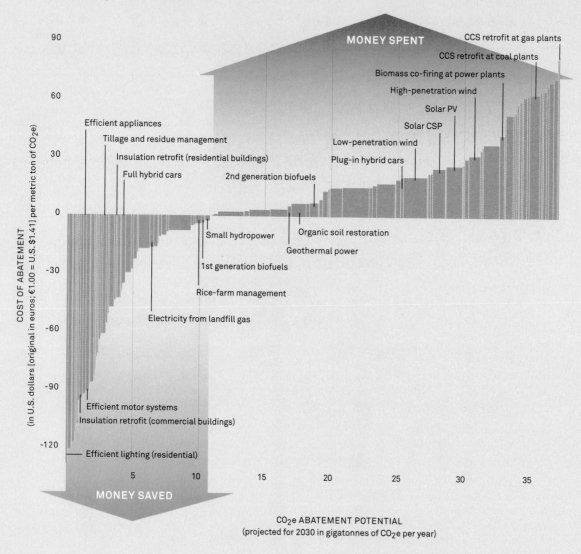

COST OF ABATEMENT (in U.S. dollars [original in euros; €1.00 = U.S. $1.41] per metric ton of CO_2e)

MONEY SPENT

CCS retrofit at gas plants
CCS retrofit at coal plants
Biomass co-firing at power plants
High-penetration wind
Solar PV
Solar CSP
Low-penetration wind
Plug-in hybrid cars

Efficient appliances
Tillage and residue management
Insulation retrofit (residential buildings)
Full hybrid cars
2nd generation biofuels

Small hydropower
Organic soil restoration
Geothermal power
1st generation biofuels
Rice-farm management
Electricity from landfill gas

Efficient motor systems
Insulation retrofit (commercial buildings)
Efficient lighting (residential)

MONEY SAVED

CO_2e ABATEMENT POTENTIAL
(projected for 2030 in gigatonnes of CO_2e per year)

SOURCE: Adapted, with changes, from McKinsey & Company, *Pathways to a Low-Carbon Economy. Version 2 of the Global Greenhouse Gas Abatement Cost Curve,* 2009

▶ The replacement of inefficient industrial electric motors with far more efficient modern motors.

▶ Adequate insulation of buildings in every sector—particularly residences.

▶ The replacement of inefficient windows, lighting systems, hot water heaters, appliances, and electronics with more efficient modern versions.

▶ Higher fuel-economy standards for cars and trucks and the greater use of mass transit.

The favorable economics of introducing efficiency gains are clear and compelling. Any nation adopting a determined and persistent national strategy for implementing pervasive improvements in the efficiency with which energy is converted to useful work will quickly find that this is, by far, the most effective way to save energy and reduce global warming pollution.

This is true in rich and poor countries alike. McKinsey & Company found in a report published in July 2009 that the United States could reduce its projected energy consumption 23 percent by 2030 simply by making economically beneficial investments in energy efficiency. Earlier, a report by the McKinsey Global Institute found that developing countries "could reduce their energy demand growth by more than half…and reduce their energy consumption in 2020 by 22 percent from the projected levels.…Boosting the energy productivity of developing countries' economies alone has the potential to reduce global CO$_2$ emissions by 15 percent in 2020, making it critical from the global climate change perspective."

For example, Johnson Controls, one of the leaders in industrial energy efficiency, points out that while Japanese factories run at an efficiency level of 85 percent or better, Chinese factories run at only 50 percent efficiency. Since each 10 percent reduction in factory efficiency represents a doubling of

energy consumption, the result is that Chinese factories use an average of 350 percent more energy than Japanese factories for each unit of output.

The United States went through a brief period of careful attention to efficiency gains, from 1977 through 1985—in the first serious effort to reduce oil consumption, after the oil embargoes of the 1970s—and the results were stunning. In response to intelligent and focused policies pursued by President Jimmy Carter and continued through the first years of Ronald Reagan's presidency, the United States cut oil use by 17 percent while increasing economic output by 27 percent. Our dependence on imported oil dropped by half

CALIFORNIA LEADS THE WAY

Since the energy crises of the 1970s, California's laws have encouraged efficiency. Over the past 25 years, the state's per capita electricity use has remained flat, while that of the rest of the U.S. has jumped by 60 percent.

SOURCE: California Energy Commission

NEW STEELMAKING TECHNOLOGIES, SUCH AS
THIN-SLAB CASTING, CAN SAVE LARGE AMOUNTS
OF ENERGY, LOWER CO_2 EMISSIONS, AND REDUCE
PRODUCTION COSTS BY UP TO 20 PERCENT.

and, in response, OPEC was forced to cut oil prices significantly. Yet when the emphasis on conservation was allowed to lapse, we slowed the rate of efficiency gains. There were continued improvements, and our energy use per unit of GDP decreased by a third between 1985 and 2008. Nevertheless, our dependence on imported oil skyrocketed once more, and oil prices started climbing rapidly again.

One state that did not lose its focus on efficiency was California. Art Rosenfeld, the state energy commissioner who designed California's efficiency initiative, points out that energy use in his state had increased rapidly from the end of World War II until the first oil embargo in 1973, just as it had in the rest of the country. However, in the past three decades, California's total per capita electricity consumption has not increased at all, even though its per capita economic output almost doubled. Meanwhile, in the rest of the nation, per capita electricity use increased by more than 60 percent over the same period, with virtually the same economic-output gains.

Compared with 1993, the ratio of energy use to each dollar of GDP in California has been cut in half, and the total savings for the state from energy efficiency have been roughly $1 trillion. The experience in California and a few other states has proved that the cost of efficiency measures that avoid electricity use is typically far less than the cost of building new generating capacity. These impressive results were achieved with the decoupling of utility incentives from increased energy use, higher standards for efficiency in appliances and buildings, and higher mileage requirements for automobiles than in the rest of the United States.

In the global economy as a whole, using fuel twice by simultaneously generating electricity and thermal energy offers the largest opportunities for energy efficiency savings. Although many of the most important opportunities for large savings are specific to particular industries—and there are literally hundreds of examples—there are four that are frequently mentioned by energy-efficiency experts as being among the most valuable.

First, one of the largest opportunities for efficiency gains throughout the industrial sector of the global economy is through the replacement of older inefficient electric motors with much more efficient modern motors that pay for themselves in a short period of time—and then achieve further efficiency gains through the optimization of their motor-driven systems.

According to the Department of Energy's Office of Energy Efficiency and Renewable Energy, "Motor-driven equipment accounts for 64 percent of the electricity consumed in the U.S. industrial sector." Indeed, electric motors are ubiquitous throughout global industry and, for the most part, are extremely inefficient. Electric motors play a similarly prominent role in industrial activity around the world.

According to Amory Lovins, perhaps the world's leading expert on energy efficiency, "Motors use three fourths of industrial electricity, three fifths of all electricity, and more primary energy than highway vehicles. This use is highly concentrated: about half of all motor electricity is used in the million largest motors, three fourths in the three million largest.... A comprehensive retrofit of the whole motor system typically saves about half its energy and pays back in around 16 months." Moreover, Lovins adds, the new, more efficient industrial motors are almost always quieter, more reliable, and easier to operate.

The larger electric motors can often be replaced with new, more efficient motors that produce savings that exceed their purchase price in only a few weeks of electricity usage. Even when factoring in

the cost of lost production time for the period necessary to replace these engines, most factories can
save money and reduce pollution by choosing the
more efficient motor design.

Second, steel mills have traditionally required
enormous amounts of energy to first melt the metals into a liquid, in order to cast thick sheets—
which are then heated a second time in order to
recast the steel into the particular form desired,
whether I-beams, slabs, sheets, or some other
form. Recent advances known as "thin-slab casting" and "direct casting" now make it possible to
dispense with the second heating and recasting
step—saving prodigious amounts of energy and
CO_2 emissions. In the new, more efficient process, the steel is cast directly from its liquid state
into the final form desired. Many experts claim the
resulting product is not only 20 percent cheaper
to produce but also both lighter and stronger than
steel made by the older process. Although this new
technology was first employed in "mini-mills," it
has now advanced to the point where it can be used
in large mills as well.

Third, fluid handling—with pumps and piping
systems—is common in most manufacturing
plants and commercial buildings. Indeed, more
than a quarter of the electricity used by industrial
systems in the manufacturing sector is used to
power pumps. In addition, municipalities use lots
of pumps for water and wastewater transfer and
treatment. However, engineers have traditionally
erred on the side of oversizing pumps and have
often failed to optimize the interaction of the overall system of pumps and pipes. By replacing older
inefficient pumps with modern versions, redesigning piping systems to optimize the efficient flow
of fluids, and choosing the most efficient size of
pump for each task, many companies have found
enormous energy savings, increased productivity,

In just 10 years, citizens of the U.S. wasted enough aluminum cans to reproduce the world's entire commercial air fleet 25 times.

CONTAINER RECYCLING INSTITUTE

BALES OF METAL AWAIT RECYCLING IN
SEATTLE. METAL RECYCLING CAN REDUCE
ENERGY NEEDS BY 95 PERCENT, COMPARED
WITH PRODUCING NEW METAL FROM ORE.

lower production and maintenance costs, and better reliability and product quality.

Fourth, most oil refineries and many chemical processors rely on a very energy-intensive first step called "distillation" that many experts say can be profitably replaced by a much more energy-efficient technique known as "membrane separation." The rapidly advancing design and manufacture of highly specialized membranes has recently found new applications in many fields. In just the past 10 years, specialty membranes have been designed for gas separation in these two industries, requiring a small fraction of the energy used in the traditional distillation process.

In some industries, recycling of widely used materials can sharply reduce the energy requirements for reprocessing. For example, increased use of paper recycling could also save enormous amounts of energy and help avoid increased deforestation in the world. (This book is printed on 100 percent recycled paper manufactured in a mill certified by the Sustainable Forestry Initiative and the Forest Stewardship Council operated by Newton Falls Fine Paper, which has purchased carbon offsets for the production of the paper.)

Recycling offers even larger gains in the aluminum industry. The production of aluminum from bauxite ore is one of the most energy-intensive processes in the global economy. However, 95 percent of the energy can be eliminated by processing recycled aluminum instead of producing fresh quantities from bauxite. Yet vast quantities of aluminum are discarded annually, filling up landfills unnecessarily.

In the United States alone, more than 50 billion aluminum cans are thrown away each year—more than half of the 100 billion aluminum cans sold each year. The Container Recycling Institute reported that in just the 10 years between 1990 and 2000, citizens of the United States wasted enough aluminum cans "to reproduce the world's entire commercial air fleet 25 times." Large quantities of aluminum are also discarded—instead of being recycled—in the form of appliances and other durable goods made of aluminum.

Another 50 billion containers—in the form of plastic bottles made of polyethylene terephthalate (PET)—are thrown away by Americans each year instead of being recycled. The phthalates in these bottles, by the way, have been identified by scientists as "endocrine disruptors," and have been linked by many medical experts to sexual deformities in babies, lower sperm counts, diminished sperm quality, and other health problems. The Endocrine Society has warned that phthalates and other endocrine disruptors are "a significant concern to public health."

When you add the 29 billion glass bottles and seven billion higher-density plastic bottles and jugs that are also thrown into landfills each year instead of being recycled, it means that there are nine aluminum, plastic, or glass containers thrown away every week on average by every man, woman, and child in the United States. If they were routinely recycled instead, the energy saved in making new containers would equal the energy equivalent of 53.5 million barrels of imported crude oil—which would be the equivalent of eliminating all the gasoline used by two million cars.

One of the largest opportunities for efficiency gains—an opportunity that is present in electricity generation and in many industrial sectors—involves the capture of waste heat. Indeed, most industrial facilities using large amounts of heat can profitably capture their wasted thermal energy and reuse it in their own processes—or sell it for use in nearby buildings for space heating and cooling. They can also simultaneously use their waste heat to generate electricity on-site—thereby significantly reducing their purchases of electricity from utilities, and thus sharply reducing CO_2 emissions.

Using energy sequentially for two productive purposes is called cogeneration, or combined heat and power (CHP). Entrepreneurs like Tom Casten, chairman of Recycled Energy Development, have demonstrated time and again how industrial companies that are willing to invest in cogeneration technologies can become more efficient and more profitable in a short period of time.

The Oak Ridge National Laboratory (ORNL) concluded in a major study that CHP is "a proven and effective energy option, deployable in the near term, that can help address current and future U.S. energy needs," adding, "Energy efficiency, including CHP, is the least expensive and most rapidly deployable energy resource available today."

In the United States, a Department of Energy study calculated in 2007 that the potential for CHP generation from industrial operations currently emitting recoverable waste heat is equivalent to the output of 40 percent of the

HOW COGENERATION WORKS

Cogeneration, or combined heat and power (CHP), systems use a single fuel source to create and capture both electricity and heat. (In conventional power plants, two thirds of the heat is wasted.) The captured heat can be used in various ways, including the creation of steam for electricity and direct space heating. The efficiency of a CHP plant can reach 80 to 90 percent.

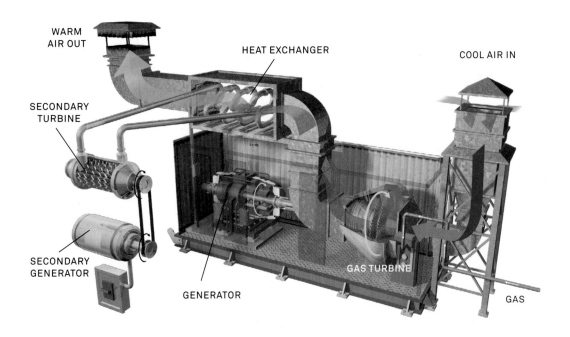

WARM AIR OUT

HEAT EXCHANGER

COOL AIR IN

SECONDARY TURBINE

SECONDARY GENERATOR

GENERATOR

GAS TURBINE

GAS

coal-fired generating plants now producing electricity in the United States.

ORNL also calculates that by 2030, the wide use of CHP in settings where it is profitable would "create nearly one million new highly skilled, technical jobs throughout the United States. CO_2 emissions could be reduced by more than 800 million metric tons per year, the equivalent of taking more than half of the current passenger vehicles in the United States off the road." According to ORNL's analysis, if only 20 percent of generation capacity came from CHP in the United States, "over 60 percent of the projected increase in CO_2 emissions between now and 2030 could be avoided." And 20 percent is not a very aggressive target.

a short period of time and still reduce the overall cost of energy.

A recent study by experts at the American Council for an Energy-Efficient Economy (ACEEE) and the International District Energy Association (IDEA) found that the industrial sector of the economy has "the greatest potential for near-term growth. The majority of this capacity exists in industrial sites with large steam loads."

On a global basis, the International Energy Agency reports that five industrial sectors that use large amounts of heat in their processing—food processing, pulp and paper, chemicals, metals, and oil refining—"represent more than 80 percent of the total global electric CHP capacities."

"Energy efficiency, including CHP, is the least expensive and most rapidly deployable energy resource available today."

OAK RIDGE NATIONAL LABORATORY

A few countries have already expanded their use of CHP, with five countries now getting between 30 and 50 percent of their total power generation from CHP: Denmark, Finland, Russia, Latvia, and the Netherlands.

In fact, if all of the wasted energy from U.S. factories was captured and recycled, the amount of energy saved could reduce fossil fuel use and CO_2 emissions in the United States by approximately 20 percent. Moreover, the true market cost of these investments would pay for themselves in

Unfortunately, these industries are still exceptions. In fact, almost every industry that uses large amounts of heat energy could profitably employ cogeneration technologies and sharply reduce its energy bills and global warming pollution. However, the IEA concluded, "Despite increased policy attention in Europe, the United States, Japan, and other countries, the share of CHP in global power generation has remained stagnant for the past several years at around 9 percent."

Moreover, the United States is well below the average for major industrial countries, and the ACEEE notes that one of the reasons is "many facility managers are unaware of technology developments that have expanded the potential for cost-effective CHP."

But a wide variety of experts agree that the *principal* barrier to the wider use of CHP in the United States is that electric utilities actively block its use through a variety of discriminatory practices designed to maximize their profits and avoid competition from lower-cost energy generated by their customers on-site.

As the Oak Ridge National Laboratory puts it, "Many current U.S. rate structures that link utility revenues and returns to the number of kilowatt-hours sold are a disincentive for utilities to encourage customer-owned CHP and other forms of onsite generation." As Tom Casten puts it, "Electricity is a cost-plus world." And most people are familiar with some of the gross abuses caused by the "cost-plus" formula in U.S. defense contracting.

Electric utilities in most states are using a variety of techniques to block CHP use by their largest customers. Often, they petition regulators to require industrial facilities desiring to use CHP to pay for expensive backup generating capacity to protect the utility against a sudden surge in demand in case the CHP system fails. Although this sounds reasonable in principle, the tactic is routinely used in ways that raise the price of CHP to cover wildly improbable contingencies. It is not unusual for utilities to try to force industrial customers to make large unprofitable investments to protect against a contingency that statistical analysis shows has a one in six million chance of occurring.

The ACEEE notes that many utilities also charge "prohibitive 'exit fees' to customers that build CHP facilities." The organization lists among other roadblocks to CHP:

▶   The failure of current regulations to recognize or give credit for the superior energy efficiency of CHP and the lower emissions resulting from the displacement of electricity generation no longer needed.

▶   Distorted depreciation schedules for CHP investments that can run as long as 39 years—when the true depreciation schedule should be seven years.

▶   The lack of national standards for the interconnection of CHP to the electric grid—opening the door for some utilities to require prohibitively costly studies and unnecessarily expensive equipment as a means of discouraging the use of CHP.

There are generous investment tax credits and production tax credits for renewable energy, but until 2009, no credits for CHP. Put another way, governments provide the power industry with almost no incentives to deploy efficient CHP. Most states do not recognize CHP as part of the Renewable Portfolio Standards that many states have adopted. Another major barrier is treating power from a local CHP plant as equivalent to power from a remote central generation plant, completely ignoring the value of local generation in avoiding investment in transmission and distribution wires, and avoiding the line losses of moving power from remote generation to users. If CHP plants were paid fairly for avoiding these costs, entrepreneurs would rush to build more efficient local generation.

Moreover, the political power and influence of utilities has frequently been used to block changes in obsolete legal and regulatory provisions. For example, 49 states ban use of a private wire to move electricity across a public street. A university that installs a CHP plant can move the steam

IN SOME TOKYO NEIGHBORHOODS, SKYSCRAPERS GET THEIR HEATING AND COOLING FROM COMBINED HEAT AND POWER COGENERATION SYSTEMS (LIKE THE ONE BELOW). THESE SYSTEMS CAN BE 30 TO 40 PERCENT MORE EFFICIENT THAN CONVENTIONAL ONES.

across public streets to all of its facilities, but must pay exorbitant rates to the local utility to move the electricity across public streets.

California and a few other states have decoupled the profit opportunities for utilities from the constant pressure to sell more energy. California's plan enables energy utilities to share in the savings CHP and other energy-efficiency programs make possible for residential and commercial customers. If all states adopted a similar approach, the need for new generating capacity would fall dramatically.

Even more outrageous than the practice by utilities of blocking the use of CO_2-reducing cogeneration technologies by their largest customers is the fact that the utilities have failed to build their own

The volume of energy wasted is enormous. According to ORNL, "The energy lost in the United States from wasted heat in the utility sector is greater than the total energy use of Japan." The Center for Building Performance and Diagnostics at Carnegie Mellon University reports that 65 percent of all the energy consumed to generate electricity in the United States each year is lost. (The Electric Power Research Institute found that roughly 10 percent of the electricity coming out of the generators is also lost during transmission and distribution.) The McKinsey Global Institute adds, "With conversion efficiencies ranging from less than 30 percent in older coal generators to 60 percent in advanced combined-cycle gas turbines,

Sixty-five percent of all the energy consumed to generate electricity in the United States each year is lost.

CARNEGIE MELLON UNIVERSITY

new generation using waste-energy streams from their customers' factories and have also failed to develop new appropriately sized generation near customers that could use the by-product heat. Indeed, the absurdly wasteful way in which we now generate most of our electricity needlessly doubles fuel use, cost, and CO_2 emissions. These old electricity-only plants are long beyond their planned lives, propped up by their right to pollute. It is time to replace them with distributed CHP plants. This is one of the single largest sources of huge, profitable, near-term reductions in CO_2 emissions.

there are large opportunities for reducing losses in both new and existing generation plants."

Nevertheless, there are hundreds of plants proposed for construction in the world today that will generate electricity only, wasting 40 to 65 percent of the fuel instead of generating heat and power and achieving 65 to 95 percent efficiency.

The potential savings are so large that, according to McKinsey, "in many regions, it makes economic sense to replace inefficient, old power capacity because future energy savings pay for the investment cost of new, more efficient equipment." According to the National Science Foundation, the typical large power plant has an average efficiency of 30 percent, with additional losses of electricity in transmission. By shifting to CHP, coupled with distributed generation, we could achieve an average efficiency of 80 percent, with virtually no transmission losses.

Unfortunately, most utility regulators continue to use outdated approaches that make it more profitable for utilities selling electricity to waste two thirds of the energy in the fuel they burn. The regulatory barriers preventing the avoidance of this inefficiency have held improvements in the efficiency to power generation of less than

1 percent in the past 50 years—since Dwight David Eisenhower was president.

Ironically, CHP was first used more than 125 years ago. On September 4, 1882, at precisely 3 p.m., Thomas Edison flipped a switch to begin operating the world's first electricity generating plant, in lower Manhattan. It turned on the new electric lights he had installed in the offices of *The New York Times* and other leading newspapers of the day, Wall Street brokerage houses, the Stock Exchange, and several major banks.

The *Times* reported the following day that the new lights were "soft, mellow, and grateful to the eye … with no nauseous smell, no flicker, and no glare." The reporter, referring to his colleagues at the paper, said that the new lights were "tested by men who have battered their eyes sufficiently by years of night work to know the good and bad points of a lamp, and the decision was unanimously in favor of the Edison electric lamp as against gas."

Significantly, this same observer noted that "there was a very slight amount of heat from each lamp, but not nearly as much as from a gas-burner—one fifteenth as much as from gas, the inventor says." Indeed, the wasted heat from Edison's incandescent bulbs was only a fraction of the wasted heat from old gas lamps. But it still gave off only 1.4 lumens (a lumen is the standard unit of measurement for visible light) for each watt of electricity it consumed.

Modern incandescent bulbs are more efficient still, yielding 15 to 20 lumens per watt, with the latest versions at or above the high end of that range. The new compact fluorescent lightbulbs (CFLs) have received so much attention because they are four times as efficient in the use of electricity and last four times as long before they have to be replaced. While early versions of CFLs

A BETTER LIGHTBULB

While Thomas Edison's incandescent bulbs were more efficient than gaslight, advances since then have increased bulb efficiency more than 70 times, as measured by the amount of light per watt of electricity.

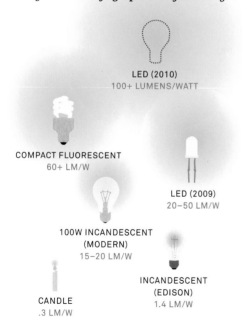

LED (2010)
100+ LUMENS/WATT

COMPACT FLUORESCENT
60+ LM/W

LED (2009)
20–50 LM/W

100W INCANDESCENT
(MODERN)
15–20 LM/W

CANDLE
.3 LM/W

INCANDESCENT
(EDISON)
1.4 LM/W

SOURCE: Philips; U.S. Department of Energy

AT THE NONPROFIT ELECTRIC POWER RESEARCH
INSTITUTE, RESEARCHERS TEST THE EFFICIENCY AND
PERFORMANCE OF ELECTRICAL EQUIPMENT, FROM
GENERATORS TO LIGHTBULBS AND APPLIANCES.

irritated many users with their lower output of
light compared with the incandescent bulbs that
most people are still used to, newer versions are
now brighter.

And an even newer form of lighting, light-
emitting diodes (LEDs), has already achieved 100
lumens per watt of electricity and is doubling in
efficiency every 18 to 24 months. Newer consumer
versions of these LED bulbs are expected to be
available in designs that fit standard light sockets
within one year.

The potential savings from the wide substitution of new LED lights for incandescent lights are
astonishing. Approximately 12.5 percent of global
electricity production is used for lighting. Better
and more efficient street lighting can bring especially large savings for cities.

Just as Edison's first lightbulb was far less efficient than modern incandescent bulbs, his first

generator—named after P.T. Barnum's famous
African elephant, Jumbo—was far less efficient
than those of today. Jumbo captured only 3 to
4 percent of the energy in the coal Edison burned
to produce electricity, while today's generators
are 10 times as efficient.

Yet they still dump two thirds of the heat
into the atmosphere instead of reusing it to
produce more electricity, or recapturing it for
other purposes.

Even though Edison's first electricity generator
was less efficient than today's generating facilities,
he was nevertheless careful to capture the waste
heat and put it to productive use. A few years after
starting his first generator in Manhattan, Edison
personally inspected another new electrical system that had been installed to power the Hotel Del
Coronado in San Diego. This hotel was the first to
get district heating from combined heat and power.

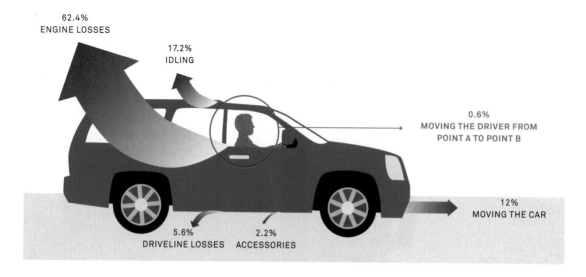

62.4%
ENGINE LOSSES

17.2%
IDLING

0.6%
MOVING THE DRIVER FROM
POINT A TO POINT B

12%
MOVING THE CAR

5.6%   **2.2%**
DRIVELINE LOSSES   ACCESSORIES

WHERE OUR GASOLINE'S ENERGY GOES

As a result of the energy losses related to internal-combustion engines and other energy-using systems in conventional cars, less than 13 percent of the energy in a gallon of gasoline actually moves a typical car. Most of that energy moves the car itself, with less than 1 percent going to move the person inside the car from point A to point B.

District heating and district heating and cooling are the terms used to describe systems that can transfer otherwise wasted heat from electric generation to surrounding buildings. In typical European systems, hot water from the CHP system is piped under the streets throughout a city and into heat exchangers installed in the buildings where it is used for space heating—and, with absorption chillers, for space cooling. This thermal energy is a by-product of electricity generation and displaces boiler fuel throughout the city.

Indeed, one of the most important guidelines for experts in energy efficiency is to design energy systems that avoid the transformation of energy from one form into another. Any such transformation inevitably results in the loss of large amounts of overall energy. For example, when natural gas is burned to produce electricity, which is then used in electric stoves to produce heat for cooking, the transformation of the energy in the gas is reduced by 65 percent in the first transformation; another 10 percent of what remains is lost during the transmission and distribution of the electricity; further losses occur when the voltage is stepped down for distribution into the home; and yet more is lost when the electricity is transformed back into heat at the stove.

The most efficient applications for district heating thus far have been supplying thermal energy to large institutions such as hospitals, universities, and military bases, where the institution is able to heat all of its buildings with a single installation and a single contract.

In recent years, the fastest growth for district heating and cooling systems has been in urban

SOURCE: U.S. Department of Energy; Amory Lovins, Rocky Mountain Institute

downtown areas—with particular emphasis on cooling systems. For example, Helsinki, the capital of Finland, has long received more than 92 percent of the heat for its buildings from CHP—and has generated so much electricity at the same time that it sells the excess to other Nordic countries. According to a report from Helsinki Energy in 2008, the entire nation of Finland receives almost 50 percent of its space heating from district heating systems, and 75 percent of that heat is supplied by CHP plants.

One of the surprising leaders in the use of district heating is Russia, which receives more than 30 percent of its total power generation from combined heat and power and uses hot water for district heating. Germany and China both get more than 10 percent. But other countries, including the United States, have not yet exploited even a fraction of the potential of this energy-efficient approach.

Smaller-scale CHP units have been successful in Europe and Japan, providing heating and cooling for individual buildings. As an example, there are now more than 50,000 homes in Japan with a CHP device—either a small piston engine or a small turbine—that operates only when the home requires heat or hot water, and then runs just enough to produce the heat. The by-product electricity winds the meter backward. The systems are 90 percent efficient, and the community wins. However, such systems and larger CHP plants have not yet been widely used in the United States and elsewhere. Still, the growing interest in "self-powered buildings" is beginning to create increased demand for these smaller-scale applications.

In 1882, Edison sold electricity as direct current—which, at the low voltage then used, could be profitably transmitted only half a mile. Six years later, Nikola Tesla, a brilliant Serbian immigrant briefly employed by Edison, invented alternating current and then—with financing from George Westinghouse—started his own business competing with Edison. The radius over which alternating current could be transmitted gave it a decisive technology advantage, and AC power quickly became the new standard.

Because utilities using the new AC generators could service many more customers, they quickly began to build larger generators in an effort to capture economies of scale. However, the large amounts of coal these generators burned produced unacceptable quantities of air pollution, and by the 1930s, these generators began to be located some distance from downtown areas. As a result, they were no longer close to the large buildings that could easily use the otherwise wasted thermal energy they created. The buildings installed boilers and burned more fuel to create the heat the electricity-generation plants just threw away. In an era of cheap fossil fuel and little appreciation for the problems of air pollution and global warming, we stopped using energy twice. The electricity industry in the United States and elsewhere adopted the habit of simply dumping all of its waste heat into the atmosphere.

Over time, this inefficient technique became accepted as normal, and utility regulations developed in ways that rewarded only capital investment, requiring utilities to pass on all efficiency gains to customers. As a result, in most jurisdictions, utilities receive no incentive for increasing efficiency by capturing and recycling their waste heat. Incredibly, this distorted use of regulatory power makes it more profitable for most electric utilities to completely waste two thirds of the energy in the coal, gas, and other fuels they burn. Virtually all of this wasted heat could be economically recaptured at the point of generation and used efficiently for additional electricity—thus displacing the need for more coal and more generating capacity.

The Clean Air Act of 1970 grandfathered old, inefficient plants on the assumption that these older facilities would eventually wear out and be replaced, but the economic value of this right to continue polluting at high levels has given old generation plants virtual immortality. The average age of coal-fired generating plants has continued to increase because this legal distortion gives them such a large economic advantage over new, more efficient plants.

Since the wasted heat energy is free, and the CO_2 produced in creating it is emitted whether the waste heat is recovered or not, this energy is effectively CO_2-free. Capturing this energy to displace the burning of yet more carbon fuel would simultaneously reduce CO_2 emissions and the cost of energy—thereby increasing the efficiency and competitiveness of industries and businesses that would gain the benefit of reduced electricity costs.

Any serious effort to recapture the enormous amounts of energy wasted due to inefficiency each year must also focus on buildings. The United Nations Environment Program estimates that 30 to 40 percent of CO_2 emissions worldwide are from heating, cooling, and lighting buildings that leak energy like sieves. The United States is at the top of the building inefficiency range, with almost 40 percent of CO_2 emissions coming from buildings.

The largest energy savings in the built environment can come from retrofitting homes with adequate amounts of insulation and proper weather stripping. Unfortunately, as noted in Chapter 15, builders and developers of homes have different incentives from owners; in an effort to minimize the first purchase price of homes, they often install significantly less insulation than is necessary to ensure that the home is energy efficient.

Although building practices and technology have improved in recent years, most building codes are still based on outdated knowledge and do not reflect most of the new opportunities that are now available. As just one example, many local codes require constant lighting of certain common spaces and emergency exits, even though modern sensors can detect when people are present and turn on lights. National standards for building codes can rectify this problem, though political resistance to change is, as always, difficult to overcome.

Most efficiency improvements require an up-front investment that pays for itself over time, which means that homeowners who do not have the cash for the initial investments must borrow the money for the initial expense and then subtract the cost of the capital from the amount paid back in the form of reduced energy bills. What homeowners need is a source of low-interest loans to finance efficiency upgrades, a readily available source of reliable knowledge about what works best, and a network of one-stop service providers who compete to conduct inexpensive or free energy audits as the first step in identifying the most cost-effective efficiency upgrades for each individual home. As mentioned in Chapter 15, the current obsessive focus on short time horizons further discourages large investments in the present that are repaid in savings that stretch out over a period of years.

When this bias in our habitual way of thinking is added to the split incentives between builders and owners of structures, and the unfamiliarity that many builders as well as homeowners have with the new opportunities for saving, the result is huge and unnecessary energy losses from most buildings in the U.S. and many other countries.

Legislation now pending in the U.S. Congress would require energy audits and establish minimum efficiency standards when new homes are

THERMAL IMAGING CAN REVEAL EXACTLY WHERE HEAT ENERGY LEAKS OUT OF OUR HOMES. ROOFS AND WINDOW FRAMES ARE OFTEN THE MOST WASTEFUL AREAS.

HOT

COOL

The cumulative effect of efficiency improvements in homes can offer significant reductions in greenhouse gas emissions. Improvements range from replacing old windows (top left) and installing better insulation (top right) to switching to efficient appliances (middle left) and compact fluorescent lightbulbs (middle right). Further home retrofits include rooftop solar water heaters (bottom left) and green roofs (bottom right).

built or when old ones are sold. Other countries—Germany, Sweden, and Japan, for example—have already made good use of such laws.

In recent years, there have been efforts to establish green building standards, such as the LEED (Leadership in Energy and Environmental Design) certification. The energy savings—and consequent CO_2 reductions—from using new, energy-efficient forms of construction are so great that most homeowners can benefit significantly by retrofitting their homes with more insulation, better lighting, and better windows.

Older homes are, of course, usually the most inefficient but can be upgraded with investments that typically pay for themselves through saved energy bills in less than three years. Moreover, the retrofitting of inefficient residential and commercial buildings to make them more energy efficient would create millions of jobs in the United States that cannot be outsourced. And of course, similar efforts in other countries would also create employment and strengthen the global economy.

Modern furnaces, air conditioners, and heat pumps usually save so much energy compared with older versions that they too recover their purchase price in a short period of time. In addition, the wider use of ground source heat pumps—particularly with new construction—(as discussed in Chapter 5) can utilize the natural thermal energy of the earth itself to sharply reduce heating bills in winter and cooling bills in summer.

One of the biggest sources of wasted energy in buildings is leaky ductwork. Since the leaks, and the wasted energy they cause, are hidden in walls, attics, and basements, these losses are usually invisible—except when the utility bill comes each month. And since there is no division between energy used and energy wasted when you get the bill, it doesn't occur to most people to fix the leaks in their ducts.

When the state of California began to focus on energy efficiency, it found that residential air ducts, on average, were leaking 20 to 30 percent of the air for heat and air-conditioning that flowed through them. The state now requires leakage rates of less than 6 percent and inspects every seventh new house to enforce the standard.

Most windows in buildings are so inefficient that they leak huge amounts of heat to the outside air during winter and allow almost as much heat from the outside into the building during summer. The new, much more efficient windows on the market today are more expensive but pay for themselves many times over in the form of significantly lower energy bills for the owner of the building—by cutting heat loss in winter by two thirds and reducing unwanted heat gain during summer by one half. In commercial buildings, many architects have rediscovered the virtues of allowing people to actually open windows at times when doing so can save energy.

Architects and builders are also paying new attention to the design and construction of roofs in order to minimize heat absorbed from the sun by buildings during summer and to reduce heat loss during winter. Highly reflective white roofs are particularly efficient in areas that use air-conditioning heavily.

With the expectation of higher energy prices, and with increased pressure to reduce CO_2, the options for upgrading building efficiency are receiving much more attention. Architects have demonstrated sharply higher interest in recent years in investigating and using designs that minimize heat loss and energy consumption, and many purchasers of buildings are now beginning to pay much more attention to features that reduce their annual electricity, heating, and cooling expenses.

A DEDICATED LANE FOR BUSES IN JAKARTA, INDONESIA, ALLOWS MASS TRANSIT RIDERS TO SPEED PAST AUTOMOBILES DURING RUSH HOUR.

TESTING IS UNDER WAY ON LIGHTER, MORE EFFICIENT JET AIRPLANE ENGINES. THIS ENGINE PRODUCES 20 PERCENT LESS CO_2 THAN ITS PREDECESSORS.

As discussed in Chapter 3, what is called "passive solar" design in new construction can make a dramatic difference in the amount of energy needed to heat in winter and cool in summer. Roof gardens can also insulate structures and lower the temperature of buildings during summer, while making them warmer in winter.

On the inside of most residential buildings, further significant gains in energy efficiency are available with the upgrading of appliances and electronic systems with more efficient, modern versions. The Department of Energy's Energy Star program gives useful information to consumers about which appliances can save the most money with lower energy consumption. Older hot water heaters can profitably be replaced—or wrapped with insulation to reduce wasted heat. In most latitudes, roof-mounted solar hot water heaters pay back their purchase price in a short period of time and produce further savings for as long as they're used.

Most television sets, DVD players, and other electronic devices in homes waste prodigious amounts of energy even when they are turned off. Indeed, so much energy is used by American television sets while they are in standby mode, it takes the full output of a 1,000-megawatt coal-fired generating plant just to supply energy to television sets that are *turned off*.

Virtually all products and systems can be made far more efficient with the use of modern computer-aided design and computer-aided manufacturing (CAD/CAM) approaches. New manufacturing technologies now make it possible to use new materials and new design approaches that in many cases result in revolutionary improvements to products and processes that have followed the same basic design for many generations.

The opportunity for reducing global warming pollution in the transportation sector is enormous. The internal-combustion engine, for example, is only 20 percent efficient, while a switch to electric vehicles would instantly improve that percentage to 75 percent—although overall efficiency gains depend upon the source of the electricity; coal-fired generating plants are, after all, only 35 percent efficient.

As we have long known, comfortable modern mass transit, such as light rail systems in urban areas, can sharply reduce both energy used and CO_2 produced in transportation. New engine technologies, hybrids, and plug-in electric vehicles produce huge energy and pollution savings in personal vehicles. New lightweight materials offer the promise of much larger gains in vehicle efficiency. Mandated improvements in mileage efficiency have been responsible for most of the gains experienced thus far in the vehicle fleet. The new increased standards adopted in the U.S. in 2008 will achieve even greater savings, although U.S. standards are still far below those in China, Europe, and Japan.

A number of airline companies are actively investigating the use of alternative sources of fuel, though the lower energy content of many biofuels has convinced many analysts that this transition will be difficult. However, Boeing Aircraft is making progress in the development of much lighter and much stronger materials in the manufacturing of airplanes that are already achieving gains in energy efficiency.

The obstacles confronting increased efficiency include, most prominently, a lack of awareness at the national, regional, local, and individual levels of how large the opportunities are. Since we are creatures of habit, the prevailing unfamiliarity with the new efficiency gains now possible is compounded by the natural inertia that prevents many from taking the initiative to capture the energy savings—

even though they can save money in the process.

Moreover, the lack of "systems thinking" has inhibited the introduction of efficiency improvements that come from the redesign and reengineering of large systems. And the lack of high-quality, easy-to-use information systems that identify the opportunities for efficiency improvements has contributed to the failures to adopt them.

Ultimately, the redesign of larger systems— like cities—will make it possible to achieve much larger savings in efficiency. And while that may sound too visionary or unachievable, it is important to remember that in nations like China, India, Nigeria, and other fast-growing countries, new cities are being laid out for construction every year. Incorporating these new energy-efficiency design principles at the beginning can produce massive savings.

Most of the solutions to the climate crisis will become much easier to adopt when the consequences of global warming pollution are reflected in the cost of the choices we make. But most energy-efficiency experts agree that factors other than price are also extremely important to address in order to seize opportunities for large gains from efficiency improvements.

Because these opportunities are so varied and appear in so many different settings, there is not a single body of knowledge that is universally applicable to all of the steps that should be taken. Virtually the only common feature is that all of these steps end up saving money as well as reducing CO_2 pollution. Moreover, there is not a single service provider capable of eliminating the "hassle factor" that faces those who undertake efficiency improvements.

Energy utilities would seem to be the obvious providers of a soup-to-nuts package of efficiency improvements, but the perverse incentives in the utility industry reward them for selling more energy and penalize them for reducing consumption by their customers. As noted earlier, the state of California and a few other jurisdictions have redesigned the incentives for utilities to allow them to share in the savings they make possible for their customers. This one change was responsible for a significant amount of the energy-efficiency success story in California. The enormity of the efficiency opportunity is a powerful argument for a national program based on the experience of California and the few other states that have taken the lead in this area. However, the roller coaster of world oil prices, which is always mirrored in the prices for energy of all kinds, has weakened national stamina and frustrated the sustained effort necessary to introduce pervasive efficiency improvements throughout the national and global economy. Partly because the global warming consequences of fossil energy are not included in the price of electricity or oil, the illusory low prices have discouraged investments in efficiency. But the distorted price signal is only a small part of the reason for our failure to use efficiency gains, because they are almost always profitable, no matter the price of energy.

What is most needed is the kind of vision, focus, and determination at the national leadership level that we have seen produce revolutionary improvements in companies and in government policy when visionary leaders were in charge of driving efficiency improvements.

ENERGY EFFICIENCY DEPENDS UPON LEADERSHIP AND VISION

Frito-Lay CEO Al Carey and Governor Arnold Schwarzenegger tour Sun Chips' solar plant in Modesto, California.

Among businesses that have been most successful in adopting efficiency improvements, the primary reason for their success is that, in almost every instance, a determined and visionary leader has driven the organization's commitment to efficiency. And in too many other businesses, organizational failures and a lack of leadership have prevented the recognition and adoption of efficiency improvements.

Frito-Lay, a division of PepsiCo, is led by its CEO, Al Carey, who has acheived remarkable savings in energy by focusing on how to reduce his company's CO_2 emissions.

Under Carey's leadership, Frito-Lay has looked at every component of its corporate operations to find new efficiencies and savings. The company recently installed the largest corporate solar-power system in Arizona, and last year it began using solar power to help make its Sun Chips in California. (In this system, solar-produced steam heats the oil in which the chips are cooked.) The company has also instituted a "zero landfill" program in four of its plants in pursuit of the goal of reducing waste to less than 1 percent.

The company has also made a number of changes that reduced its CO_2 emissions by almost 50,000 tons from 2006 to 2007. Additional reductions of greenhouse gas emissions of another 14 percent have been identified, and Frito-Lay has set a goal of achieving a 50 percent reduction in global warming pollution by 2017.

Another strategy Frito-Lay is adopting to reach that goal is reusing shipping cartons. By reusing each carton an average of five to six times, the company has reduced its consumption of paperboard by 120,000 tons per year. When the cartons finally wear out, they are recycled.

Frito-Lay has also set a goal of making its fleet of trucks the most fuel efficient of any in the country, converting many to hybrid vehicles and training drivers to reduce fuel consumption. By changing the materials it uses and redesigning its processes, the company is making a huge difference and demonstrating what is possible for others with a similarly sustained commitment.

HOW WE USE ENERGY

CHAPTER THIRTEEN

THE SUPER GRID

IN SHANGHAI, CHINA, AND ELSEWHERE AROUND
THE WORLD, A SURGE IN ELECTRICITY DEMAND HAS
TAXED EXISTING POWER GRIDS WITH OUTDATED
TRANSMISSION TECHNOLOGY.

For more than a hundred years, our way of thinking about electricity has been shaped by the dominance of large, centralized electricity-generating plants—coal-fired, hydroelectric, nuclear, and gas-fired—always connected to customers in real time by electricity transmission and distribution networks. When the present electrical system was first put in place, it was a marvel to the world. Indeed, the National Academy of Engineering described it as "the most significant engineering achievement of the 20th century" in the year 2000.

However, growing problems and several significant new developments in technology have begun to shatter the usefulness of the old way of thinking. And it is now painfully clear that our present electrical grid has become outdated and inefficient.

Utilities, regulatories, and federal and state legislators are trying hard to understand how they can best adapt to and take advantage of the new technologies and new patterns of generation, transmission, distribution, storage, and use of electricity.

But there are numerous reasons to speed up the introduction of these modern advances in technology and construct new continent-wide unified smart grids—or super grids—in the United States and in other countries. Doing so will significantly reduce the huge and unnecessary emissions of global warming pollution caused by inefficiencies and failures in the transmission, distribution, and storage of energy.

The technologies necessary to build a super grid are all fully developed and available now. The only missing ingredient is political will. The first steps toward building a U.S. super grid were taken in early 2009 when President Obama formally proposed such a project and included the initial stage of funding in the economic stimulus bill.

Just as the United States benefited from the national vision of an Interstate Highway System, and, later, the "Information Superhighway" that became the Internet, the development of a unified national smart grid would create millions of new jobs and sharply reduce its CO_2 emissions.

Moreover, with a large enough vision and adequate planning, the U.S. and other nations could simultaneously lay high-capacity optic fibers for an expanded and much more efficient national broadband network in the same trenches dug for the high-voltage transmission lines making up the backbone of the new electricity grid.

The National Energy Technology Laboratory (NETL) for the U.S. Department of Energy proposed a vision statement to guide the development of a modern U.S. grid, "to revolutionize the electric system by integrating 21st century technology to achieve seamless generation, delivery, and end use that benefits the nation."

It is important to be clear about the four interconnected elements that make up a unified national smart grid, or super grid:

▶ Much more efficient, higher-voltage long-distance transmission lines, connected to all generators of electricity, including the new intermittent sources, solar and wind.

▶ "Smart" distribution networks connected by the Internet to smart meters at homes, substations, transformers, and every other element of the transmission and distribution grid.

▶ Modern, dynamic, and efficient electric-energy storage units placed throughout the transmission and distribution networks, with most storage devices placed near or in the facilities owned by end users. Full interconnection of all transmission and distribution grids.

▶ Distributed intelligence with robust, information-rich, two-way communication throughout the grid.

A smart-grid distribution system will allow consumers to use time-of-day pricing to reduce their energy costs and could automate the process based on customer preferences. One of the key components of a smart grid, smart meters, will help consumers take control of their energy-use patterns. Electricity consumers will be able to determine in advance how much they want to pay for their power each month and then make choices as to what devices, appliances, lighting, and other electricity-using features they are willing to have turned off for several hours each day in order to meet the price target they themselves have set. The days of meter readers coming to the corner of your house will be over.

The entire grid will be digital, instead of relying on old electromechanical and analog features. It will be self-monitoring and, to a large extent, self-healing. The smart grid will eliminate many power outages, minimize others, and assist utilities by instantly informing them of the exact location

The technologies necessary to build a super grid are all fully developed and available now. The only missing ingredient is political will.

where emergency repairs are needed. It will also provide utilities with far more accurate information with which to plan future investment and construction. Modernizing the grid can also make it far less vulnerable to what national security experts say is an increasing threat of digital terrorist acts aimed at intentionally blacking out large areas.

Properly designed, a smart grid will be more reliable, more secure, more efficient, less costly to operate, and far less harmful to the environment. Older grids like the one in the United States are also vulnerable to outages that interrupt the electrical supply periodically for millions of people. As well, our system is plagued by "congestion" problems; because of too little transmission capacity

The estimated annual cost to society of the outdated U.S. grid is $206 billion per year.

IN AUGUST 2003, THE LARGEST BLACKOUT IN NORTH AMERICAN HISTORY AFFECTED MORE THAN 50 MILLION PEOPLE, INCLUDING ALL OF NEW YORK CITY.

and too few "smart" features, it is ill-equipped to efficiently manage the flows of energy from multiple generating sources that all stream through bottlenecks throughout the grid.

A modern grid would fix these problems. In the United States alone, according to the Lawrence Berkeley National Laboratory, electric power outages and blackouts—lasting from seconds to days—cost the United States approximately $80 billion every year from businesses and industries. Other estimates that include residential customers and "power quality events" that suddenly change the voltage supplied to business users of sensitive electrical equipment put the value of annual losses at even higher levels. In the United States, some manufacturers in businesses particularly sensitive to fluctuations in voltage and even very brief power outages have deferred construction of new facilities in locations where the electricity supply has poor quality.

In 2007 NETL estimated that the annual cost to society of the outdated U.S. grid is $206 billion per year. A more reliable, efficient, and smart electrical network will foster improved competitiveness and greater job creation. The Galvin Electricity Initiative, which studied the impact of the outmoded power system on productivity and competitiveness, concluded, "At least a trillion dollars in gross domestic product is already being lost each year as a result, and that cost is growing rapidly as the digital economy expands."

Modern, highly efficient long-distance transmission lines must be put in place in order to unleash the potential of these CO_2-free sources of energy. Transmission lines that use a much higher voltage than is common in today's grid can transmit large amounts of electricity over long distances with extremely low losses. As noted in Chapter 12, the low voltage used by Thomas Edison to transmit direct current sharply limited the distance over which he could transmit electricity to roughly half a mile. The success of alternating current, invented by Nikola Tesla in the late 1880s, was based primarily on its much lower losses when transmitted many miles.

Today, however, both DC and AC electricity can be transmitted over modern lines with very low losses. High-voltage DC current (HVDC) is actually somewhat more efficient than HVAC as a long-distance transmission technology, but that advantage disappears if converter stations are placed along the line to draw electricity supplies for use along the route of the line. That is the principal reason why approximately 98 percent of transmission worldwide is now high-voltage AC.

HVDC lines are probably the best choice for the transmission of electricity from remote solar and wind locations—whether from the southwestern desert of the United States, from Mongolia to Eastern China, the desert in northwestern India to Delhi, or elsewhere. One advantage of HVDC lines is the ease with which they can be cost-efficiently buried instead of placed on overhead transmission towers, thus avoiding much of the public opposition to new transmission lines. They may also be used for intercontinental links like the ones proposed between North Africa and Europe.

Transmission systems at even higher and more efficient voltage levels already exist elsewhere in the world and more are planned in North and South America, South Africa, Scandinavia, Western Europe, and Asia. China, which has announced its intention to interconnect its three regional electricity networks in north, central, and south China by next year, is hard at work building an 800-kilovolt national super grid that it says will be the most advanced in the world by 2020.

In Europe, many policy makers are now moving

THE EUROPEAN AND NORTH AFRICAN SUPER GRID

A new, large-scale electric grid has been proposed for Europe and North Africa, which would—once built—supply power to the European Union, the Middle East, and North Africa. The DESERTEC Concept is designed to collect clean, renewable energy in Saharan North Africa and the Middle East for transmission throughout the area connected by the super grid. The plan predicts that by 2050, 100 gigawatts of electricity will be generated within the system.

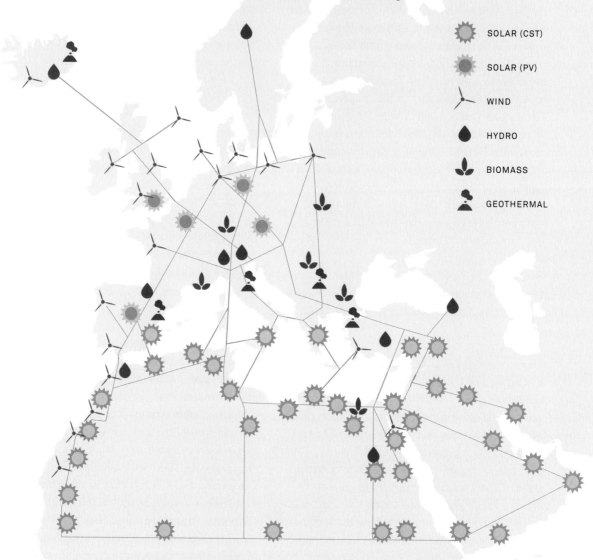

SOLAR (CST)

SOLAR (PV)

WIND

HYDRO

BIOMASS

GEOTHERMAL

SOURCE: DESERTEC Foundation

forward with the development of a super grid linking Europe with North Africa and the Middle East—where the potential for solar and wind electricity generation is virtually limitless. Similar ideas are being discussed in South America, Asia, and Australia—all of which have areas with great sun and wind potential far from their cities.

The present U.S. grid is older and less efficient than newer grids that exploit the most modern transmission technologies and equipment. For example, the average age of a substation transformer is 42 years—older than its projected useful life. The highest voltage used in the United States

grid modernization will exceed the costs by four to one. Energy experts estimate that a U.S. grid similar to the one now being built in China would reduce peak-load losses by more than 10 gigawatts (enough to power over 2.5 million homes) and reduce CO_2 emissions by millions of metric tons each year.

There are actually two important reasons why the electricity transmission and distribution grid must be upgraded in order to use far larger amounts of renewable electricity. Not only are the renewable sources generally found in remote locations that require new high-voltage

"The benefits to society from grid modernization will exceed the costs by four to one."

ELECTRIC POWER RESEARCH INSTITUTE

today is 765 kilovolts, although there are few such lines. Even when it is operating properly, the U.S. grid loses more electricity during transmission than modern state-of-the-art electrical grids.

Building a modern super grid will allow the deferral or elimination of tens of billions of dollars for new centralized generating plants, transmission lines, substations, and other distribution assets. The combination of money saved from the avoidance of outages and expensive additions to the old patchwork system will more than pay for the cost of a modern grid.

According to a study by the Electric Power Research Institute, the benefits to society from

transmission lines but the grid must also be equipped with smart features linked by the Internet to widely distributed dynamic electric energy storage devices in order to accommodate the intermittent nature of the electricity generated by solar and wind.

Indeed, the growing interest in using carbon-free sources of electricity is one of the principal drivers of the burgeoning interest in energy storage by utilities that need a way to smooth out or "firm" the electricity streams produced intermittently by solar and wind generators. Both solar and wind, as noted in Chapters 3 and 4, have an enormous potential for supplying virtually all of the electricity the United States consumes—and a very high percentage of that in most other countries.

Currently, wind and solar electricity represent relatively small percentages of the total amount of electricity flowing through the grid, and therefore utilities can manage the consequences of intermittency without harming the reliability of the grid.

However, the flow of electricity is uneven because passing clouds and the setting of the sun interrupt the flow of solar electricity, and variations in wind speed, and periods of no wind, interrupt the flow of wind electricity. As the percentage of electricity from these sources increases, intermittency becomes a more serious challenge.

SOLVING WIND'S INTERMITTENCY

With their very fast response times, new batteries, such as NAS cells, add precise amounts of electricity to a transmission source when input from a generating source declines. The combination produces a steady electric current.

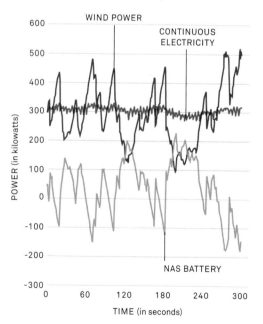

SOURCE: NGK Insulators, Inc.

Already, the Bonneville Power Administration (BPA) in the Pacific Northwest is struggling to integrate more than 2,000 megawatts of intermittent wind electricity (enough for two Seattles) with all of the other electricity it receives from its 31 hydroelectric dams and one nuclear plant. After growing over the past 10 years from 25 megawatts to its current level, the amount of wind electricity in the BPA's 10,500-megawatt peak-load balancing system is expected to triple within the next three years.

Several of the new batteries providing electric energy storage are capable of providing large surges of electricity in as little as two milliseconds for meaningful durations of time. These systems can be linked to the transmission and distribution grids receiving intermittent supplies of electricity from wind and solar generators in ways that automatically trigger releases of electricity whenever the flow from the generators fluctuates. These continual, episodic surges, when combined with intermittent streams of electricity in a mirroring pattern, can produce an even, continuous flow of electricity with high reliability and consistent and predictable voltages. (See "Solving Wind's Intermittency," at left.) By contrast, even the fastest gas turbine generator requires at least 15 seconds to ramp up its power output to the four-megawatt level— a response time that is way too slow to firm the intermittent flows from solar and wind.

There are actually a lot of gas-fired (and some coal-fired) generators that are held in reserve by utilities to meet longer and more predictable periods when they must meet a surge in electricity demand. In most countries, including the United States, the difference between the amount of electricity used during the hours of peak demand and the amount of electricity used, on average, during the rest of the day has been growing—largely because of increased use of air-conditioning, lighting, and larger television sets when people come home from work and before they turn in for the night. Moreover, the highest peaks occur on the hottest summer days, for only a small number of hours in each year.

These brief periods of peak demand are now usually met by revving up gas-fired (and a few coal-fired) generating units that are held in reserve for this purpose. However, when fossil fuel generators are kept running—even at a low level of output, in a "spinning reserve" mode, they are producing CO_2 all the while.

One analyst compared the inefficiency of this technique to the amount of gasoline someone wastes by constantly stepping on the gas for full acceleration and then stepping on the brake to idle the motor again. Anyone who drives a car like that is getting the worst fuel economy possible, but that's essentially how many electric utilities must operate their fossil fuel reserve plants, given the existing load patterns and the technologies and assets they have.

As a result, the power produced by these plants is the dirtiest and most inefficient power in the system. Replacing it with storage would therefore bring multiple benefits. Chris Shelton, the vice chairman of the Electricity Storage Association, refers to the combination of efficient storage with fossil fuel power plants as "hybridizing the grid."

One study of the CO_2 reductions that could be achieved by substituting efficient energy storage for standby spinning-reserve coal-fired generators operating during peak usage periods could eliminate between 76 and 85 percent of the CO_2 emitted by those generators.

Moreover, while energy use is growing at a rate of 1.5 to 2 percent annually, peak load demand is growing at a rate of 5 to 7 percent annually. In the United States 50 years ago, the average use of electricity was approximately 67 percent of the peak usage of electricity. Today, it is approximately 50 percent of the peak. The Department of Energy reports that on a nationwide basis, "Ten percent of all generation assets and 25 percent of distribution infrastructure are required less than 400 hours per year, roughly 5 percent of the time."

The growing disparity between average electricity use on the one hand and peak use for a three- to four-hour period each weekday evening, when almost twice as much electricity is needed in many developed countries, has driven new attention by utilities to the competitive benefits of storing electricity for use during peak load periods

ELECTRICITY DEMAND IS VARIABLE

In the course of a year, electricity usage varies considerably from month to month and even hour to hour. Summertime power loads are the highest because of air-conditioning. This chart shows demand in 2006 at PG&E, a California utility. Usage peaked on July 25 at 5 p.m. during a heat wave.

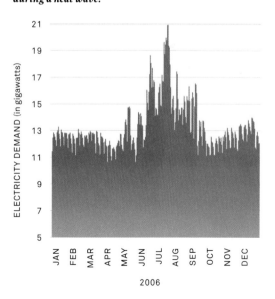

SOURCE: Pacific Gas & Electric

NEARLY ALL OF THE WORLD'S CURRENT CAPACITY
FOR STORING ELECTRICAL ENERGY COMES
FROM PUMPED HYDRO. THESE LARGE PIPES CARRY
WATER AT VATTENFALL'S HYDROELECTRIC PLANT
IN WENDEFURTH, GERMANY.

in order to defer or avoid costly investments in new generating capacity.

Unfortunately, electricity is difficult to store in large amounts. It constantly wants to be in motion. "Lightning in a bottle" is a metaphor based on the extreme difficulty of holding electricity in place until you are ready to unleash it again.

The overwhelmingly dominant energy storage technology used by utilities in the world today doesn't store energy itself, but the potential to create it. First introduced in America in 1929, the "pumped hydro" system is based on pumping water up to a reservoir on top of a mountain during the night, when there is excess generating capacity. Then, when the peak demand hours begin, the water is allowed to fall back down through hydroelectric turbines that generate electricity during the hours when it is most needed—recovering between 70 and 80 percent of the electricity used to pump the water up the mountain the night before. More than 150 pumped-hydro sites are now operating in the United States and worldwide; this mature technology now provides more than 99 percent of the world's total storage capacity for electrical energy.

Many additional pumped-hydro projects are now in the planning stages around the world. However, the future of this resource is limited for many of the same reasons that hydroelectric dams—which accounted for 40 percent of American electricity production in the early 1900s—will not play a significantly larger role in the future: Most of the best sites have already been developed; permits for such large energy installations are difficult to get approved. And of course, this option is not available in areas without mountains and/or adequate sources of water.

The second largest electrical energy storage technology is called compressed-air energy storage (CAES), which involves compressing air with a gas-fired generator and storing it in a geologically appropriate underground cavern, like a salt dome, then recapturing the energy when the air is released and expands, reproducing most of the energy used to compress it, when it is needed during peak usage hours. Although this option is almost as efficient as pumped hydro, its reliance on gas-fired generators makes it less attractive when its overall energy efficiency and CO_2 emissions are taken into account. Moreover, its use is limited by the availability of underground caverns with all the needed characteristics. New work is now under way to develop CAES systems that store the compressed gas in large pools of water and aboveground containers. There are only two large-scale compressed-air systems utilizing underground caverns, one in Alabama and one in Germany, both using salt domes.

The older methods of storing electric energy—pumped hydro and compressed air—have characteristics similar to the old centralized generating facilities: they handle large amounts of electricity generation quite well, but they are neither nimble nor smart; the electricity flowing from them cannot be easily and instantly turned on and off.

Other forms of energy are much easier to store. Coal and oil, for example, have been our energy sources of choice, not only because of their high energy density but also because they efficiently store energy with no losses for millions of years. It is worth noting, however, that when U.S. oil production peaked around 1970 and the Middle East oil embargo occurred a few years later, the United States found it necessary to establish a very large Strategic Petroleum Reserve, in order to store enough oil to supply the economy's needs during a future interruption of supply.

Biomass resources burned to produce electricity

A PROTOTYPE CHEVROLET VOLT IS FITTED WITH ITS LITHIUM-ION BATTERY, WHICH CAN POWER THE CAR FOR ABOUT 40 MILES. BEYOND THAT POINT, A SMALL GASOLINE MOTOR GENERATES FURTHER ELECTRICITY, EXTENDING THE CAR'S RANGE TO MORE THAN 300 MILES.

can rot if they are improperly stored, but newly applied techniques like torrefaction can extend their storage lifetimes considerably. Concentrating solar thermal (CST) plants produce large amounts of heat from the sun to boil water in order to fire a steam electric generator. Some of that heat can be stored in tanks of synthetic oil or molten salt and then the heat can be withdrawn when clouds pass over the solar collectors to continue running the generators for a few hours at a time. These thermal storage tanks can also be used to continue producing solar electricity after the sun goes down—for at least a few hours.

Mechanical flywheels, which store energy kinetically in the motion of a massive rotating cylinder, are also used for electrical energy storage in some applications, but the size and expense of these units has limited their attractiveness for storing large amounts of energy.

Some experts believe that the application of nanotechnology to solid-state electrical energy storage devices will lead to a development pathway that takes advantage of the well-known rapid improvements in solid-state chips, which are reducing costs by 50 percent every 18 to 24 months. Others believe that revolutionary developments in magnetic storage of electricity are now visible on the horizon. However, for now, efficient and competitive storage of electric energy remains a difficult challenge.

Indeed, for many years, there was a widely shared belief that electricity was so difficult to store in large amounts that storage would never play a big role in the interaction between electric utilities and their customers. In recent years, however, that myth has been shattered.

Most of the new excitement in the field of energy storage is now focused on developments in batteries of various designs. The competition to develop more efficient batteries is now raging in multiple countries, with many entrepreneurs and established companies throwing millions of dollars at the task. Scores of new battery companies are now feverishly developing and demonstrating innovative ways to store larger amounts of electric energy in smaller containers at less cost. Government R&D funding in numerous countries has been focused on this challenge, and venture capitalists and other investors are competing ferociously to come up with cost-effective solutions. Many innovative breakthroughs are reported regularly, and it is at present impossible to predict which new designs—or dramatically improved old designs—will emerge as winners.

Lead acid batteries—invented in 1859—are still the most commonly used rechargeable batteries. They are not considered practical for higher-volume grid-storage systems because of their short lifetimes and high maintenance costs—though some entrepreneurs have recently improved this old technology in ways that may make it a contender for many new applications.

However, the growing need for electricity storage has driven the investment of large amounts into the research and development of brand-new designs for much larger and ever more efficient batteries that are now capturing a lot of attention.

A Japanese ceramics company, NGK, now sells a room-size sodium-sulfur battery that has high energy and power density and relatively high efficiency. Originally developed by Ford Motor Company for an early attempt to build electric cars, this design was adapted for use in the electric utility market in the 1990s by NGK and the Tokyo Electric Power Company, which has already deployed them at almost 100 locations throughout its grid and at more than 200 locations around the world, including nine megawatts of capacity currently operating in three U.S. utilities.

Even though each of these batteries is expensive ($2,500 per kilowatt), the cost of the electricity they provide during peak usage is far cheaper than the peak electricity from new generating capacity used only a few hours each day. Because it can provide one megawatt of power for six hours—and can be clustered in groups that provide significant amounts of energy on short notice—the NGK battery is now regarded by many as the most effective design available at a large scale.

Its size and expense make it appropriate for some settings but not for others. And in addition to its size and expense, it suffers from one other disadvantage: only one company in the world makes it, and that company has only a limited number available each year. Utilities in Abu Dhabi and France have already purchased the company's entire output for the next four years.

Nevertheless, at present, this sodium-sulfur battery beats all competitors in the large-scale utility energy storage marketplace, though General Electric is now preparing to introduce a technology it first developed for hybrid locomotive engines, based on a sodium-metal-halide design, that it says will have performance capabilities comparable to those of the NGK battery.

Part of the challenge in evaluating which form of energy storage is the most efficient and is best to use in the generation, transmission, or distribution part of the electricity industry is predicting future developments that are likely to bring sharp cost reductions and efficiency improvements.

One of the reasons why several market leaders are betting on lithium-ion batteries stacked together for storage options of up to tens of megawatts is the fact that, all over the world, there is a growing competition to provide more efficient lithium-ion batteries for plug-in hybrid electric vehicles.

The safety and environmental requirements for automobile batteries also give utility planners increased comfort that successive generations of these batteries will be not only progressively more cost-effective but also safe, environmentally responsible, and widely accepted by customers. These factors have influenced a number of storage specialists in the energy and utility industries to piggyback on the cost-reduction curve they see emerging in batteries for electric cars; many are working to essentially stack these car batteries and use them as a coordinated fleet of batteries to meet many of the industries' storage needs.

This fast-growing global competition to replace heavily polluting and grossly inefficient internal-combustion vehicles has driven an enormous amount of research-and-development funding toward the rapid improvement of batteries powerful enough (and safe enough and small enough) to power automobiles and light trucks with the same speed and performance that drivers are used to.

Indeed, the impending widespread introduction of plug-in hybrid electric vehicles and, soon thereafter, all electric vehicles, will dramatically change the ability to store vast amounts of electrical energy during off-peak hours for use during the one sixth of each day when electricity usage is so much higher than the daily average.

Since the average vehicle is used for transportation only 4 percent of the time, the new plug-in electric hybrids and all electric vehicles will arguably have a bigger impact on energy storage than on transportation. Since each one stores only small amounts of electricity and remains available for transportation at all times, these batteries are not useful for baseload power, but they have enormous potential value for peak management.

As the global market for all-electric vehicles begins to expand, the growing demand for lithium-ion batteries will further drive economies of scale for the enormous automobile marketplace. Already, the demand for lithium-ion batteries is increasing rapidly, because of their high energy and power density, high efficiency, and ongoing and projected cost reductions.

The first mass-market electric hybrid, Toyota's Prius, still employs a nickel–metal hydride battery. However, both the Chevrolet Volt plug-in hybrid and the new all-electric Prius expected to be sold to fleet owners in 2010 are based on a lithium-ion design. The Great Wall Motor Company, in China, is one of a few Chinese automobile manufacturers planning to introduce electric vehicles with lithium-ion batteries next year. Nissan Motor Company announced in 2009 that it will sell all-electric vehicles with lithium-ion batteries in 2010 and will manufacture the Leaf for the U.S. market in its Smyrna, Tennessee, plant. And the Tesla, an American all-electric vehicle already on the market, uses several thousand lithium-ion cells stacked together in the back of the car.

In China, Japan, the United States, and the European Union, the race is on for market share in the anticipated mass conversion of the automobile fleet to electric vehicles. The dramatic advantages in efficiency from electric motors compared with the internal-combustion engine have combined with the growth in carbon-free renewable electricity sources and the societal advantages provided by cheap and efficient electricity storage in PHEVs, which can discharge electricity to

THE ELECTRIC TESLA ROADSTER HAS A RANGE OF MORE THAN 200 MILES. ELECTRIC CAR BATTERIES CAN SERVE AS WIDELY DISPERSED STORAGE SYSTEMS FOR ELECTRICITY—ENHANCING THE PERFORMANCE OF A SUPER GRID.

the grid during peak generating hours, to produce a surge in research-and-development spending to develop cheaper and better PHEV batteries.

The AES Corporation is installing a 12 megawatt system in northern Chile by stacking together lithium-ion batteries made for hybrid buses. Ironically, these lithium batteries are being deployed in an electrical system that provides power to the lithium mining operations in Chile. (Chile is the second largest source of lithium after neighboring Bolivia.)

One of the traditional strategies employed by electric utilities to cut power demand during periods of peak usage has been to negotiate electricity supply contracts with large users that require them to shut off power for a few hours at a time when peak usage strains the generating capacity of the system. This is sometimes referred to as "shedding load." For that reason, the lithium mining

operations in Chile have, in the past, been forced to shut down frequently during peaks. One of the improvements in efficiency that will be provided by the new lithium-ion battery stacks will be to allow the lithium mining operations to avoid shutting down during peak usage periods.

American Electric Power (AEP) is using lithium-ion batteries in a storage system called Community Energy Storage (CES), which pushes electricity storage out into the distribution grid with small boxes of multiple lithium-ion batteries operating in tandem immediately adjacent to the transformer boxes that typically serve every four to five houses in a neighborhood (with hundreds of transformer boxes connected to each utility substation). All of these small storage units would be operated as a fleet, with the electronic brains located in each substation. The advantages include fewer power outages, greater system reliability, and the installation of a storage infrastructure that could accommodate the shift to plug-in electric vehicles.

Ali Nourai, whose day job is with AEP, is also chairman of the Electricity Storage Association and an enthusiastic supporter of moving energy storage "toward the customer," where the selling of storage services could be a new revenue source for utilities adapting to a modern distributed energy marketplace for electricity and provide very high service reliability levels. He advocates open standards and robust competition and envisions energy storage eventually becoming a commodity for sale throughout the world's electricity distribution infrastructure. Eventually, he believes, "utilities will sell the service of storage."

While competition in the market for wholesale electricity takes place throughout the United States, only some of the states allowing retail competition have fully separated the markets for electricity generation from the transmission and distribution functions. In some of those areas, utilities are now attempting to persuade their regulators that energy storage ought to be operated in support of the transmission and distribution grid. But rules established to protect competitive generation make it more difficult to justify direct utility ownership of energy storage. Additionally, allowing regulated utilities to own or control access to and rules for distributed energy storage could run the risk of stifling competition in energy storage just at the time when many entrepreneurs are coming up with exciting ideas for much lower-cost, more efficient energy storage options.

The ability on the part of customers to own and operate their own devices for both generating and storing electricity is now clearly growing at a rate that will soon undermine the monopoly model and begin to shift the grid toward a "widely distributed" model often labeled "micro-power."

The combination of more efficient storage and smarter distribution grids will clearly accelerate the expansion of small-scale generating capacity throughout the grid, with everything from photovoltaic panels on the roofs of homes to small windmills in areas where they are profitable.

In the United States, one of the policy questions raised by micro-power is whether, and on what terms, households and small businesses can sell some of the electricity they generate—and energy that they store during off-peak hours—back into the grid. Most utilities have fought hard against liberal provisions for "net metering" that would encourage and accelerate the spread of widely distributed micro-power.

The utilities' argument is that, in many cases, the rate they would have to pay for electricity generated by their customers is not fair to the utilities. They fear being stuck with all of the fixed costs of their transmission and distribution infrastructure and centralized generating plants and yet be progressively deprived of the revenue from electricity sales that they need to pay for it and remain profitable. The idea of not only losing electricity sales but also competing with new electricity generating sources spread throughout their service area has created a sense of foreboding in parts of the industry—and has led some utilities to work overtime to convince state regulatory commissions and state and federal legislators to put up roadblocks intended to slow down the spread of micro-power.

Although rooftop photovoltaic cells still play only a minor role, in many areas the number of installations is doubling every year. American Electric Power reports that renewable sources of electricity generation owned and operated by their customers have grown "by a factor of 1,000 in a decade." California, New Jersey, and Arizona have been leaders in encouraging and financing distributed PV electricity. And in Germany, one of the world's leaders in solar energy, rooftop solar units accounted for 90 percent of the solar energy produced in 2008.

With projected cost reductions and efficiency gains, it is now clear that we are not far from the day when a large percentage of electricity will be generated on-site. Some analysts have predicted that up to half of all households in the U.S. will be generating at least some of their own electricity from renewable sources within 10 years.

If that prediction turns out to be true, it is

Internet. The super grid—which some refer to as "the Electranet"—will create the kinds of markets for electricity generation, distribution, and storage that the Internet created for small devices that process, transmit, and store information. In fact, the economic benefits of widely distributed generation and storage will, within a decade, completely transform the nature of the electricity marketplace.

In developing countries, where electricity grids are still uncommon, small rooftop photovoltaic systems are proliferating rapidly in many

> The super grid will create the kinds of markets for electricity generation, distribution, and storage that the Internet created for small devices that process, transmit, and store information.

difficult to imagine the competitive dynamism and economic growth that will be unleashed by the competition to supply distributed energy systems to tens of millions of American homes and households throughout the world. The only analogy that comes close is the surge in economic growth and productivity that was unleashed when the Internet led to the development and sale of hundreds of millions of laptop computers, smart phones, and other new electronic devices now widely distributed globally and connected to the

areas—leapfrogging the older electricity architecture of the developed world just as cellular telephones found huge markets in developing nations with very few fixed-line telephone grids. Harish Hande, a photovoltaic entrepreneur in India, described his experience in learning from a housewife in Mumbai who told him, "Three hundred rupees a month is impossible, but 10 rupees a day I can afford." (Ten rupees is equal to approximately 25 cents in the United States—not bad for a day's worth of electricity.)

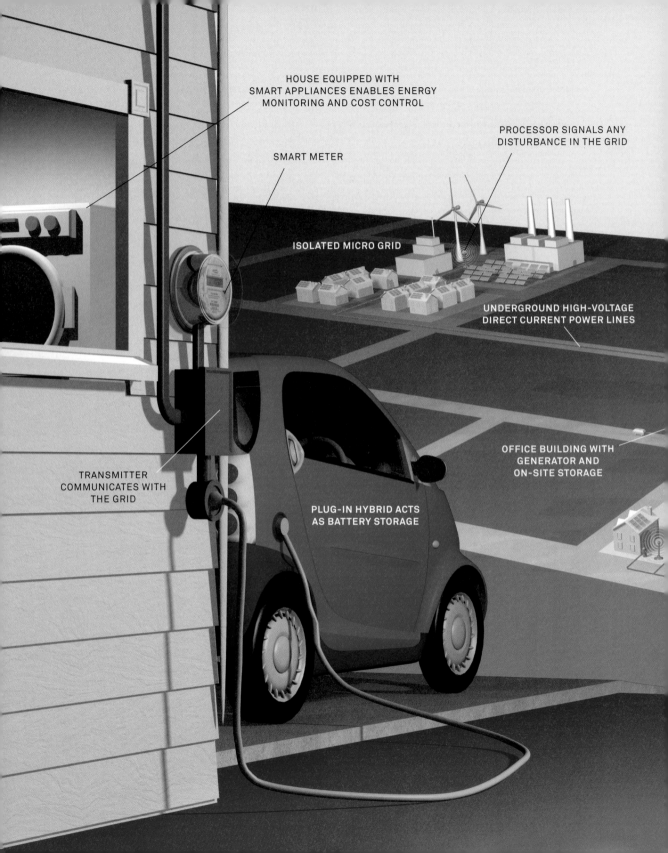

HOUSE EQUIPPED WITH
SMART APPLIANCES ENABLES ENERGY
MONITORING AND COST CONTROL

SMART METER

PROCESSOR SIGNALS ANY
DISTURBANCE IN THE GRID

ISOLATED MICRO GRID

UNDERGROUND HIGH-VOLTAGE
DIRECT CURRENT POWER LINES

OFFICE BUILDING WITH
GENERATOR AND
ON-SITE STORAGE

TRANSMITTER
COMMUNICATES WITH
THE GRID

PLUG-IN HYBRID ACTS
AS BATTERY STORAGE

HOW A SUPER GRID WILL WORK

A unified national super grid will use microprocessors and sensors to distribute information and constantly balance supply and demand throughout the system. The super grid is made up of many different technologies, including improved transmission lines, large- and small-scale batteries that smooth out intermittent power sources like wind and solar, micro-power installations such as rooftop solar panels, and smart meters and appliances that can adjust power usage depending on supply and pricing. Throughout the super grid, computers and the Internet facilitate feedback between components and end users, allowing for conservation, much higher efficiency, and sales of excess electricity by customers.

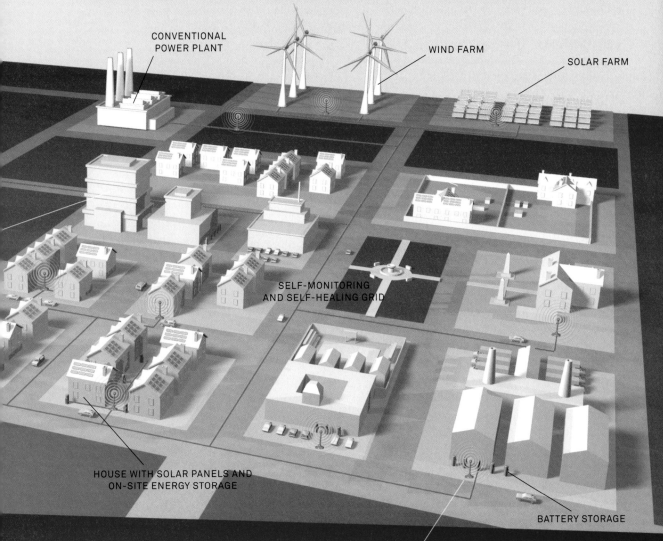

CONVENTIONAL
POWER PLANT

WIND FARM

SOLAR FARM

SELF-MONITORING
AND SELF-HEALING GRID

HOUSE WITH SOLAR PANELS AND
ON-SITE ENERGY STORAGE

BATTERY STORAGE

PROCESSOR AND
DATA TRANSMITTER

The National Renewable Energy Laboratory describes the new architecture of distributed energy as, "a variety of small, modular power-generating technologies that can be combined with energy management and storage systems and used to improve the operation of the electricity delivery system, whether or not those technologies are connected to an electricity grid." These energy generation and storage systems are "placed at or near the point of use" and include "fuel cells, microturbines, reciprocating engines, load reduction, and other energy management technologies. Combined heat and power systems provide electricity, hot water, heat for industrial processes, space heating and cooling, refrigeration, and humidity control to improve indoor air quality and comfort."

For example, there are two options for thermal storage located in the buildings owned by users of electricity. One of the elements common in "passive solar" designs is the strategic placement of masonry or some other heat-absorbing mass where it will absorb heat from the sun during the day and then give it back to the interior of the home at night. Overhangs above the sun-facing windows can be designed to block the sun's rays during summer when they are arriving at a higher angle, and admit the sunlight during winter when the sun is lower in the sky and shines beneath the overhang.

In commercial buildings, there has recently been a surge of interest in a form of highly efficient thermal storage that sharply reduces cooling costs for large buildings. By making large quantities of ice at night when electricity costs are lowest, the owner or manager of the building can reduce air-conditioning costs the following day by using the cooling power of the ice collected the night before. Even though this technology is highly cost-effective and is commercially available,

its usefulness has been limited by the well-known mismatch between the incentives of those who construct buildings and those who pay the operating costs. Moreover, if the customer pays a flat rate for electricity that obscures the cost differential between peak electricity and off-peak electricity, there is no incentive for end users to take advantage of the enormous benefits that energy storage technologies can provide.

The increased prominence of energy efficiency in the real-estate market has caused a renewed focus on these options. However, it is all too common for lessors of buildings to pass through utility costs, and force of habit has slowed recognition of these new savings opportunities. When the owners and operators of large buildings pay attention to the design and construction process, they can insist upon inclusion of energy saving and energy storage technologies in order to sharply reduce the daily and annual operating costs of the building. For example, the new highly efficient Bank of America Tower in New York City makes more than half a million pounds of ice every night, which enables the shifting of 1,000 tons of air-conditioning to off-peak hours. The developer of the system, Mark MacCracken, says that his company's experience proves that it is "dramatically less expensive to store the cooling than the electron to make it."

In addition, the economic incentives that shape the behavior of electric utilities distort any effort by utilities to rationally and objectively compare the cost of storage and the cost of new generating capacity. They get reimbursed for the latter, which profits based on a percentage of their cost. They don't get reimbursed for the former in the same way and therefore have invested almost nothing in storage or R&D into better storage. If the utility is uncertain that its regulator will approve the recovery of investments in storage, and if the regulator

"It is dramatically less expensive to store the cooling than the electron to make it."

MARK MACCRACKEN

BANK OF AMERICA'S NEW HEADQUARTERS IN
NEW YORK IS EXPECTED TO BE THE FIRST SKY-
SCRAPER TO EARN LEED PLATINUM CERTIFICATION.
AMONG ITS GREEN ADVANCES IS AN ICE STORAGE
SYSTEM THAT HELPS COOL THE BUILDING
DURING PEAK HOURS.

AT NIGHT, DETROIT (TOP) AND WINDSOR, ONTARIO (BOTTOM), DISPLAY DIFFERENT LEVELS OF POWER DEMAND.

is confused about how to classify the storage asset, the result is that nothing happens. Some regulators, however—in Connecticut and New Hampshire, for example—have successfully persuaded legislators to allow inclusion of energy storage devices in their rate bases.

In many jurisdictions, there are large economic implications that flow from the classification of electrical energy storage in one or the other of these categories. Depending on the decision, the utility may or may not be allowed to recover its investments in storage as part of the "recoverable costs" that are allowed to be added to the rates paid by customers, with profits added to compensate for the risk taking.

of using storage enhance the efficiency of three different parts of the electricity system: generation, transmission, and distribution—all at once. In Texas, sales of the NGK battery were blocked because of disputes over who should "take ownership of the energy while it is stored in the battery."

In California, the company was told that in order to sell its battery, it had to meet "the challenge of establishing the precedent for Battery Energy Storage Systems as a Transmission Asset recoverable in the Transmission Access Charge." The state of California, in turn, was following a ruling by the Federal Energy Regulatory Commission requiring the rigid separation of assets in the

It is past time for a complete revision of the national rules and regulations that have guided the electricity industry for more than a century.

In some jurisdictions, utilities primarily engaged in the business of generating electricity can be prohibited from owning storage assets if they are classified as part of the transmission and distribution network. Conversely, in other jurisdictions, state laws say that anyone other than the power-generating utility may be prohibited from owning storage assets if the benefits they add are seen as part of the generating function.

For example, some U.S. utilities have been temporarily blocked by regulators from deploying sodium-sulfur batteries because the benefits

generation part of the system from assets in the transmission part of the system.

Understandably, the California investor-owned utility attempting to use this efficient storage system concluded that the regulators were placing at risk the ability of the utility to recover the costs of the energy storage assets.

For these and other reasons, it is past time for a complete revision of the national rules and regulations that have guided the electricity industry for more than a century. The national and state legal and regulatory frameworks surrounding electricity generation, storage, transmission, and distribution are even more antiquated than the equipment.

Too often, current regulations and prevailing assumptions dissuade utilities and electricity customers from investing in energy storage and efficiency. It is time to develop a set of new ideas, statutes, and rules that recognize, value, and facilitate the interplay between generation, storage, transmission, distribution, and customer-owned assets—to produce a more efficient, functional, and environmentally acceptable electric system.

The parallel development of the electric utility industry over the past 120 years and the development of regulations governing every aspect of the investments, expenditures, and profits of the industry have both been driven by a way of thinking about electricity generation and use that is now hopelessly outdated. The common assumption built into this obsolete model is that there is always a real-time connection between the electricity that is being generated and the electricity that is being used. Moreover, because most utility regulation is the responsibility of states, regulators have not had an incentive to focus on benefits that extend beyond the borders of the small area that occupies their attention.

More than 10 years ago, there was widespread recognition in the United States that electric utilities should no longer be allowed to continue as vertically integrated monopolies controlling generation, transmission, distribution, and storage. The electricity industry has been partially deregulated nationwide in order to create a competitive market for the generation of electricity. But most states and most other countries do not yet allow competition in the full range of services that are needed to generate, transmit, store, distribute, and sell electricity in the most efficient manner possible.

The Federal Energy Regulatory Commission has begun to approve cost allocation formulas that spread the cost of new transmission capacity. However, in many jurisdictions, regulators still use "license plate pricing" to charge the full costs of transmission assets built within state boundaries to rate base that determines the cost of electricity to the citizens of the state where it is located.

In the words of energy expert Bruce Radford, "If you're building transmission from North Dakota to bring energy to Chicago, it means the price of all the transmission infrastructure in North Dakota gets passed under license plate pricing to North Dakotans." Alison Silverstein, former senior energy policy adviser at the Federal Energy Regulatory Commission, says, "If we are still going to have a utility system that is going to be reimbursed by 'cost plus a return,' then you are not incentivizing anybody to save money."

There are several ways that regulators could align utility incentives with the public interest goal of building a unified national smart grid, or super grid. They could:

▶ Explicitly break the link between electricity sales and utility profits through what is called "decoupling."

▶ Offer performance-based rate-making incentives linked to smart grid investments, renewable generation interconnection, additions of energy storage, and distributed generation.

▶ Eliminate the up-front capital risks for utilities desiring to invest in transmission, end-use efficiency, distributed generation, and on-site storage by guaranteeing the recovery of costs for these investments.

▶ Mandate that a certain percentage of utilities' resource portfolios be dedicated to efficiency improvements and demand response options.

AT THE DEPARTMENT OF ENERGY'S ROCKY MOUNTAIN OFFICE, POWER SYSTEM DISPATCHERS DISTRIBUTE HYDROELECTRICITY THROUGHOUT THE WESTERN UNITED STATES.

▶ Deny full-cost recovery for investments in facilities that do not include smart features.

▶ Develop national interconnection standards and unifying protocols to help eliminate distortions in the way electricity-generating capacity is now developed, and facilitate distributed generation.

▶ Enact new laws and regulations allowing the rational and most efficient allocation of the costs and benefits of the new super grid among all commercial entities involved and society as a whole.

Unfortunately, the partial deregulation of utilities was complicated and frustrated by the continuing political power of the utilities in their dealings with state regulators. And the obscene greed and corruption of Enron created a bad taste in the mouths of citizens who had been persuaded that intelligent deregulation was in their interests.

As a result, the inevitable movement toward a widely distributed electricity infrastructure has been only partly completed. The electric utilities continue to play a crucial role, and even advocates of micro-power usually acknowledge the need to make this transition in a way that protects the overall energy economy from the consequences of large-scale utility bankruptcies and the burdens of enormous stranded costs.

Nevertheless, most experts agree that it is now only a matter of time before the electricity grid is completely redesigned to fully integrate widely distributed generation and storage in a unified smart grid that is far more efficient, far less costly, and far more environmentally responsible than today's grid.

THE OBSTACLES WE NEED TO OVERCOME

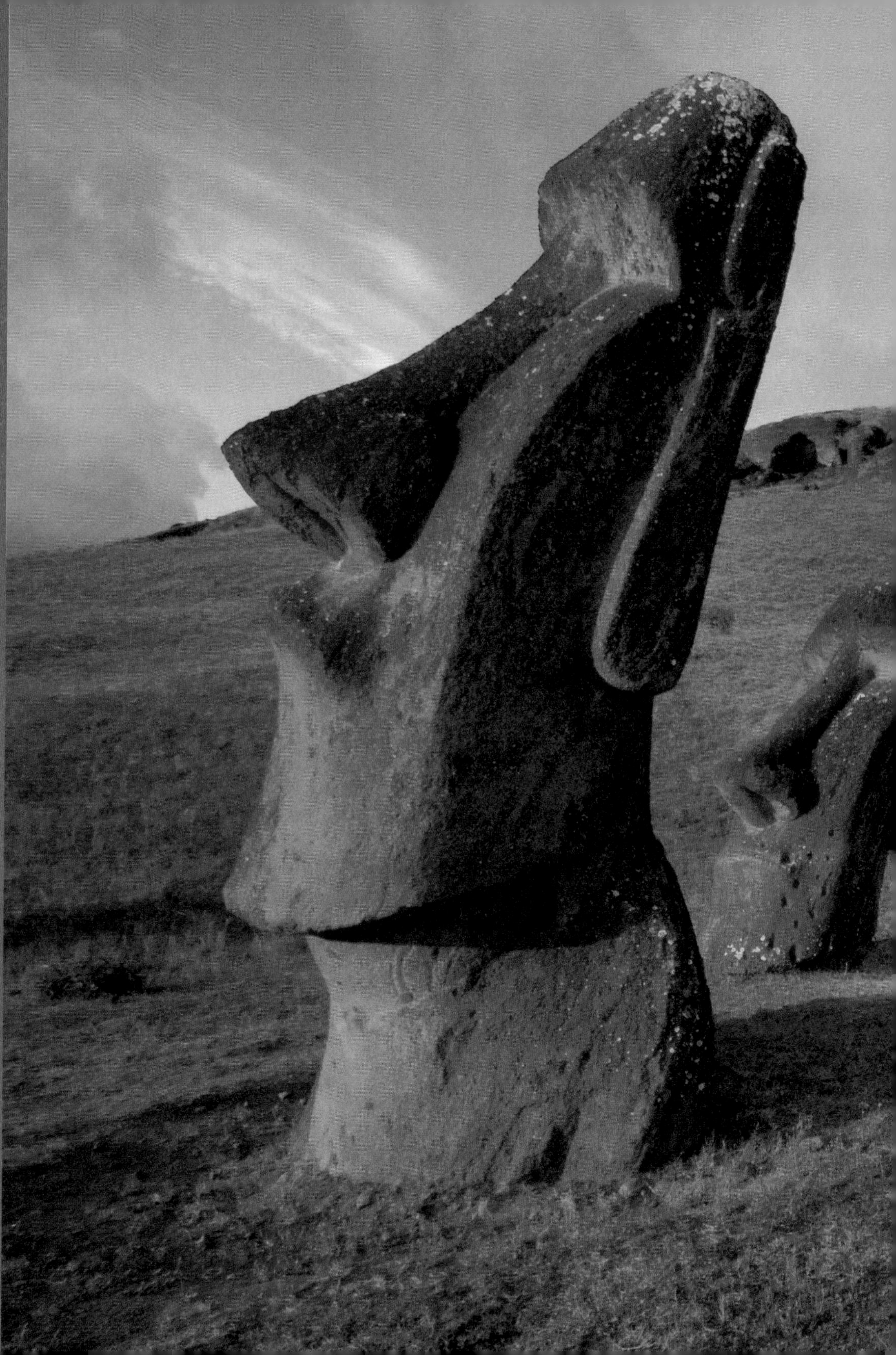

CHAPTER FOURTEEN

CHANGING THE WAY WE THINK

SOME OF THE MOAI—GIANT STONE HEADS—
ON EASTER ISLAND. THE REMOTE ISLAND
HAS BECOME A SYMBOL OF THE DANGERS OF
UNSUSTAINABLE PRACTICES.

It's increasingly clear that part of the challenge we face in solving the climate crisis stems from the way we think about it, both individually and collectively.

Why is it that humanity is failing to confront this unprecedented mortal threat? What is it about the way we human beings process information and make choices that promotes global procrastination?

In fact, there are striking differences between the commonly accepted mythology about the way we make decisions and the way we *actually* make decisions, according to psychologists, neuroscientists, and some economists. In many ways, most of modern civilization is based on decision-making structures that assume the existence of an archetypal "reasonable person," one who takes in all available information about decisions to be made, selects the evidence most relevant to the decision, discusses that evidence with other reasonable persons, and then makes a rational decision and sticks to it.

This prototype originated in late 18th century England, Scotland, and Europe as the culmination of a philosophical movement called the Enlightenment, which enshrined the "rule of reason" as a new sovereign source of authority to take the place of monarchs, the medieval church, and the feudal structure.

It is not coincidental that Adam Smith's *The Wealth of Nations,* Thomas Jefferson's Declaration of Independence, and the first volume of Edward Gibbon's *The History of the Decline and Fall of the Roman Empire* were all published in the same year, 1776. The philosophical architects of the modern world were filled with optimism, a belief in progress, and a keen sense that they were building something new from the remnants of an old order that was collapsing around them.

The design of representative democracy and market capitalism were both based on the assumption that Reason could be made paramount in the ongoing decisions that determined human affairs. The collective judgments of reasonable people were to be found in the results of democratic elections, which aggregated all of the individual decisions by voters in order to steer the ship of state. And the best guidance for the economy was to be found in the "invisible hand" of the marketplace, which aggregated the net result of millions of individual economic decisions to balance the supply and demand for goods and services.

This overarching belief in the rule of reason was enhanced by the nature of the information ecosystem on which 18th century inhabitants depended. Thanks to the development of the printing press three centuries earlier, new information spread rapidly to lay publics, leading to widespread literacy. There was a growing optimism that individual citizens, armed with knowledge previously available only to elites, could make decisions for themselves in ways that produced better political, social, and economic results than mere acquiescence to the dictates of the few who ruled by divine right.

The printed word was equally accessible to anyone who learned to read and write. For the first time in history, any literate individual, without regard to wealth or force of arms, could use knowledge and ideas as a source of power. Reason was the principle governing the aggregation of individual judgments and choices into a single net result. While emotions and feelings were recognized as powerful motivators, it was assumed that rationality would determine the ultimate outcomes. As Benjamin Franklin wrote in 1749, "If Passion drives you, let Reason hold the Reins."

According to this view, reason could also be employed to cleverly design safeguards protecting without governmental interference—were added to the text, along with the rest of the individual protections included in the Bill of Rights.

The United States of America's stunning success over the last 200 years (emulated by aspiring democracies on every continent) and the dominance of market capitalism in most of the world (especially after its philosophical victory over communism in the late 20th century) both serve as evidence of the unprecedented power and vitality of these two designs based on the assumed primacy of reason in human affairs.

Both systems were traditionally seen as self-correcting. Market failures would be sorted out with the benefit of new information from the failures themselves, with said information subjected

Why is it that humanity is failing to confront this unprecedented mortal threat?

the operations of the rule of reason against well-understood threats inherent in human nature. For example, since the accumulation of too much power in the hands of one person (or one small group of people) could unbalance the operations of reason, America's founders divided power between state governments and the federal government and divided power within the federal system among three coequal branches of government. These checks and balances were woven into the fabric of the U.S. Constitution. It is worth remembering that the states refused to ratify the Constitution until individual freedoms provided by the First Amendment—which, among other things, guarantees citizens access to the free flow of information

to the rule of reason in order to find progressively better solutions to newly discovered problems. For example, when reformers pointed out the unhealthy economic consequences of concentrated economic power, Congress responded with anti-trust laws and other protections. When the Great Depression of the 1930s shook public confidence in market economics, the U.S. government adopted vast new powers of regulation to protect against a repeat of that massive market failure.

Similarly, it was assumed that voters would respond to political failures by correcting them over time, ideally at the next election. The key to continued smooth functioning of both intertwined systems was freedom of information. So long as

Global warming has often been described as the greatest market failure in history.

TRASH DUMPED ON THE TUNDRA OUTSIDE
ILULISSAT, GREENLAND

reasonable people were guaranteed access to a free flow of information in both politics and commerce, they could be trusted to fix any problem and continue the march of progress.

It is now apparent that the climate crisis is posing an unprecedented threat not only to the future livability of the planet but also to our assumptions about the ability of democracy and capitalism to recognize this threat for what it is and respond with appropriate boldness, scope, and urgency. Global warming has been described as the greatest market failure in history. It is also—so far—the biggest failure of democratic governance in history.

In searching for the underlying reasons behind these twin historic failures, psychologists and neuroscientists have begun to argue that the climate crisis poses a unique and unprecedented challenge to our ability to use the rule of reason as a basis for urgent response.

The specific flaws in market capitalism highlighted by the climate crisis are discussed in more detail in Chapter 15, while the political obstacles to an effective solution are discussed in Chapter 16. But there is something more basic in our relationship to the climate crisis that reveals the fundamental flaws in the way we have thus far been able to think about it collectively.

Our capacity to respond quickly when our survival is at stake is often limited to the kinds of threats our ancestors survived: snakes, fires, attacks by other humans, and other tangible dangers in the here and now. Global warming does not trigger those kinds of automatic responses.

We also have the demonstrated ability to respond urgently to indicators associated by repeated experience with harmful consequences: the smell of leaking natural gas or a run on the bank, for instance. We may be slow to learn these kinds of habitual responses, but once we learn them, we gain the ability to respond to an appropriate stimulus almost automatically, with very little need for thinking.

The phenomena that alert scientists to the onset of the climate crisis are, by contrast, unfamiliar, because they are unprecedented in human experience and seem slow-moving due to the vast global scale of the ecological systems under siege. In other words, because of its planetary scope, this crisis masquerades as an abstraction.

As a result, the automatic and semiautomatic brain responses that have ensured our survival over the millennia are uniquely unsuited to the role of motivating new behaviors and patterns necessary to solve the climate crisis.

The impact of global warming seems remote. Its effects are distributed throughout the earth in a pattern that makes it difficult to ascribe an unambiguous cause-and-effect relationship between what is happening to the earth as a whole and what is happening to a single individual in a given time and place. Because local consequences are still difficult to attribute to the global catastrophe, we are slow to perceive its immediate and growing effects.

The necessary perception of local impacts may finally be changing, however, given the incredible number of record floods, record droughts, record fires, and record storms. Unprecedented changes in the distribution of living species important to people in specific areas—whether the disappearance of salmon off the coast of California, the loss of specific songbird species in multiple areas, or the radical changes in duck populations in places where duck hunting is popular—have also raised awareness. Increasingly, scientists have grown confident in saying we have crossed a threshold beyond which it is irresponsible not to point out the cause-and-effect relationship between global warming and the exact kinds of consequences so long predicted.

The scientific evidence that forms the basis of our understanding of this looming catastrophe is unequivocal, but as alarming as that statement is, it lacks emotional impact, because its conclusions point toward "probabilities" and unfamiliar "nonlinear effects." Moreover, our species has no historical memory of any comparable catastrophe in the past, and thus no point of emotional reference. And so we are putting more weight on the rule of reason than we ever have in the past.

Because the benefits of solving the crisis lie in the future, while the solutions must be undertaken now, reason-based analysis has thus far proved insufficient to motivate action. The common behaviors that cause the climate crisis—particularly the massive burning of coal and oil worldwide—are ingrained in our civilization. Since changes in behavior by individuals seem so unlikely to impact the global crisis, it's exceedingly difficult for reason to challenge the powerful forces of habit.

Scientists who study human behavior, the nature of the brain, and the way we make decisions have developed a sophisticated understanding of the limits to our ability to rely on the rule of reason. Specifically, they describe with increasing precision the limitations that circumscribe our ability to rely on rationality, to devote sustained attention to one particular problem, and to apply limited reserves of willpower to solving a problem that persists over decades or centuries.

Fortunately, there is a third brain system that can guide us in making the crucial decisions necessary to safeguard the future of our civilization. Neuroscientists and behavioral psychologists have long understood the process by which we as human beings—individually and collectively—set long-term goals based on shared values and continue pursuing them in a dedicated way over decades, generations, even centuries.

The great cathedrals of Europe were built by human beings just as vulnerable to distraction and just as desirous of short-term gratification as we are. The pyramids of Egypt, Angkor Wat in Cambodia, and the Palace of Knossos in Minoan Crete are among the many examples of multigenerational achievements throughout our history. The Marshall Plan, NATO, and the unification of Europe were pursued in a focused manner over a long period of time.

Scientists can now monitor how the human brain makes decisions that motivate changes in behavior. They have learned that values-based, goal-directed decisions sustained over long periods of time require a great deal of thought and are made slowly but deeply. The prefrontal cortex is the part of the human brain that enables us to focus in a sustained way on the future reality we intend to shape with long-term efforts beginning in the present moment. Dr. Greg Berns of Emory University notes: "Some researchers have speculated that the difference between humans and other animals lies in our ability to form a mental image of, and care about, delayed outcomes, and there is widespread agreement that the prefrontal cortex, which is disproportionately large in humans relative to other species, has an important role in this capability.... Humans

THE LONG BATTLE AGAINST SMOKING

In 1964, when a landmark Surgeon General's report called "Smoking and Health" officially linked smoking with public health problems for the first time, smoking was generally treated as a pleasurable vice, not a public-health emergency. The 50 percent decline in U.S. smoking rates since that report is a long-term success story illustrating how changes in attitude and policy can change a nearly suicidal habit.

The rate's decline was kick-started by the Surgeon General's warning placed on cigarette packs in 1966, and by the Public Health Cigarette Smoking Act of 1969, which banned tobacco advertising from TV and radio and strengthened the wording of the pack warnings.

Even as their customer base declined, cigarette makers maintained that no proof existed that their products were harmful or addictive. The PR backlash and settlements of hundreds of lawsuits provided impetus and funding for effective public education campaigns and subsidized addiction treatment, adding further positive momentum to the anti-smoking movement.

Cigarette use, responsible for 90 percent of all lung cancers, is now largely unwelcome in public. Through a combination of government action and

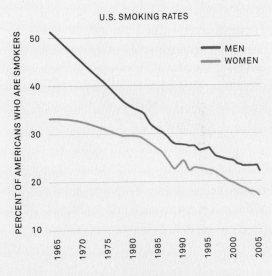

Smoking rates have dropped significantly since 1965.

societal progress, smoking has become stigmatized. The most recent studies show that fewer than 20 percent of American adults smoke; the Centers for Disease Control and Prevention credit the decline in smoking with saving thousands of lives every year.

SOURCE: U.S. Centers for Disease Control and Prevention

CHARTRES CATHEDRAL TOOK MORE THAN A
CENTURY TO BUILD, WITH SEVERAL GENERATIONS
OF PEOPLE WORKING TOWARD—AND REMAINING
FOCUSED ON—A COMMON LONG-TERM GOAL.

undoubtedly share with other animals the mechanisms that produce rapid hyperbolic time discounting, but we also have the capacity, seemingly enabled by the prefrontal cortex, to make decisions that take account of a much longer span of time."

Once they are made, such decisions can produce powerful commitments to sustained behavioral change and allow enormous flexibility in pursuit of the goals that have been set.

Over the millennia of human civilization, we have often formalized the process of making long-term, values-based decisions. For example, the great faith traditions on our planet—including Christianity, Islam, Hinduism, Judaism, Buddhism, Taoism, and others—all embody important and largely consistent teachings about the value of protecting and preserving the health of our environment and serving as good stewards of the bounty it provides us. Among the many roles they play in people's lives, these faith traditions have sought to elevate important shared values as factors that guide the decisions of adherents who wish to live ethical lives. These values could, if people chose to make them a priority, play a crucial role in reinforcing the ability of humankind to sustain a multigenerational commitment to the changes now necessary to fulfill our roles as good stewards of planet Earth.

Brain scientists have identified the specific parts of the brain within the prefrontal cortex that keep us on course once we decide to pursue a value-based goal over a long period of time. Significantly, they also note that the particular part of the human brain—the dorsolateral prefrontal cortex (or DLPFC)—just above the temples, may weaken in the presence of high stress and that the mental effort necessary to maintain its crucial function may be exhausted by excessive distraction and anxiety. The DLPFC keeps us on track by coordinating our ability to recall things from

Do not dump waste in any place from which it could be scattered by the wind or spread by flooding.

JUDAISM (Maimonides' Mishneh Torah)

DO NOT CUT TREES, BECAUSE THEY REMOVE POLLUTION

HINDUISM (Rig Veda, 6:48:17)

This is what should be done by those who are skilled in goodness, and who know the path of peace: let them be able and upright, straightforward and gentle in speech, humble and not conceited, contented and easily satisfied, unburdened with duties and frugal in their ways, peaceful and calm, wise and skillful, not proud and demanding in nature. Let them not do the slightest thing that the wise would later reprove.

BUDDHISM
(The Metta Sutta, The Buddha's Teaching on Loving-kindness)

The world is beautiful and verdant, and verily God, be He exalted, has made you His stewards in it, and He sees how you acquit yourselves.

ISLAM (Hadith of sound authority, related by Muslim on the authority of Abu Sa'id al-Khudri)

You should not burn [the vegetation of] uncultivated or cultivated fields, nor of mountains and forests.
You should not wantonly fell trees.
You should not throw poisonous substances into lakes, rivers, and seas. You should not wantonly dig holes in the ground and thereby destroy the earth.

TAOISM (180 Precepts of Lord Lao)

Do not disturb the sky and do not pollute the atmosphere.

HINDUISM (Yajur Veda, 5:43)

THEN THE LORD GOD TOOK THE MAN AND PUT HIM IN THE GARDEN OF EDEN TO TEND AND KEEP IT.

CHRISTIANITY
(Genesis 2:15, New King James Version)

memory, plan for the future, and juggle all of the things competing for our attention. Take the DLPFC offline—with constant, excessive levels of stress, for example—and we become locked in the here and now, with little care for the past and an ambivalence for the future. Unsurprisingly, there is overwhelming evidence that modern societies routinely generate much higher levels of stress than was common in previous centuries.

One big source of the new, extraordinarily high levels of ambient stress is the information environment in which most of us now live. Because the evolution of modern culture and ubiquitous electronic media serve up a constant diet of distraction, we may be less able to rely on a sustained, decades-long focus of collective willpower just when we need it more than we ever have.

And part of the reason is that—ironically—the primary users of the new brain research are the marketers and advertisers of goods and services. The advertising industry fuels the massive, expensive, and omnipresent electronic media that are underwritten by the constant push for more and more consumption. The average American now sees an average of 3,000 advertising messages per day.

Virtually every Pavlovian trigger discovered in the human brain is now pulled by advertisers. Partly as a result, material consumption in our society has reached absurd levels, declining slightly only in the teeth of the worst economic downturn since the Great Depression.

Per capita purchases of clothing in the United States doubled from 1991 through 2005. In the first seven years of this decade, U.S. household debt—

THE BRAIN'S SELF-CONTROL CENTER

Scientists at the California Institute of Technology studied how we make diet choices to pinpoint the parts of the brain that help us consider long-term risks and values. In fMRI images, regions of the ventromedial prefrontal cortex (vmPFC) were active in every decision, but the dorsolateral prefrontal cortex (DLPFC) became active only when self-control was used. When the vmPFC was tempted by a candy bar, the DLPFC prompted consideration of the benefits of an apple.

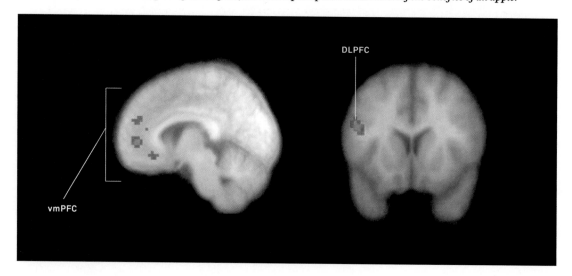

driven by unprecedented levels of consumption—grew to 138 percent of disposable income. Per capita production of waste in this frenzy of production and consumption in the United States has now reached an astonishing 141 pounds per day for every man, woman, and child in the nation. (This figure includes a per capita attribution of the combined total of individual, household, commercial, and industrial waste—though it excludes most of the waste associated with the high volume of products made in China and elsewhere and then sold in the

agree with the statement "I feel left out of things going on around me" has quadrupled.

Increased consumption of goods and services has come to be equated with the pursuit of happiness. Yet the level of happiness in modern U.S. society—by any measure—has not increased with the level of consumption. The results are similar in other high-consumption countries. Numerous studies have found significantly greater levels of well-being and happiness in some societies with far lower standards of living as measured by per capita

Virtually every Pavlovian trigger discovered in the human brain is now pulled by advertisers.

U.S. It also excludes the weight of the fuel used to generate electricity.)

Indeed, the new combination of electronic media and mass advertising has produced a culture of constant mass consumption that is far different from anything seen previously in human history. Quite apart from the economic and environmental consequences of our consumption binge, the psychological consequences for civilization are profound. The average American now spends five hours per day watching television. The average home, according to Nielsen, now has "more TVs per home than people." Most of that time has been subtracted from opportunities to talk with family, friends, and neighbors, and from participation in community affairs and the democratic life of the nation. No wonder the number of Americans who say they

production and consumption.

Voluminous studies in the relatively new field of "happiness research" show that, after achieving the ability to meet basic needs for food, shelter, transportation, and health care, individuals and families stop making tangible gains in their sense of well-being with further increases in consumption. In the aggregate, the United States has tripled its economic output over the last half-century, with absolutely no gain in the general sense of well-being.

The distortion of what we deem "valuable" and the confusion about what is likely to make us happy is driven partly by our obsession with material goods. "He who dies with the most toys wins" is intended as a humorous comment on our current behavior, but as a diagnosis of our current beliefs about the purpose of life, it contains more than a grain of truth. Indeed, consumption has become a goal in and of itself.

Some behavioral psychologists and brain researchers have recently begun an effort to discern how we might develop a system of common valuation for psychological rewards and improvements in our sense of well-being that can be quantified in ways that reflect how much they are worth to us in comparison with cash income, goods, or services. In many ways, the work of these researchers represents what a previous generation might have regarded as simple common sense, but it is a measure of how much we have been immersed in the culture of consumption that we find new meaning in the attribution of monetary value to things such as time with loved ones, enjoyment of a beautiful environment, and the ability to have a voice when community decisions are made.

Recently, a few innovative policy makers have attempted to identify what they call "psychological subsidies" that can encourage the positive changes needed to save energy and promote efficiency, conservation, and renewable energy. These same innovators have tried to identify "psychological taxes" that serve as barriers to new approaches that would move policy in the right direction. By removing these barriers, they hope to make more progress.

Business managers have long known that social rewards can be more important in motivating positive change than material rewards. Now there is a focused effort to use that knowledge in the shaping of policy. Such rewards are sometimes called "nudges" and, in the context of climate change, can be as simple as providing clear information to people about the consequences—positive and negative—of the choices they are making.

However, when it comes to convincing whole societies to make the transformative changes

Material consumption in our society has reached absurd levels.

THE AVERAGE AMERICAN SUPERMARKET, LIKE THIS
ONE IN PORTLAND, OREGON, CARRIES MORE THAN
45,000 DIFFERENT ITEMS.

necessary to stabilize Earth's climate, behavioral and brain researchers recommend completely new approaches to designing and communicating policy. Simply laying out the facts won't work, they say. The barrage of negative, even terrifying, information can trigger denial or paralysis or, at the very least, procrastination.

It can also trigger what psychologists call "single-action bias." This deeply ingrained brain pattern, combined with our often unrealistic confidence that technology alone will save us, has led some to assume that once we make up our minds to act, we can simply choose yet another single technological solution to "fix" the problem in short order.

Some of the more bizarre manifestations of this disorder in our thinking include a proposal from a distinguished nuclear physicist, Edward Teller, to put billions of tinfoil strips in orbit around the Earth in order to reflect up to 2 percent of the incoming sunlight and cool down the planet. An even more extreme proposal involves placing a giant parasol in orbit around the sun at a point where it will partly shade the earth and reduce the amount of incoming sunlight. Both proposals overlook the important fact that sunlight in adequate amounts is necessary for growing food and sustaining the health of plants and animals. Fortunately, neither of these proposals is being taken seriously.

However, some other proposals for blocking incoming sunlight have, unfortunately, begun to attract a modicum of support from a few scientists who ought to know better but who have grown increasingly desperate about the failure of the world's political systems to respond to the climate crisis. Some distinguished scientific groups have now begun serious discussion of a few of these ideas. One such proposal is to inject huge quantities of sulfur dioxide into the atmosphere in order to block a portion of the incoming sunlight.

Advocates of this idea acknowledge that it might have serious unintended consequences, such as changing the chemistry of our atmosphere in ways that could lead to irreversible effects we don't understand. They point out that volcanoes inject about the same amount of SO_2 every couple of decades, apparently without any irreversible harm. However, the intermittency of volcanic SO_2 is very different from maintaining a constant, artificially high level of SO_2 in the atmosphere. Moreover, so long as the amount of CO_2 and other greenhouse gases continued to increase in the atmosphere, we would have to constantly replenish and add to the amounts of sulfur dioxide that we put into the sky. If we ever stopped doing this, we would face the sudden acceleration of global warming at a rate of 2 to 4°C per decade (compared with 0.2°C now)— just as the protagonist in *The Picture of Dorian Gray* suffered a terrifying acceleration of the aging process when the artificial but temporary suspension of his aging was suddenly no longer functioning.

Moreover, the sulfur dioxide cloud circling the earth would partially negate the effectiveness of the effort now beginning to shift electricity production to solar panels. And in any event, the effort to counter the global warming impact of carbon dioxide by blocking sunlight would do nothing to stop the other consequences of CO_2 buildup, such as seriously harming the world's oceans through acidification.

A more benign proposal involves painting roofs white all around the world in order to increase the amount of sunlight reflected off the surface. If this idea worked, its proponents add, we could paint parking lots and highways and portions of deserts white. And while this particular proposal for white roofs is a good one and should receive serious consideration, the benefits would quickly be wiped out if we continued increasing the accumulation of CO_2 and the other air pollutants that cause global warming.

Another family of technological fixes—or "geoengineering," as all of these proposals are sometimes called—involves efforts to change the chemistry of the oceans in ways that might increase their absorption of CO_2 by stimulating larger plankton blooms. Early experiments involving enrichment of some areas of the ocean with iron have failed to absorb carbon at anywhere near the levels predicted, because the process by which the ocean absorbs CO_2 is vastly more complex than this

of civilization. We should not begin yet another planetary experiment in the hope that it will somehow magically cancel out the effects of the one we already have.

The only meaningful and effective solutions to the climate crisis involve massive changes in human behavior and thinking—changes that lead, in turn, to the widespread use of efficiency and conservation, a shift from fossil fuels to solar and wind and other renewable forms of energy, and an end to the burning of forests and croplands and the depletion of carbon-rich soils.

In order to bring about these changes,

The only meaningful and effective solutions to the climate crisis involve massive changes in human behavior and thinking.

theory originally suggested. Massive tree-planting programs, on the other hand, make excellent sense and, as described in Chapter 9, such programs can—if carried out on a sufficiently large scale—measurably increase the amount of CO_2 taken out of the air by trees and forests. Similarly, efforts to enhance the ability of soils to absorb CO_2 (as described in Chapter 10) can be valuable tools in a multipronged strategy to solve this crisis.

In any case, we are already involved in a massive, unplanned planetary experiment. We already have all the evidence we need to know that human interference in the natural climate balance and in the relationship between the earth and the sun carries with it enormous risk of harming the health of the ecosystem in ways that can threaten the future

scientists who study behavior and thinking advise us to strengthen the linkage between solutions to global warming and solutions to other challenges (economic, strategic, and social) that seem more immediate and are more likely to induce a desire to make the necessary changes.

In communicating the urgency of the climate crisis, it is important to use relevant, everyday language and to link the virtue of solving the crisis to the shared values and aspirational goals that have proved in our past to sustain long-term collective commitments. We are inherently social animals, and our survival as a species has depended not only on the survival of the fittest individuals but also upon our ability to cooperate with one another and to strengthen the social bonds that make such cooperation possible.

Moreover, there is also evidence that the legacy gifts of one generation to the next carry with them a felt obligation to reciprocate by doing well and passing on the legacy to the next generation. Even though we are inherently vulnerable to the desire for short-term gratification and even though we usually have an ingrained preference for short-term actions, those preferences can be and often are overridden by an innate and powerful desire to do right by those to whom we feel some connection.

The strategy we follow must give people an active role in helping to solve the crisis and connect the value of what we're encouraging them to do to personal experiences that carry emotional meaning. Rather than dwelling on the elimination of all remaining uncertainty, we need to communicate the essence of why we already know more than enough to begin acting urgently. We also need to structure the choices that lie before us in ways that enable the necessary changes to seem easier and more automatic.

Once we find ways to embody these shared values in new social norms, we will benefit from the natural desire of people to follow the lead of others in circumstances similar to their own who are actively becoming part of the solution. We must pay attention to the alignment of individual choices with the incentives that businesses have to reinforce those same choices.

It's also important to design systems that will give our society constant feedback on the progress we're making in order to continually update the strategy we decide to follow in solving the crisis. Fortunately, the new information technologies that make such communication possible can play a crucial role in keeping us on this course.

CHINESE SCHOOLCHILDREN TAKE PART IN A TREE PLANTING CAMPAIGN NORTH OF BEIJING, PART OF A NATIONAL EFFORT TO STOP DESERTIFICATION.

THE OBSTACLES WE NEED TO OVERCOME

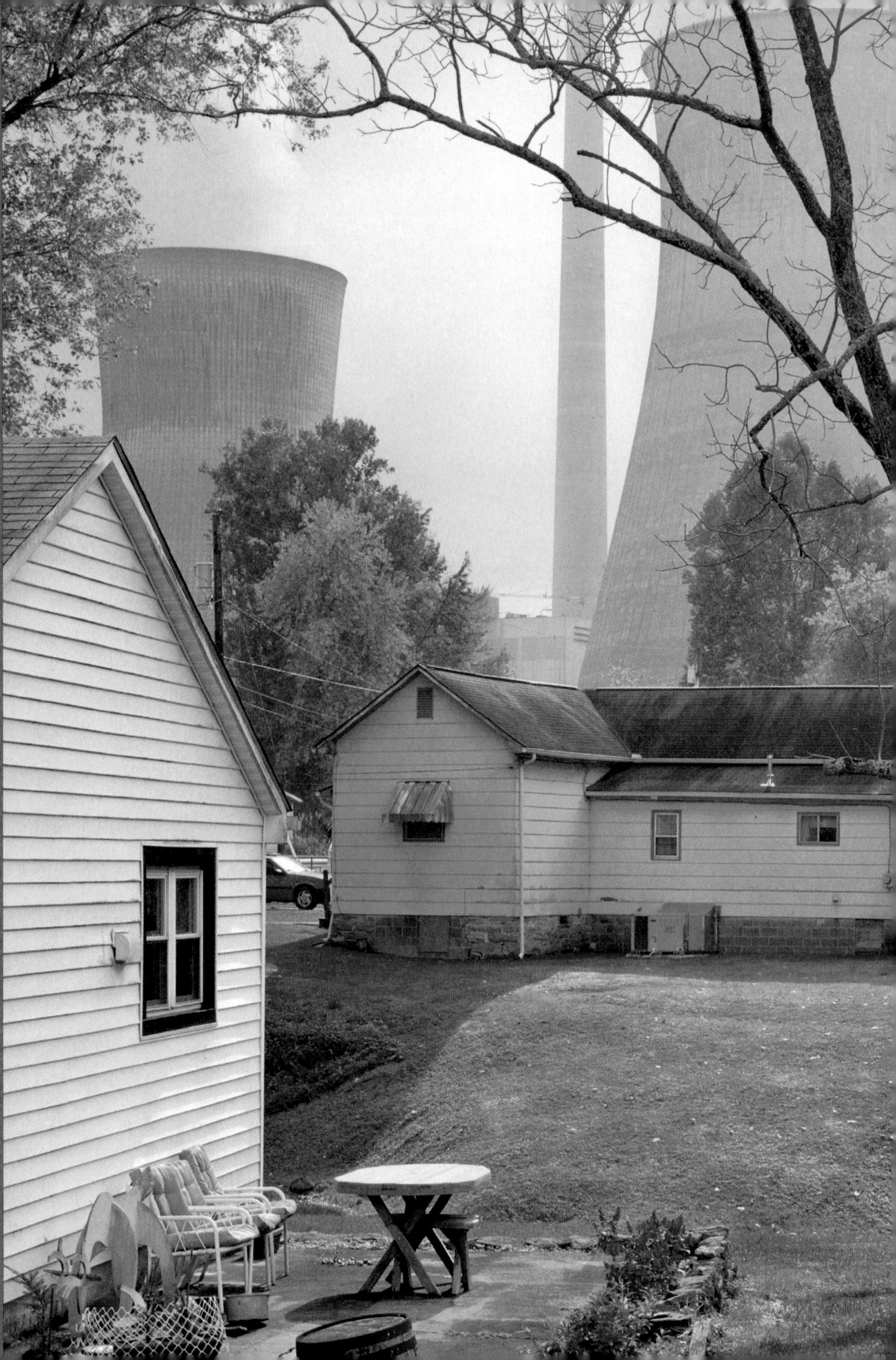

CHAPTER FIFTEEN

THE TRUE COST OF CARBON

THE AMOS COAL POWER PLANT IN WINFIELD, WEST VIRGINIA, PRODUCED MORE THAN 18 MILLION TONS OF CO_2 EMISSIONS IN 2006.

It is remarkable that at the very moment when we finally seemed ready to address the climate crisis we were hit with the worst global economic crisis since the Great Depression. Initially, many assumed this steep global financial downturn would halt progress on climate crisis solutions. But in fact, the relationship between these two monumental challenges has turned out quite differently. Economic policy experts from almost all points on the ideological spectrum acknowledged early on the need for a massive economic stimulus through government spending. The subsequent funding of large-scale projects intended to create millions of jobs has accelerated development of a green infrastructure in ways that promote solutions to the climate crisis.

Even so, we still are not using the power of the market economy to address this issue. It is deeply ironic that many of those opposing meaningful efforts to avert catastrophe raise fears of economic harm while, at the same time, absolutely refusing to allow the use of market mechanisms to help solve the crisis. Change is urgently needed because there are serious flaws in the quality and nature of the information we receive about the environment from market signals.

Our current system of measuring what is good for us and what is bad for us is deeply flawed. At present, global warming pollution—indeed, all pollution—is described by economists as a negative "externality." In public discussions, this technical economic term has come to mean: we don't want to keep track of this stuff, so let's pretend it doesn't exist.

Carbon dioxide, the most important source of global warming pollution, is invisible, tasteless, and odorless. It is largely invisible to market calculations as well. And when something's not recognized in the marketplace, it's much easier for government, business, and all the rest of us to pretend that it doesn't exist. But what we're pretending doesn't exist is destroying the habitability of the planet. We put 90 million tons of it into the atmosphere every 24 hours, and the amount is increasing decade by decade.

The easiest, most obvious, and most efficient way to employ the power of the market in solving the climate crisis is to put a price on carbon. The longer we delay, the greater the risk the economy faces from investments in high-carbon-content assets and activities. The artificial value placed on such investments ignores the reality of the climate crisis and its consequences for business. As Jonathan Lash, president of the World Resources Institute, recently said, "Nature does not do bailouts."

When we accurately acknowledge the consequences of the choices we make, our choices improve. Our market economy can help us solve the climate crisis problem if we send it the right signals. We've got to tell ourselves the truth about the economic impact of pollution, and we have to measure it. We need to internalize the externalities.

The system of "national accounts," which still serves as the backbone for determining today's

THIS COAL PLANT IN ROME, GEORGIA, INSTALLED A SCRUBBER IN 2008 TO PREVENT SULFUR DIOXIDE EMISSIONS AND TO BRING THE PLANT INTO COMPLIANCE WITH FEDERAL CLEAN AIR STANDARDS. NEW EPA REGULATIONS COULD REQUIRE LIMITATION OF CO_2 AS WELL.

"Nature does not do bailouts."

JONATHAN LASH

A WORKER HELPS IN THE CLEANUP AFTER A
SEWAGE SPILL NEAR RIO DE JANEIRO, BRAZIL, THAT
CAUSED LARGE FISH KILLS IN LOCAL WATERWAYS.

gross domestic product (GDP), is woefully incomplete in its assessment of value. Principally established in the 1930s, this system is precise in its ability to account for all produced goods and services, including capital goods, but dangerously imprecise in its ability to account for natural and human resources.

John Maynard Keynes led a group of economists who worked during the 1930s to give policy makers better tools to avoid a repeat of the Great Depression. In spite of the brilliance they brought to meeting this challenge, they were encumbered by unspoken assumptions then prevailing in a world in which industrialized countries still possessed colonies in Africa, Asia, and Latin America.

The colonial era would come to an end a few decades later, but during the time when the national accounts were being devised, it seemed easy to assume that natural resources need not be accounted for in the same way as capital resources. As a result, common accounting devices like "depreciation," which were routinely applied to man-made capital assets like equipment, buildings, factories, and other fixed assets, were simply not used in the same way for raw materials that seemed then to be available in almost limitless supplies.

In fairness to the economists who created GDP, they never intended for it to become widely used as a measure of general well-being. They were focusing, in the national accounts, on domestic production. However, others soon began using GDP as a way of measuring the overall health of a nation's economy, when that is not the purpose for which the measurement was intended.

GDP VS. GPI: "PRODUCT" VS. "PROGRESS"

Whereas GDP is the standard measure of a country's economic performance, summing the market value of all goods and services, the genuine progress indicator (GPI) is an attempt to measure the sustainability of income and the socioeconomic well-being of a nation. GPI adjusts the personal-consumption data of GDP by adding the benefits of nonmarket work, like unpaid housework and volunteering, and subtracting social costs like crime, air and water pollution, and the loss of farmland and forests. Over the past 50 years, GPI has increased at a much lower rate than GDP.

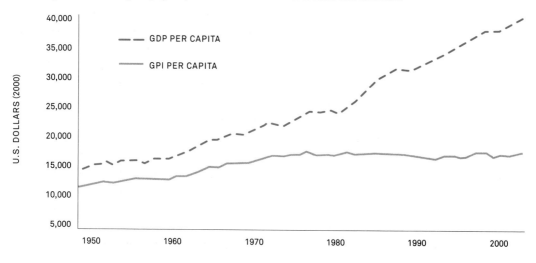

SOURCE: Robert Costanza, et. al., *The Pardee Papers*, No. 4, January 2009

It is ironic that such an elaborate system of accounting—one resting on an assumption that economic actors possess "perfect information"—would assign large categories of inputs and outputs to a netherworld that was considered safe to ignore. Most forms of pollution fell into this category because of a false assumption that the earth was so vast and resilient that dumping pollution into nature surely had no consequences worth tracking on the balance sheets.

The most serious consequence of failing to account for pollution in routine business judgments was that we made markets "blind" to the

the portion of that spectrum that's made up of light that we can see with our eyes is a very tiny slice. There's so much more out there, but human nature being what it is, we tend to assume that what we see is all that really matters. And since most everybody else does as well, we get along pretty well that way. The practice of looking solely at common financial reports means looking only at a narrow slice of that spectrum of information. Yet the information that lies in a company's environmental practice, employee practices, and the other nonfinancial factors is very important.

Physicist Werner Heisenberg discovered that

The gross national product "measures everything, in short, except that which makes life worthwhile."

SENATOR ROBERT F. KENNEDY

consequences of decisions and business plans that resulted in unrestrained pollution. This selective blindness, especially when coupled with supreme confidence that markets are routinely helping us make wise decisions in the aggregate, has made it extremely difficult to rely on market forces to solve the climate crisis. In fact, market forces cannot solve the climate crisis on their own. Markets fail to get the right outcome in the presence of externalities; policy intervention is required.

Consider this analogy that describes what we see and what exists alongside what we see, even though we are blind to it. If you look at the electromagnetic spectrum from ultraviolet to infrared,

in quantum physics, the act of observing changes what's being observed. It seems also to be true that when we as human beings observe something, the act of observing affects *us*. Information carries an imperative. Managers of investment funds who get daily financial reports begin to rely heavily on them, to the exclusion of other important information not included in those. As a result, their judgment is affected by the tool on which they come to rely.

Psychologist Abraham Maslow once said, "It is tempting, if the only tool you have is a hammer, to treat everything as if it were a nail." In the same way, if the only tool we use to analyze what's valuable is a price tag, then those things that don't have price

THE PRACTICE OF CLEAR-CUTTING, AS PICTURED HERE IN WASHINGTON STATE, CAN MAXIMIZE TIMBER REVENUE IN THE SHORT TERM, BUT AT ENORMOUS COST TO THE LOCAL ECOSYSTEM.

tags can begin to look like they have no value. And those things that are not on balance sheets can begin to look invisible and not worth taking into account.

In a market economy like ours, every one of the solutions to the climate crisis will be more effective and much easier to implement if we place a price on CO_2 and other global warming pollutants. We need to use the right tools for this job. Once we have a price on carbon, the negative externality that was invisible and not tracked by the market will become visible and will be included in the decisions of the market participants.

Forty years ago, Robert F. Kennedy reminded Americans that measures like the Dow Jones Industrial Average and gross national product fail to consider the integrity of our environment, the health of our families, and the quality of our education. As he put it, the gross national product "measures neither our wit nor our courage, neither our wisdom nor our learning, neither our compassion nor our devotion to our country. It measures everything, in short, except that which makes life worthwhile." His insightful observation about the gross national product represented a rare demarcation of the internal boundary between democracy and capitalism and invited a discussion about where to place the appropriate boundaries between decisions left to the market and decisions that ought by rights to be made in the sphere of democracy.

The philosophical system of which the United States of America has been the avatar is Democratic Capitalism. Adam Smith wrote *The Wealth of Nations* in the same year that Thomas Jefferson wrote the Declaration of Independence. The combination of free markets and self-governance by free citizens in the American republic was responsible for the rise of the United States as the leading nation of the world and for the prosperity that made it the envy of peoples throughout the world.

Throughout American history, reform movements, such as those of the progressive era, and the civil rights, women's rights, and environmental movements of the 1960s, were implicitly aimed at remedies in democratically enacted laws and regulations for perceived excesses and failures in the operations of unrestrained market forces.

The decisive victory of Democratic Capitalism over Communism in the 50-year post–World War II struggle between the United States and its allies and the Soviet Union led to a period of unquestioned philosophical dominance for market economics worldwide. The disappearance of Communism as a serious competitor to Democratic Capitalism led to the illusion of a unipolar world with one superpower. It also led, in the United States, to a hubristic bubble of "market fundamentalism" that encouraged opponents of regulatory constraints to mount an aggressive effort to shift the internal boundary between the democracy sphere and the market sphere by asserting that, over time, markets would most efficiently solve most problems and that laws and regulations interfering with the operations of the market carried a faint odor of the discredited statist adversary we had just defeated.

Simultaneously, changes in America's political system—including the replacement of newspapers and magazines by television as the dominant medium of communication—conferred powerful advantages on wealthy advocates of unrestrained markets and weakened advocates of legal and regulatory reforms.

This period of market triumphalism led to the dilution and removal of many protections against harmful pollution and, ironically, coincided with the scientific community's discovery that earlier fears about global warming were—according to the evidence that quickly mounted in the 1980s and

THE "SUPER PIT" GOLD MINE IN KALGOORLIE,
AUSTRALIA, PRODUCES 850,000 OUNCES OF
GOLD ANNUALLY. IT IS ALSO THE COUNTRY'S
TOP EMITTER OF MERCURY.

"There is something fundamentally wrong in treating the earth as if it were a business in liquidation."

HERMAN DALY

1990s—grossly understated. But the entire political context in which this debate took form was by then tilted heavily toward the views of market fundamentalists, who fought ferociously to weaken existing constraints and scoffed at the possibility that a new set of global constraints would be necessary to halt the dangerous dumping of global warming pollution into the earth's atmosphere.

Few doubt that the renewed inquiry into the structure, premises, and effects of capitalism will lead to anything more than tighter regulation of excesses and new efforts to extend some regulations beyond national boundaries to cover global financial flows. However, many have seized the opportunity to freshly raise serious questions about fundamental flaws in the way the market system, at least as we have practiced it, deals with natural resources and pollution—including global warming pollution.

Everyone knows the old saying "There is nothing so powerful as an idea whose time has come." I would like to propose a corollary to Victor Hugo's insight: the greatest source of destructive power is the sudden collapse of a widely accepted assumption that is suddenly recognized as wrong. In the case of subprime mortgages, the idea that some alchemy inherent in global financial markets would somehow eliminate the inherent risk of the bad mortgages if they were simply lumped together and sold as securities was an assumption that suddenly collapsed when the global markets realized, with a sinking feeling, that most of what had been sold with triple-A ratings was actually worthless.

In much the same way, many institutional investors are now beginning to suspect that another widely held assumption that undergirds the value of their portfolios is beginning to collapse. Several trillion dollars' worth of "subprime carbon assets" depend, for their valuation, on the belief that it's perfectly okay to put 90 million tons of CO_2 into the earth's atmosphere every 24 hours—and on a zero price for carbon emissions that reflects this assumption. The world's scientific community has presented irrefutable evidence that we must quickly stop burning carbon-based fuels in ways that destroy the future of human civilization. The owners of these assets will soon face a reckoning in the marketplace. They are in roughly the same position as the holders of subprime mortgages before they realized the awful mistake they had made.

The longer we continue making investments in subprime carbon assets, the longer we will increase the risk faced by our economy from stranded investments on a large scale. Subprime mortgages became "toxic assets." And since so much money had been invested in these suddenly worthless assets, the sheer size of the stranded investment became a powerful drag on the economy. The amount of investment sunk into high-carbon assets whose value is likely to plummet in the foreseeable future also represents a serious problem for our economy. Moreover, the owners of these high-carbon assets have an incentive to aggressively defend their value. Unfortunately, some of them have chosen to defend their value by fighting against the reforms needed to solve the climate crisis.

Economic history is full of similar mistaken assumptions that were belatedly recognized to the rue of investors, from the Dutch tulip craze in the early 17th century to Pets.com in the last months of the 20th century. But the financial crisis of 2008–09 illustrates the new global risks of spectacular bubbles when they burst. Environmental problems that used to be local or regional have become global in scale.

The longer we fail to recognize and measure the true cost of burning carbon-based fuels in the way we presently burn them, the bigger the bubble will grow and the more destructive the bursting will be. A wise economist named Herman Daly said decades ago that "there is something fundamentally wrong in treating the earth as if it were a business in liquidation."

A second serious defect in the way our markets now operate—alongside the failure to put a price on the cost of carbon emissions—is the pervasive reliance on short-term profit and earnings projections as the most important measure of whether a company is doing well or doing poorly. This is especially true in public companies that rely on institutional investors who buy and sell stocks with way too much emphasis on quarterly earnings reports as the dominant indicator of a company's value.

Numerous studies show that long-run performance determines most of a company's true value.

For years the best investors, including the legendary Warren Buffett, have understood this. Yet today the majority of the investment community (made up of investment managers, company executives, pension fund managers and trustees, investment consultants, and the research community) now acts in ways that give the impression that the long term simply does not matter anymore.

This is a relatively new phenomenon. Thirty-five years ago, the average holding period for stocks in the United States was almost seven years. But today the holding period is down to six months, and quarterly earnings targets are an obsession with research analysts.

As the evidence presented in Chapter 14 makes clear, we have a strong bias toward short-term decision-making. But this natural vulnerability

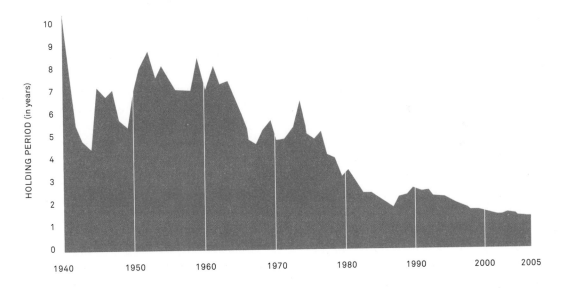

AVERAGE HOLDING TIME OF STOCKS

In the 1950s and 1960s investors held stocks for an average of seven to eight years. As recently as 1974, the holding period was almost seven years. Investment has become more like speculation as the average holding time has steadily decreased over the past four decades. In 2007 the average length of investment fell to 11 months. As of August 2009, it stood at six months.

SOURCE: James Montier, *Behavioural Investing*

has been artificially enhanced by the way in which corporations and financial markets have recently begun to make most of their decisions. Many experts in corporate finance believe that 75 percent or more of a corporation's true value is built up over a period of five to seven years—not coincidentally, the same time frame that used to characterize the average holding period for stocks. If most buyers of stock are now selling their purchases in six months, that is evidence that their decisions are being driven by something other than a desire to pick winners and invest in good companies. Strictly speaking, this short-term obsession that has taken over the financial markets has transformed the very act of investing into a behavior that, for many investors, more closely resembles speculation or gambling.

Most of the short-term metrics that are now used to drive decisions on buying and selling stocks are useful in trying to anticipate how other investors will react to similar short-term indicators. The decisions are still tethered, in theory, to an ability to project longer-term performance on the basis of short-term results. But the pressures now dominant in the market drive investors to rely increasingly on shorter holding periods and higher turnover. This, in turn, puts pressure on corporate managers—especially chief executive officers and chief financial officers—to adjust their own decisions with an eye toward maximizing good results every 90 days when the quarterly earnings per share are reported.

The market is long on short and short on long. This short-term orientation has significant negative repercussions for the global economy. If businesses forgo value-creating investments to manage short-term earnings, it damages economic vitality in the future. A short-term perspective also hinders innovation and research and development, diminishes investment in human capital, encourages financial gymnastics, and discourages leadership.

How do we reverse this trend toward the dominance of short-termism? First, the investment community should embrace genuine long-term thinking. That would mean managing portfolios with a long-term investment horizon of roughly five years, or through a business cycle. To do this, portfolio managers and analysts need to systematically take account of a number of factors that are not routinely monetized on balance sheets today—including sustainability issues—as opposed to focusing solely on short-term financial returns. As Abraham Lincoln said at the time of America's greatest danger, "We must disenthrall ourselves, and then we shall save our country."

This means analyzing the implications for shareholder value of long-term economic, environmental, and social challenges. Such challenges include future political or regulatory risks, the alignment of management and board with durable long-term company value, quality of human-resources capital management, risks associated with governance structure, the environment, corporate restructurings and mergers and acquisitions, branding, corporate ethics, and stakeholder relations. These extrafinancial issues clearly affect a company's ability to enhance shareholder value, create a competitive advantage, and generate sustainable returns over the long term.

Short-termism is not a problem that is found only in markets, by the way. My previous life was in politics. (I'm a recovering politician now.) When I first ran for office in 1976, I believe I took one poll. By the time I left politics in 2000, it was common practice to have overnight polling. Every day. And now tracking polls run continuously. Policy decisions made by politicians today are influenced by information flows derived from this never-ending polling and from its computerized data analysis.

We've heard for a long time about the dangers

AFTER THIS TENNESSEE POWER PLANT'S COAL ASH SLURRY POND COLLAPSED IN 2008, THE SLUDGE DESTROYED NEARBY HOMES, DAMAGED FARMS AND ROADS, AND TAINTED THE EMORY RIVER. SCIENTISTS HAVE WARNED OF CONTINUING HEALTH THREATS TO LOCAL RESIDENTS.

A BEEKEEPER TENDS HONEYBEES IN
WESTERN FRANCE.

of CEOs' responding to quarterly reports. The *McKinsey Quarterly* reported that a survey of business managers found "more than 80 percent of the executives responding said that they would cut expenditures on R&D and marketing to ensure that they met their quarterly earnings targets." This is not venal behavior; it is predictable behavior. If executives are evaluated on whether or not they hit the quarterly earnings mark, they're going to behave accordingly. And if they don't, they'll be replaced by someone who will. This happens on a regular basis. *McKinsey* added, "A majority of the managers polled said that they would forgo an investment offering a decent return on capital if it meant missing their quarterly earnings expectations."

There are managers who are trying to break out of this pattern. But if a company's major investors are looking for short-term rewards, then a manager who adopts a longer-term outlook will not survive.

Investors who want to adopt a longer-term horizon should not approach this issue simply as a challenge to reconcile their conscience and concern for the planet with the mathematics that come out of their investment offices. What's really required is to challenge the structure of the decision-making process. What are the incentives that drive the managers? What is the time horizon? What kind of information is taken into account?

Consider this example: Six years ago, a report from the World Resources Institute and Sustainable Asset Management titled "Changing Drivers" analyzed the carbon intensity of profits in

THE TRUE VALUE OF OUR ECOSYSTEM

In 1997 a team of researchers estimated the annual benefits of the critical "services" provided by our global ecosystem at $44 trillion. The total covered only renewable services, excluding nonrenewable fossil fuels and minerals. By comparison, the U.S. GDP was $14.3 trillion in 2008.

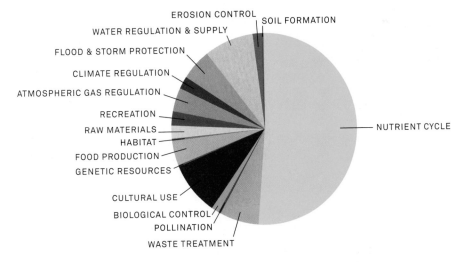

ECOSYSTEM SERVICES (global flow value)
$44 TRILLION (2008 dollars)

SOURCE: Robert Costanza, et al., *Nature*, May 15, 1997

the automobile industry. They then took that metric—carbon intensity of profits—into account and integrated it with the rest of the traditional analysis of the auto company's value, giving a more accurate picture of the sustainability of that stock's value over time.

The true price of burning carbon-based fuels like oil and coal would—in a perfect system of accounting—include additional costs that are now routinely ignored. The cost of failing to address the climate crisis is, of course, incalculable—and once this fact becomes sufficiently clear, it will have an enormous influence on our choices.

Where coal is concerned, we do not presently include in the purchase price the enormous damage being done by the common practice called

A BRIEF HISTORY OF THE PRICE OF OIL

The volatility of oil prices—which are affected by complex political issues, financial speculation, security concerns in the Persian Gulf, and the weather, among many other things—has proved to be an expensive obstacle to sustainable economic growth. The fast rise of prices in 2008 brought oil back to its highest level ever—equaled only by a peak during the Iran-Iraq War, almost 30 years earlier.

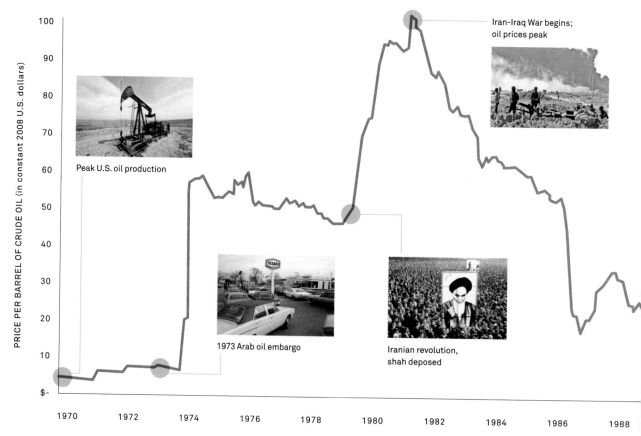

Iran-Iraq War begins; oil prices peak

Peak U.S. oil production

1973 Arab oil embargo

Iranian revolution, shah deposed

PRICE PER BARREL OF CRUDE OIL (in constant 2008 U.S. dollars)

SOURCE: U.S. Department of Energy

"mountaintop mining." Many coal deposits in the eastern U.S.—especially in West Virginia and Kentucky—are routinely mined by the systematic removal of entire mountaintops, with the toxic residue dumped into the streams of water at the base of the mountains. This despicable practice has already poisoned a number of community drinking water sources in Appalachia and has wreaked havoc on the lives of many families. The new mechanized techniques for removing the mountaintops have also destroyed many jobs for coal miners.

After the coal is mined, it typically is burned in ways that create the largest source of mercury pollution in the world. Mercury is, of course, a potent neurotoxin that accumulates in fish. It is the single largest cause of health warnings against

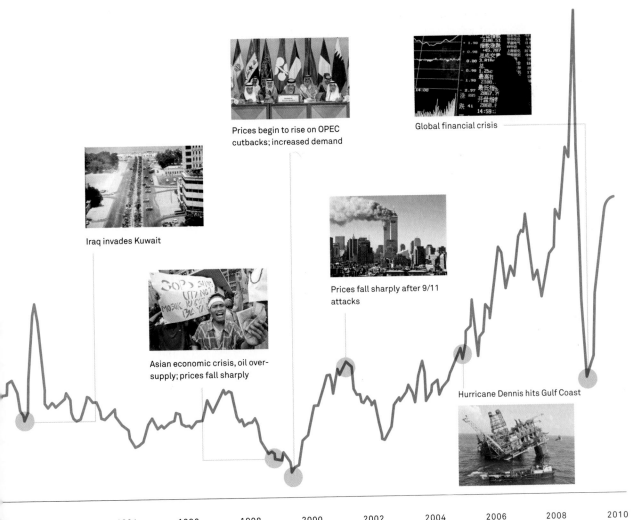

Prices begin to rise on OPEC cutbacks; increased demand

Global financial crisis

Iraq invades Kuwait

Prices fall sharply after 9/11 attacks

Asian economic crisis, oil over-supply; prices fall sharply

Hurricane Dennis hits Gulf Coast

90   1992   1994   1996   1998   2000   2002   2004   2006   2008   2010

Military expenditures by the United States could be reduced by tens of billions of dollars per year if we switched to renewable sources of energy.

A U.S. SOLDIER STANDS NEAR BURNING OIL WELLS IN KUWAIT, 1991.

eating many varieties of fish that used to be considered excellent and safe sources of protein.

Moreover, after the coal is burned, massive amounts of toxic sludge—the second largest volume of toxic waste produced annually in the U.S.—are typically dumped into holding ponds. In December 2008, more than a billion gallons broke free from a holding pond in my own state of Tennessee, forcing families from their homes and ruining farmland in the nearby community of Harriman.

Ironically, given our current system of national-income accounts, the cost of cleaning up that sludge will be added to GDP as a positive entry, while the cost to the families of having their lives disrupted and the cost to the environment of the pollution in the river and the damage to drinking water supplies will not be subtracted from our GDP.

Our current system of accounting also ignores the present value of all of the important services that healthy ecosystems provide, most of which we now take for granted. The predictable pattern of ice-pack melting in mountain regions provides clean drinking water for hundreds of millions of people, yet we take it for granted even as it is disappearing. Healthy soils recycle nutrients that make modern agriculture possible, yet we routinely ignore this function even as we contribute to its destruction with the heavy use of pesticides and oil-based fertilizer. A healthy population of bees pollinates many food crops, not to mention flowers, yet the systemic threat to bees from environmental degradation is not taken into account. Looked at in aggregate, the value of these and other ecosystem services worldwide was calculated in a 1997 study in *Nature* magazine at $44 trillion per year (in 2008 dollars)—none of which shows up in business or market accounting.

Yet another cost of our reliance on oil and coal that is not included in our routine measurements is the cyclical disruption of the global economy that comes as a result of instability in the price of oil on world markets. These fluctuations drive the price of coal up and down in tandem with oil because there is enough substitution of fuels at the margin to link the commodities together.

Most oil reserves in the world today are not owned privately by companies but are controlled by sovereign governments. The owners of the largest reserves work in concert within OPEC (the Organization of the Petroleum Exporting Countries) to pursue two strategic goals simultaneously. They wish to maximize price, of course, and thus they routinely cut back production quotas to drive the price of oil higher than it might otherwise be, just as a private cartel might do. But they also pursue a second objective that is not as well understood among the oil-consuming countries: they are highly conscious of their strategic interests in preventing the formation of political will in the West that might lead to a sustained effort—like the one I'm advocating in this book—to make the investments necessary to complete an historic shift away from oil and coal to renewable sources of energy.

There is enough oil production in non-OPEC countries to limit OPEC's ability to dictate price increases or sudden price reductions, and thus the oligopoly does not have complete control over prices on world oil markets. And the tensions within OPEC between richer members like Saudi Arabia and cash-strapped members like Iran also limit OPEC's freedom of movement. But when conditions in the markets allow them to sit in the driver's seat, OPEC has enough power to pursue its twin strategic objectives at the expense of the United States and other oil-importing countries.

The world has experienced several oil-price shocks since the first oil embargo in 1973. These price shocks triggered several efforts—soon aborted—to achieve energy independence by aggressively shifting to renewable sources of energy. During President Carter's administration, we saw an impressive beginning of a shift to renewable sources, and a sharp reduction in the amounts of oil imported by the United States. But as soon as the price of oil was driven back down by OPEC, the investments in renewable energy predictably dried up. And the Reagan administration, which took over in January 1981, systematically dismantled the remaining government programs to support renewable-energy development—even to the point of symbolically removing solar panels already bought and paid for from the roof of the White House.

In 2008, we saw this cycle repeated when the sudden rise in oil prices in the first half of the year drove the largest surge in renewable-energy investment that the U.S. has ever known. But when the price of oil went back down in the second half of the year, the investments quickly dried up.

For the reasons discussed earlier, it is a mistake to rely on market signals to produce optimal decisions in the absence of a price on carbon. Beyond that, it is self-delusional to rely on market signals when they emanate from a partially rigged market dominated by sovereign states that take a long-term strategic view while we remain hostage to fickle short-term calculations.

There are many other costs involved in relying so heavily on carbon-based fuels that are not presently included in our market valuation of the decisions we're making. The military expenditures by the U.S. alone that are attributable to the cost of protecting the Persian Gulf to prevent a global oil-price shock, along with the related U.S. military expenditures attributable to maintaining the flow of oil in the Middle East, could be reduced by tens of billions of dollars per year if we switched to renewable sources of energy. One 2008 academic study

LARGE NEW HOMES HAVE BEEN BUILT ON THE
EDGE OF WETLANDS IN GALVESTON BAY, TEXAS.

estimated the reduction at between $27 and $73 billion per year, of which $6 to $25 billion per year (or up to 15 cents per gallon of gasoline) is attributable to the demand for U.S. cars and trucks alone.

Another widely known defect in the way markets for energy work is a problem that economists call the "principal-agent problem." This phrase sounds more complicated than it really is. What it actually identifies is a conflict between the market incentives that influence the choices made by one set of decision makers compared with the economic realities that confront all those subsequently affected by those choices.

For example, many builders and developers of homes and commercial buildings are driven by short-term competition to reduce the initial purchase price or lease price of their buildings—even if that means skimping on heavier insulation, more efficient windows and lighting systems, and design features that can sharply reduce energy consumption to the benefit of those buying or leasing the finished properties. If the annual operating costs of such energy-efficient technologies were included in the calculations made by the builders, their decisions would more closely match the interests of those who end up paying the annual operating costs. As it is, however, the structural divide between builders on the one hand and owners and renters on the other creates incentives that work at cross-purposes.

Partly as a result of these principal-agent problems, most buildings are highly inefficient in the way they use energy. In fact, buildings in the United States now account for almost 40 percent of the CO_2 pollution released into the atmosphere. The measures advocated in Chapter 12 will be far easier to implement if creative solutions to the principal-agent problem are used to remedy such problems in the marketplace for buildings

and for energy-efficient technologies.

During the era of cheap oil, energy prices were a much smaller percentage of overall production costs. So energy-reduction measures had a lower priority. Everyone now knows the days of inexpensive oil are numbered, but this strategic view of our energy future is not reflected in current market signals. So it must be imposed by government action in order to prevent an economic (and, of course, environmental) catastrophe.

The most encompassing and serious principal-agent problem is that between our generation and all the future generations, who will live with the consequences of our decisions. The incentives that have been driving us toward short-term decisions based on limited information in order to maximize short-term profit now threaten catastrophic damage within our own lifetimes. Damage to the prospects of the generations following us will be greater still. We must make decisions in the next few years not only with an eye to the effect they have on us but also considering their impact on future generations. We are, in this sense, agents for our children and grandchildren, which means we must somehow bridge the divide between what seems right for us and what is right for them.

For all these reasons, an effective plan for solving the climate crisis must include aggressive remedies for our erroneous reliance on deceptive market signals in carbon-based energy—signals that are both structurally flawed and intentionally manipulated by sovereign nations seeking to control our energy future.

There are three options available to us for fixing the flawed signals in the marketplace:

▶ A CO_2 tax that internalizes the true environmental cost of coal and oil.

▶ The use of a cap and trade system, which accomplishes the same result indirectly by restricting the amount of CO_2 that can be produced and allocating it through a market-based trading system.

▶   Direct regulation of CO_2 emissions under laws such as the Clean Air Act.

I have long advocated the first option—a CO_2 tax that is offset by equal reductions in other tax burdens—as the simplest, most direct, and most efficient way of enlisting the market as an ally in saving the ecosystem of the planet. However, one of the first casualties of the ascendance of market fundamentalism in the United States was in its success in creating massive opposition in the Congress to any new taxation—even taxation offset by reductions in other tax areas. The coal and oil companies, assisted by coal-burning utilities, have provided political contributions, massive lobbying

resources, and aggressive public advertising with corporate funds to buttress opposition by many elected officials to anything that these companies feel might hurt their profits.

It is possible that these attitudes may change over time as the merits of a revenue-neutral CO_2 tax become more widely understood—and as recognition of the unthinkable consequences of failing to solve the climate crisis begins to play a bigger role in our assessment of what is right and what is wrong.

In the last few years, some who used to oppose a CO_2 tax have come out in favor of it. For example, Arthur Laffer, a conservative Republican who was one of the architects of President Reagan's initial tax-reduction plan, joined with a Republican Congressman from South Carolina, Bob Inglis, to

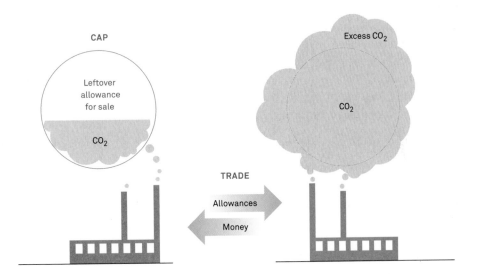

HOW CAP AND TRADE WORKS

In a cap and trade system, the law sets maximum allowable emissions—the cap—for a set of polluting industries, such as power plants. For every ton of CO_2 that a polluter reduces under the cap, it is awarded one allowance. Allowances can be bought, sold, traded, or banked for the future, and any facility that has successfully cut emissions below the mandated level can then auction their allowances to those over-polluting. This built-in cash incentive to reduce emissions encourages compliance and innovation—and maximizes the efficiency with which the market reduces pollution.

SOURCE: Patterson Clark, *The Washington Post*, February 26, 2009

HOW CAP AND TRADE HELPED REDUCE ACID RAIN

During the 1980s, acid rain—caused by sulfur dioxide (SO_2) emitted from power plants—was taking a toll on U.S. and Canadian waters, soils, and forests. The causes and effects were clear to all observers, but what was not easy to see was how the problem could be addressed. The solution, a cap and trade agreement covering sulfur emissions, is instructive now, as we must tackle the tougher global problem of climate change.

In 1988, legislators from both the Democratic and Republican sides of the aisle worked together to create a new, market-based solution to the problem. It was clear that to reduce acid rain, SO_2 emissions had to be cut by 10 million tons, halving 1980s levels. There were two approaches available at that time. Coal-fired factories and power plants could switch to low-sulfur coal from western-state suppliers. Or they could install scrubbers to catch SO_2 before it left smokestacks, a fix that industry claimed would cost $6 billion a year.

Instead of implementing strict regulations, the Acid Rain Program, which was passed as part of the Clean Air Act Amendments of 1990, set an emissions cap and allowed the coal users to decide how to meet it. The proposal built in an incentive to overcomply: for every ton of SO_2 that a polluter was under the limit, the EPA granted an allowance that could be traded, sold, or banked for the future.

Members of Congress engaged in more than 100 hours of negotiating sessions before the bills passed, with votes of 401 to 25 in the House and 89 to 10 in the Senate. President George H.W. Bush made emissions trading his achievement in market environmentalism, saying, "We should set tough standards, allow freedom of choice in how to meet them, and let the power of markets help us allocate the costs most efficiently." Evidence has shown that it has done just that. In the program's first phase, regulated plants reduced emissions 40 percent more than required. By 2004, SO_2 emissions had dropped by seven million tons, slashing 1980s levels by 40 percent. The Department of Energy estimates the emissions cuts accounted for merely 0.6 percent of the utilities' overall $151 billion operating expenses, and an M.I.T. study called the program "more successful in reducing emissions than any other regulatory program initiated during the long history of the Act."

AVERAGE ACID RAIN CONCENTRATIONS

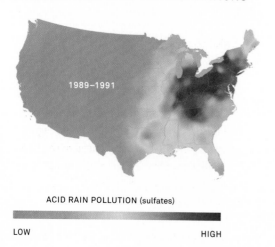

1989–1991

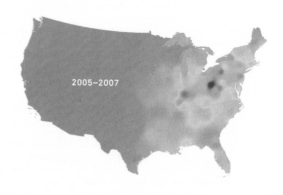

2005–2007

ACID RAIN POLLUTION (sulfates)

LOW

HIGH

Ten years after the acid rain cap and trade program was enacted to regulate sulfur-dioxide and, later, nitrogen-oxide emissions, acid rain concentrations in the eastern United States have decreased by 10 to 25 percent.

support my proposal for a CO_2 tax in a *New York Times* column they wrote in 2008.

For the foreseeable future, however, it is only prudent to assume that the U.S. political system is incapable of making such a bold and controversial decision. This could change, but I vividly remember what happened in 1993 when I persuaded President Clinton and his economic team to include a version of the CO_2 tax (at that time called a BTU—British Thermal Unit—tax) in our economic plan. With great effort, we were able to persuade the House of Representatives to adopt the measure, but the Senate refused and watered it down to the point where it was worse than nothing at all.

countries, mostly in Europe, have already enacted both approaches. Sweden, often considered the country with the most advanced strategy for reducing CO_2, has enacted both measures. Recently it increased the CO_2 tax, after an initial experience that was overwhelmingly positive.

The third option for fixing the mistaken signals in the market for carbon-based fuels involves government regulation of CO_2 emissions. In concert with a CO_2 tax and/or a cap and trade system, direct regulation of CO_2 is a very effective approach. Moreover, in early 2007, the conservative-dominated U.S. Supreme Court formally ruled that the Environmental Protection

The real solution would include both a CO_2 tax and a cap and trade system, and I believe that will eventually be our choice.

Confronted with the unlikelihood of gaining sufficient support for a CO_2 tax, most thoughtful advocates have concentrated instead on option two, the cap and trade system. Indeed, virtually all of the bills introduced in Congress by members of both political parties have featured a cap and trade system as their preferred mechanism for including the hidden costs of carbon-based fuels in our market calculations. This approach is also the centerpiece of President Obama's strategy for reducing CO_2 emissions, and the centerpiece of the global negotiation in Copenhagen at the end of 2009.

In my opinion, the real solution would include both a CO_2 tax and a cap and trade system, and I believe that will eventually be our choice. Several

Agency was required under the Clean Air Act to go forward with a formal consideration of whether or not to regulate CO_2 as an air pollutant covered by the law. Since CO_2 is obviously the most dangerous form of air pollution we face, most assumed that this court decision would inevitably lead to regulation. And early in 2009, the new head of the EPA under President Obama, Lisa Jackson, initiated formal proceedings that may result in regulation.

Another form of regulation that promises to accelerate the transition to renewable energy comes in the form of a legally required mandate to producers and sellers of electricity that they obtain a large and growing percentage of their electricity from renewable sources. This approach has already been enacted by the state of California and several other states and has already resulted in a surge of new investment for windmills and solar plants that would not have been built without the legal mandate. If this approach is codified in national law—as appears likely—this surge in renewable-energy investment will grow rapidly. Other countries, including China and the nations of the European Union, have adopted this approach as well. Several regional and provincial governments outside the United States have done so, too.

Nevertheless, regulation has its limits, and it is clear that the roller coaster of rising and falling oil prices must be addressed with more direct and accurate market signals in order to finally solve the climate crisis.

As it currently operates, the market for energy fails to integrate two important variables that enhance the attractiveness of renewable energy as a substitute for carbon-based fuels. As noted in Chapter 2, human ingenuity and innovation give us the prospect of steadily reducing the price of renewable energy in ways that simply are not available to us in dealing with limited supplies of oil and coal. By making a clear choice to shift large volumes of energy production to renewable sources, we would guarantee a major increase in research-and-development expenditures that would drive the cost of renewable energy downward. The greater the commitment we make, the greater the production we guarantee—thereby locking in further cost reductions as economies of scale reduce the price of windmills, solar panels, and other technologies that convert renewable sources into usable energy.

The slowing rate of oil discoveries, coupled with steadily growing demand in rapidly industrializing nations like China and India, guarantees that oil prices—and thus coal prices—will continue to rise over the long term, in spite of disruptive increases and decreases of oil prices in the short-term. By contrast, the long-term price of renewable energy is certain to continue to fall dramatically.

If we are truly concerned about our energy and environmental future, the choice we must make is crystal clear. But we cannot rely on market signals to make this choice for us.

Capital markets and capitalism are at a critical juncture. The domination of short-termism across our financial system will stifle innovation, damage our economies, further impair our pension systems, and ultimately erode our standard of living. The long-term investment community, which represents the significant majority of total investable assets, must adopt truly long-term thinking. Company management and the research community must also look to the long term. Our livelihoods—and more importantly, our children's and grandchildren's livelihoods—depend on it.

The financial crisis has reinforced my view that sustainable development will be the primary driver of economic and industrial change over the next 25 years. It is imperative we find new ways to use the strengths of capitalism to address this reality and, most importantly, to solve the climate crisis.

Sustainability and long-term value creation are closely linked. Business and markets cannot operate in isolation from society or the environment.

Today, the sustainability challenges the planet faces are extraordinary and completely unprecedented. Business and the capital markets are best positioned to address these issues when markets and government policies provide the right signals. And there are clearly higher expectations for businesses and more serious consequences for running afoul of the boundaries of corporate responsibility. We need to return to first principles. We need a more long-term and responsible form of capitalism. We must develop sustainable capitalism.

THE BEDDINGTON ZERO ENERGY DEVELOPMENT
IN SURREY, ENGLAND, GETS ALL OF ITS
ELECTRICITY AND HEATING FROM SOLAR PANELS,
PASSIVE SOLAR, AND TREE-WASTE-POWERED
COGENERATION.

THE OBSTACLES WE NEED TO OVERCOME

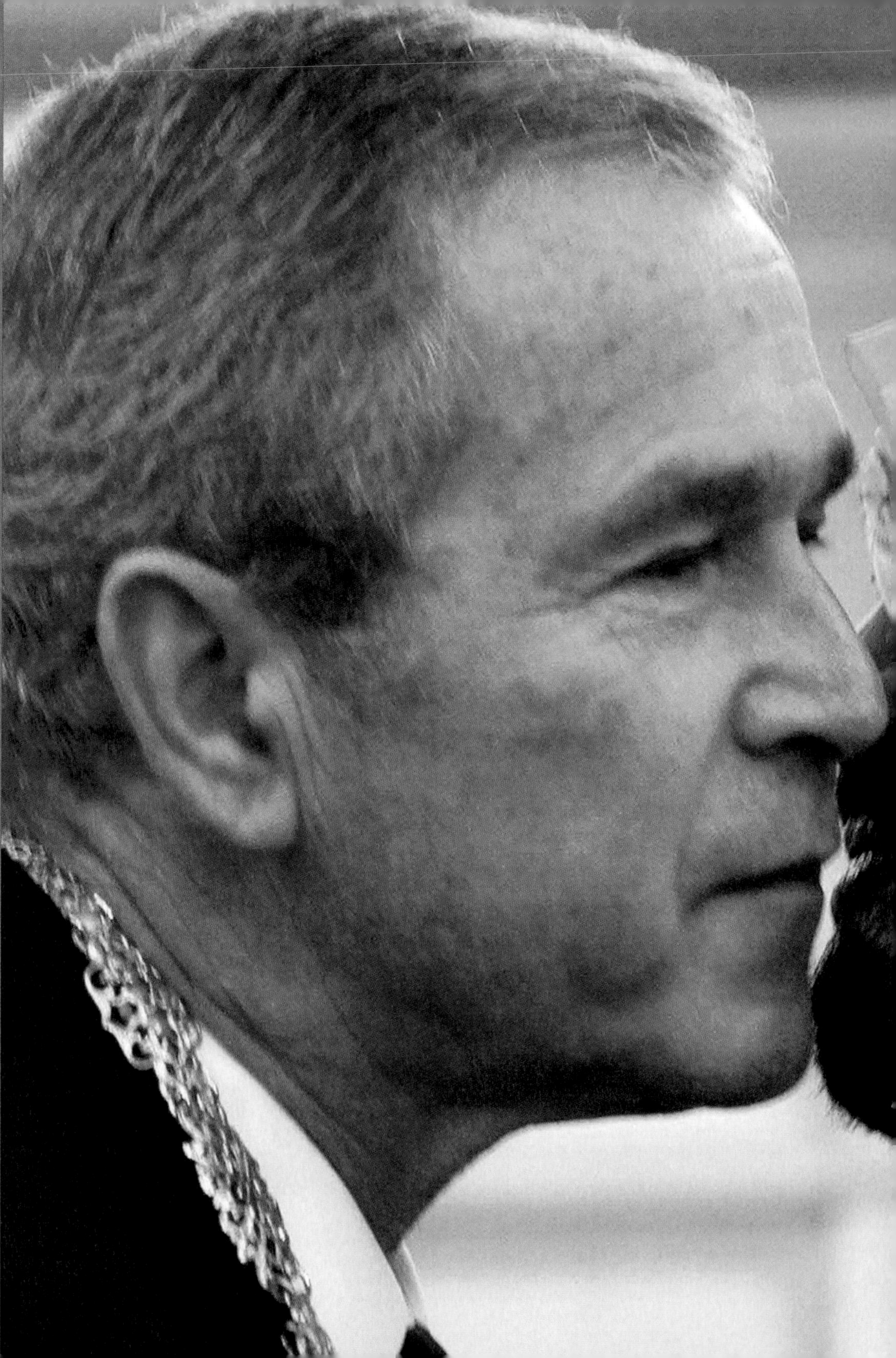

CHAPTER SIXTEEN

POLITICAL OBSTACLES

SAUDI KING ABDULLAH EMBRACES PRESIDENT
GEORGE W. BUSH AFTER AWARDING HIM THE KING
ABDUL AZIZ ORDER OF MERIT, JANUARY 2008.

We urgently need solutions to remove the political obstacles that block us from confronting the mortal threat of the climate crisis to the future of civilization. These solutions require difficult decisions that can be made only within the political systems of the United States and other nations. There is no other way. In order to do so, it is crucial to understand how and why our current politics have thus far failed us, after which we can apply the lessons learned to making political changes that will embolden elected officials to implement the solutions that will save us.

John Kenneth Galbraith once quipped, "Politics is not the art of the possible; it is the art of choosing between the disastrous and the unpalatable." In the case of the climate crisis, our choice has been clouded by confusion, generated in large part by a massive political campaign of intentional deception on the part of many corporate carbon polluters.

In the first place, the disaster we are facing has not yet been fully and clearly recognized by voters. In part, this is because its vast, global scale has disguised its role in helping to cause specific catastrophes that would, if definitively linked to global warming, motivate impacted electorates to demand action. Even Hurricane Katrina—which was exactly the kind of Category 5 storm that scientists have long warned us will become far more common as a result of global warming—did not cause many Louisiana politicians to change their position on the issue.

Second, the measures required to solve the climate crisis seem unpalatable because the CO_2 emissions that must be reduced have been an integral part of our coal-and-oil-fueled economic activity for more than 150 years. The sheer scale and pervasiveness of the policies necessary to de-carbonize the world's activities represent a unique and unprecedented challenge for the political process (even if millions of good jobs are created in the process).

In other words, the changes required are simply not the sort of incremental course corrections with which our politics usually deals.

Moreover, two other contextual factors have heightened political opposition to these large-scale changes in the United States and other industrial countries. The increased globalization of the world economy in recent decades and the new ease with which advanced manufacturing technologies travel across national borders have led to a massive migration of industrial jobs from developed nations to countries with lower wage rates. This trend has heightened the fear that new measures affecting business in any one nation might lead to additional job losses if other countries are not also required to participate in sharing the burden of change.

In this regard, one of the trickiest global political challenges is to deal with the industrialized nations' job-loss fears while simultaneously responding to the less-developed countries' argument that, since they did not create the climate crisis in the first place and since their per capita incomes are only a tiny fraction of those enjoyed by wealthier nations, they simply cannot be expected to share the same burdens.

The debate on both sides of this rich-poor divide has been sharpened by the global recession, which hit the world economy just as active consideration

AMERICAN STUDENTS RALLY AGAINST THE
CLIMATE SUMMIT IN BONN, GERMANY, IN 2001.

of global warming solutions was beginning. Fortunately, however, even though many expected the economic downturn to further delay action on the climate crisis, the prospect of millions of new green jobs has actually led to progress in addressing both the climate and economic challenges simultaneously. Still, the struggle to pass global warming legislation and enact a global treaty to reduce greenhouse gas pollution has turned into an epic political battle lasting two decades—during which time the pace of change in the natural world has accelerated dramatically.

Powerful industries affected by proposed climate crisis solutions have used all the political tools at their disposal in opposition. For example, in 2009, a lobbying firm working for coal companies and coal-burning utilities forged letters opposing climate legislation to members of Congress to create the false impression that the letters had come from citizens and nonprofit organizations. Regions especially dependent on coal have opposed any

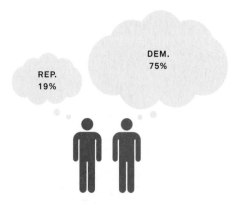

WHO BELIEVES HUMANS ARE CAUSING GLOBAL WARMING?

A 2008 survey revealed that 75 percent of college-educated Democrats believe global warming is happening because of human activity, while only 19 percent of college-educated Republicans agree.

SOURCE: Pew Research Center for People & the Press, "A Deeper Partisan Divide Over Global Warming"

measures they fear might disproportionately impact them. And ideological opponents of any enhanced role for government in managing and implementing the needed solutions have organized in opposition to the proposed policy changes.

In the United States, even though global warming should not be a partisan issue, the Republican party has aligned with opponents of government expansion, and in recent election cycles has counted oil and coal companies as key members of its political coalition. As a result, Republican elected officials have, with some distinguished exceptions, fought against taking any meaningful action on global warming. All too many Democrats have also opposed action—especially in regions with a heavy dependence on coal.

But the disproportionate and overwhelming opposition of Republican officeholders has often given the misleading impression that the issue is somehow a partisan argument. And since so many partisans on both sides tend to follow their perceived leaders, the political divide in the United States has grown deeper. By 2008, when Americans with a college education were asked in a poll whether human activities are responsible for global warming, 75 percent of Democrats said yes, but only 19 percent of Republicans said yes.

Campaign contributions from special interests have always played a significant role in politics, of course, and special interests connected to oil and coal have traditionally been among the largest contributors. Moreover, the outsize influence of campaign contributions has been greatly enhanced by the still-growing dominance of television advertising at the expense of reasoned discussion. Indeed, the historic shift from print publications to television has radically changed the balance of power in politics—particularly in the U.S.—by magnifying the importance politicians place on raising enough

"DRILL NOW" AND "DRILL BABY DRILL" WERE POPULAR SLOGANS AT THE 2008 REPUBLICAN NATIONAL CONVENTION.

Denial and ridicule of global warming science is a frequent feature of news programs seeking to appeal to right-wing voters—for some because of political bias, and also because of pressure on the media to portray a false "balance" between differing points of view. Among the most vocal skeptics have been (clockwise from top left) Rush Limbaugh, James Inhofe, Pat Buchanan, Glenn Beck, John Stossel, Lou Dobbs, and Sean Hannity.

money to buy the expensive TV ads that now determine most election outcomes.

Nowhere is this more evident than in the politics of climate. Oil and coal companies and coal-burning utilities were among the largest contributors in the 2008 election cycle and the largest television advertisers during the campaign. In just the first three months after President Obama took office, interest groups and corporations spent $200 million to influence U.S. energy policy and build opposition to action on global warming.

Since, on average, Americans now watch television five hours per day (which amounts to an incredible 17 years of television-watching in the average lifetime), they are bombarded with cleverly designed advertisements at an unprecedented rate. As noted in Chapter 14, this is partly responsible for the orgy of excessive consumption, but it is also responsible for a gross and continuing distortion of the way political decisions are now made by elected representatives.

The political model conceived by America's founders presumed that reasoned debate based on the best available evidence would play the central role in our nation's political decision-making process. Now, however, the views of voters on issues important to special interests are primarily shaped by expensive advertising campaigns that seek to manufacture "the consent of the governed" as if it were another product line.

Moreover, these same companies and their allies have flooded the U.S. Capitol with paid lobbyists in unprecedented numbers, dominating the $90 million spent on climate lobbying in 2008 alone. According to one study by the Center for Public Integrity, for every single member of the House and Senate there are now more than four lobbyists working on climate issues—an increase of more than 300 percent since the last climate legislation was brought before Congress a few years ago. And lobbyists opposed to climate legislation outnumber supporters by more than eight to one!

Even more insidious, the integrity of our democracy has been poisoned by a new kind of sophisticated, well-planned, and lavishly financed campaign aimed at actively misleading the public about what science actually tells us concerning the nature and severity of the climate crisis.

This new technique—designed to actively deceive people by intentionally distorting the science—was actually pioneered decades ago by tobacco companies. They systematically created confusion about the medical consensus linking cigarette smoke to lung cancer, emphysema, heart disease, and other deadly health threats. One tobacco-company memo from that era (recently uncovered in a lawsuit) outlined the purpose of their new approach to issue campaigning: "Doubt is our product, since it is the best means of competing with the 'body of fact' that exists in the mind of the general public. It is also the means of establishing a controversy."

Now this same unethical practice has been perfected and vastly scaled up by large carbon polluters, who have not only employed the same strategy used by the tobacco industry but have also hired some of the veteran operatives from that earlier effort in order to systematically create doubt and confusion about the scientific consensus concerning the threat of global warming.

In the late 1980s, just as the scientific consensus on global warming reached a tipping point and began to capture the attention of voters, several large oil companies, auto companies, coal companies, and coal-burning utilities joined forces to launch what can only be called a propaganda campaign designed to undermine the integrity

"Doubt is our product, since it is the best means of competing with the 'body of fact.'"

CIGARETTE EXECUTIVES TESTIFY TO CONGRESS IN 1994 THAT NICOTINE IS NOT ADDICTIVE.

"Reposition global warming as theory rather than fact."

INTERNAL FOSSIL FUEL INDUSTRY MEMO

OIL EXECUTIVES TESTIFY TO THE SENATE IN 2008
THAT OIL PRICES ARE SET BY THE MARKET.

of the scientific evidence itself. Utilizing sophisticated psychological and marketing research, these companies set out to pursue a single-minded goal that was stated in their initial strategy (a memo uncovered by investigative journalist Ross Gelbspan): "Reposition global warming as theory rather than fact."

From its beginnings in the years immediately prior to the June 1992 Earth Summit in Rio de Janeiro, this campaign of deception by the biggest carbon polluters grew to become the largest effort of its kind the world has ever seen. And although some of its largest corporate supporters have been forced by public pressure to publicly withdraw from the campaign, it continues today in full force. Indeed, it has taken on a malignant life of its own; not a day passes without right-wing commentators on radio and cable TV talk shows loudly promoting the pseudo-science of the deniers.

The creation of an entire disinformation network has been one of the most disturbing elements of this propaganda campaign. The large carbon polluters set up and funded dozens of front groups. They paid little-known and thinly credentialed "scientists" to crank out thousands of pseudo-studies, letters, books, pamphlets, and videos, all purposefully designed to raise false doubts about virtually every aspect of the emerging scientific consensus. Another internal memo stated that one of their principal goals was to "develop a message and strategy for shaping public opinion on a national scale." This manufactured campaign found a receptive audience among many citizens who would understandably prefer not to acknowledge the existence of such a frightening and potentially overwhelming threat as global climate change.

Some of the contrarians, of course, are neither insincere nor recipients of the carbon polluters' largesse. However, virtually none of them have published their views in refereed, peer-reviewed journals, and their arguments would likely be simply ignored except for the giant megaphone and echo chamber afforded them by the deniers' disinformation network.

According to one journalistic investigation, ExxonMobil—the largest and wealthiest of these companies—provided such funding to almost 40 front groups active in efforts to pervert the public's understanding of global warming science. In only one example of this phenomenon, just prior to the release of the fourth unanimous "Assessment" by the Intergovernmental Panel on Climate Change in January 2007 (which further strengthened the scientific consensus that the world must sharply reduce greenhouse gas pollution), one of the front groups financed by ExxonMobil offered $10,000 for every paper disputing the consensus findings of the world's scientific community.

Another tactic has been to routinely question the integrity of respected scientists by claiming that it's somehow in their financial interest to invent the climate crisis out of whole cloth and thereby build support for more research funding. This charge is deeply ironic on two counts. First, it is opponents themselves who are open to such accusations, while their targets are simply reporting legitimate research findings. Second, if a legitimate mainstream scientist were able definitively to disprove the reality of man-made global warming to the satisfaction of the scientific community, that person would likely become one of the most celebrated—and wealthiest—scientists of the century.

The global warming deniers' arguments are fraudulent and often nonsensical. This entire disinformation campaign is so offensive to the very nature of the democratic process, it is tempting simply to ignore it as beneath contempt. But because the evidence that has so galvanized scientific experts is still largely unfamiliar to the public, these professional climate skeptics have had an outsize impact on the politics of the issue.

The reason for their devastating impact was clearly understood by the denier network from the outset of the campaign. One of their advisers described the strategic rationale in these words: "Should the public come to believe that the scientific issues are settled, their views about global warming will change accordingly. Therefore, you need to make the lack of scientific certainty a

temperature that they claimed would prove that the earth was actually cooling. Upon closer examination, it turned out that the authors of this report had made errors in mathematical calculations and had also failed to take into account decay in the orbit of the satellites that grossly distorted the temperature readings. These errors completely negated the results the deniers had so heavily promoted, and the recalculated findings ended up confirming the consensus findings. Eventually, one of the authors was forced to acknowledge the mistakes and publicly withdraw his argument. But—

"Should the public come to believe that the scientific issues are settled, their views about global warming will change accordingly."

MEMO FOR BUSH WHITE HOUSE BY POLITICAL CONSULTANT

primary issue in the debate."

The specious and cynical arguments they put forward can be organized in terms of the stages of denial. Initially, their main message was that global warming isn't real, doesn't exist. They ridiculed the scientific consensus and cherry-picked any scrap of information they could use to undermine it. When their arguments were meticulously disproved, they refused to acknowledge the facts and continued their assertions anyway.

To choose only one among many examples: the deniers enthusiastically promoted a particular series of satellite measurements of the earth's

seemingly without embarrassment—he returned to repeating the same argument in public anyway.

After several years of record-breaking global temperatures and the rapid melting of glaciers all over the world undermined their ability to convince people that the earth was not warming, the deniers shifted to a second line of argument. While global warming may be occurring, they said, it is a purely natural phenomenon that has no relationship to the 90 million tons of global warming pollution we put into the atmosphere every day.

As the central role of man-made global warming pollution became more widely understood and accepted, the deniers shifted to a third line of argument: humans may be a factor, but global warming is predominantly a natural trend. None of their arguments were accepted for publication in refereed, peer-reviewed scientific journals, and some assertions were so extreme as to be laughable. For example, one denier argued that two observations of Pluto, 14 years apart, provided evidence of warming on Pluto. Their point was that if Pluto was warming at the same time as the Earth, that would imply that changes in the Sun's output of radiation was the common cause of warming throughout our solar system—and that we could absolve man-made greenhouse gas pollution of any responsibility for the increasing temperatures on our planet. But since it takes 248 Earth years for Pluto to trace its erratic orbit around the farthest reaches of our solar system—between 2.7 billion miles and 4.6 billion miles away from the sun—it's hard to imagine that anyone was fooled by such a ridiculous argument.

Eventually, Dr. Naomi Oreskes at the University of California–San Diego led a team of researchers who took a statistically meaningful sample of 10 percent of all the peer-reviewed scientific papers on global warming that had been published over a 10-year period. She found that, among the 928 papers in the random sample, not a single one differed with the consensus scientific view on global warming. Characteristically, some deniers then sought to raise doubts about the Oreskes study, and one claimed that there were actually 34 papers that rejected or doubted the consensus view. Even if this claim had been correct, it would have meant that less than 4 percent of the global studies disagreed with the consensus. However, upon closer examination of this denier's arguments, he was forced to acknowledge that only one of the "studies" he cited had disagreed with the consensus, and that, in fact, the item was not a study at all but rather an opinion piece, written in *The American Association of Petroleum Geologists Bulletin* by two men in the oil business, one of whom held positions at ExxonMobil. Their article concluded that, "there is no discernible human influence on global climate at this time."

The deniers' next argument makes the claim that, even if man is making significant contributions to global warming, well, that's okay because global warming is probably good for us. This assertion, like all the others, contradicted the unanimous view of the Intergovernmental Panel on Climate Change. But then, that was the whole point: to sow doubt.

A more common variation on this same theme is that, even if it's not a blessing, global warming is certainly not a serious matter; all we have to do is to make some minor adjustments to adapt to hotter weather. This claim is also specious, of course, but again, the point is not to win an argument but to create public confusion and, in so doing, to paralyze the political process in the face of what is an already difficult decision.

Global warming deniers pursue another parallel argument, stating that any effort to solve the climate crisis would be worse than the crisis itself. This is a more traditional approach, one commonly taken by polluters who don't want to pay for the cost of reducing emissions. But a number of economic analyses, including two definitive studies by Sir Nicholas Stern in the United Kingdom, have shown that the cost of cleaning up the pollution is actually minimal, compared with the devastating economic cost of allowing the damage to take place.

The last refuge of these deniers is their claim that, if global warming is real, if it is predominantly caused by man-made pollution, and if it is, in fact, very bad for us, then it's too late to address it anyway. This last argument shares one feature

How much are you willing to pay to solve a problem that may not exist?

If the Earth is getting warmer, why is the frost line moving south?

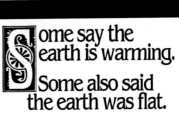

Who told you the earth was warming... Chicken Little?

Some say the earth is warming. Some also said the earth was flat.

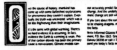

with all the other arguments: it concludes that the right course of action is to do nothing at all to stop the release of global warming pollution into the atmosphere. That is, of course, the whole reason for the campaign—and the reason that large carbon polluters are spending so much money on it.

The carbon polluters' propaganda campaign has had such success in paralyzing the political process in part because the news media have abandoned one of their traditional roles: that of refereeing important arguments in the public

employed in the coverage of differing political points of view based on subjective opinions. And conflict, after all, is often an inherently interesting factor in any news story—if it is legitimate. It is grossly inappropriate, however, to focus on bogus and artificial conflicts as a way of reporting on the current state of scientific knowledge concerning points that have been thoroughly investigated through the scientific process—especially when voluminous numbers of peer-reviewed articles result in a consensus view of what the science shows.

"What they've done is try to take scientific understanding and put it on the same level with political opinion."

MICHAEL OPPENHEIMER

domain. The decline of newspapers has precipitated layoffs of experienced reporters with the time and resources to investigate large scams like this one. As a result, the public is more vulnerable than in decades past to campaigns of deception organized by wealthy special interests.

With budgets shrinking, many news organizations have relied on shortcuts in covering disputes. Rather than devoting the time and resources necessary to investigate competing claims, they commonly use an "on-the-one-hand, on-the-other-hand" approach that can suggest a false symmetry between the merits of differing viewpoints.

This approach may be justifiable when

As distinguished climate scientist Michael Oppenheimer said in 1994: "What they've done is try to take scientific understanding and put it on the same level with political opinion. After all, if scientific understanding is the same as political opinion, then everybody's opinion is equally valid. There are no facts. And if there are no facts, there is no extra validity to acting on environmental problems than not acting."

For example, the existence of gravity and the roundness of the earth are well established, and if some group with a lot of money published a thousand pseudo-scientific papers expressing the view that the earth is actually flat and that gravity is completely misunderstood, it is hard to imagine that any serious journalist would give those views equal billing with the scientific consensus. Yet that is exactly what the news media in the United States and in many other countries have done for many years where global warming science is concerned. On the one hand is the global scientific consensus, and on the other—given equal weight—are the crackpot theories of industry-financed deniers.

An extensive study in 2004 of all newspaper articles about global warming written over 14 years in the *The New York Times*, *The Washington Post*, *Los*

Ford, General Motors, Shell Oil, Texaco, and the U.S. Chamber of Commerce formed the Global Climate Coalition (GCC). Thus began a massive disinformation campaign to deny the reality of global warming and to deny any connection between global warming and human activities. Following the onset of this campaign, press coverage changed to give equal billing to the synthetic pseudo-science purchased by the carbon polluters.

In the spring of 2009, documents surfaced (in a lawsuit) revealing that these carbon polluters had authorized and paid for an internal review of the

On the one hand is the global scientific consensus, and on the other—given equal weight—are the crackpot theories of industry-financed deniers.

Angeles Times, and *The Wall Street Journal* showed that 52.65 percent of the stories gave roughly equal billing to the scientific consensus and the views of deniers who claim that global warming has absolutely no connection to human activities. Significantly, the study's authors, Maxwell Boykoff and Jules Boykoff, found that during the first two years covered by their study, 1988–1989 (before the large carbon polluters organized their massive propaganda campaign), press coverage represented the scientific consensus fairly accurately.

In 1989, Amoco, the American Forest & Paper Association, the American Petroleum Institute, Chevron, Chrysler, Cyprus AMAX Minerals, Exxon,

scientific evidence in 1995, and that their own scientific and technical experts advised them that the consensus view was "well established and cannot be denied." Their own experts also told them that the contrarian theories of the deniers "do not offer convincing arguments against the conventional model of greenhouse gas emissions–induced climate change." Yet the recently released documents show that the "operating committee" of this coalition of carbon polluters forced the removal of that section of the report and continued to present to the public—including potential purchasers of their stocks—a view that their own censored scientific review had told them was not correct.

AS PART OF PRESIDENT GEORGE W. BUSH'S STAFF, PHILIP COONEY MADE INDUSTRY-FRIENDLY EDITS TO GOVERNMENT SCIENTIFIC REPORTS ON GLOBAL WARMING. HE NOW WORKS FOR EXXONMOBIL.

Instead, this coalition publicly distributed a scientific "backgrounder" to journalists and legislators throughout the world that claimed "the role of greenhouse gases in climate change is not well understood." In a later version, distributed in 1998, they changed the wording to focus uncertainty on whether the results of any warming would justify sharp cuts in emissions.

Normally, the best practice for publicly traded corporations is to promptly disclose important facts that have a material impact on the value of their stocks. Yet these companies concealed information created at the company's request about the single most important body of science relevant to the future value of their businesses. Since this occurred while the world was attempting to conclude a global treaty to limit the continued dumping of gaseous waste produced in the burning of oil and coal, they may have felt that, with so much money at stake, it was worth it for them to be less than candid.

By the middle of 2005, the national academies of science in the United States, the U.K., China, India, Russia, Brazil, France, Italy, Canada, Germany, and Japan had all formally endorsed the consensus view as stated by the IPCC. Yet the large carbon polluters continued their fraudulent campaign to convince the news media and the public throughout the world that the science was in dispute. Astonishingly, the news media continued, by and large, to give equal billing to the presentations of the industry-financed deniers.

The radical change in news coverage that followed the launch of the massive industry propaganda campaign led to a dramatic weakening of political support for measures to reduce global warming pollution. The peak of public concern was in the late 1980s, just before the propaganda campaign was launched. But as a result of this massive campaign to inculcate the news media and the public with a patently false view of the science and the seriousness of global warming, it is little wonder that public-opinion polls soon began showing the issue dropping further and further down the list of action priorities. This was the political result the polluters bought and paid for.

Beginning in January 2001, then-President George W. Bush appointed several of the deniers who were part of the industry disinformation campaign to key positions in his administration. One of them, Philip A. Cooney, who had led the disinformation program for the American Petroleum Institute, was put in charge of environmental policy in the White House. Cooney routinely censored the views of government scientists in official reports and substituted the point of view shared by oil and coal companies, thus making the executive branch an active participant for eight years in the effort to mislead the American people about the seriousness and urgency of global warming.

In a lengthy investigative report by Sharon Begley in the summer of 2007, *Newsweek* reported the results of this campaign: "Since the late 1980s, this well-coordinated, well-funded campaign by contrarian scientists, free-market think tanks, and industry has created a paralyzing fog of doubt around climate change."

There is, of course, a long list of fraudulent efforts to present public information that misleads people and markets about the true value of enterprises. Disgraced financier Bernie Madoff, for example, led investors in his hedge fund to believe that he was putting their money into stocks, when, in fact, he was operating the world's largest Ponzi scheme. A large number of banks convinced millions of people that so-called subprime mortgages were among the safest investments—even though the holders of those mortgages were not required to make down payments or to present the traditional financial evidence that they could make their regular monthly payments. When the awful truth about the real value of these mortgages came to light, it triggered the bursting of the housing bubble in the second half of 2007 and caused the worst financial crisis and global downturn since the Great Depression of the 1930s.

We now have several trillion dollars' worth of subprime carbon assets owned by individuals, pension funds, and other institutional investors in the form of companies whose value is artificially inflated by dishonest misrepresentations concerning the need to sharply curtail burning of carbon fuels in order to preserve a coherent civilization.

Leaving aside the environmental consequences, the financial risk of this massive fraudulent treatment of the science of global warming—financed in large part by shareholder money deployed to produce misleading representations of global warming science—is extremely high. When the truth is widely known and the appropriate actions to curtail emissions have begun, the oil and coal "bubbles" are likely to burst. The longer we wait, the bigger those bubbles will grow, and the bigger the bursting will be. Those who suffer financial harm will be put in a position not unlike the victims of Bernie Madoff, who trusted and relied on the information he gave them—to their everlasting regret.

The companies that combined to form the GCC have continued to make record profits, and the executives who were in charge when the deception was authorized have continued to receive large bonuses. (One of them received a compensation package of $400 million in his final year as CEO.)

In 2006, the Royal Society of London (the U.K.'s equivalent of the U.S. National Academy of Science) publicly asked ExxonMobil to stop misrepresenting the science on global warming and expressed "disappointment at the inaccurate and misleading view of the science of climate change" that the company continued to put before the public. The Royal Society also provided an analysis showing that ExxonMobil was giving millions of dollars to 39 groups "which misinformed the public about climate change...." Last year, company's own published report.

While the United States has been the primary target of this propaganda campaign, it is far from the only target. Throughout the world, the contrarian views of deniers who question the scientific consensus on global warming are given prominent display on a regular basis. News articles, columns, editorials, television documentaries, and advertisements have been appearing with regularity in almost every country that might have a role to play in forming a global consensus. And again, because the science of global warming has been relatively

ExxonMobil was giving millions of dollars to 39 groups "which misinformed the public about climate change...."

finally, the company was pressured to take action and announced that it would "discontinue contributions to several public-policy research groups whose position on climate change could divert attention from the important discussion" on how to produce energy without contributing to global warming. However, a scholar at the London School of Economics found in 2009 that notwithstanding ExxonMobil's pledge, it has continued this practice. "If the world's largest oil company wants to fund climate change denial then it should be up-front about it, and not tell people it has stopped," the scholar, Bob Ward, said. While ExxonMobil did cut funding to nine such disinformation outlets, it continued to fund more than two dozen other institutions engaged in climate denial, according to the

unfamiliar to most people and is more challenging to absorb than the information surrounding most issues, the deniers have had a very big impact in frustrating and delaying the world's efforts to reduce deadly pollution.

Ironically, the decline of newspapers has been accompanied by the emergence of new forms of media—especially on the Internet—that have helped to reinvigorate the political forces pushing for action against global warming.

Throughout the world, new grassroots organizations are using the World Wide Web to communicate the truth about the climate crisis and to organize in support of action to solve it before it's too late. In the United States, I founded one such organization—the Alliance for Climate

THE ALLIANCE FOR CLIMATE PROTECTION

Repower America's campaigns include TV spots that point out the benefits of a new energy plan and encourage citizen action.

The Alliance for Climate Protection, founded in 2006, is a nonprofit organization dedicated to changing the way people think about the climate crisis and catalyzing the solution.

Through its campaigns, the Alliance's message is that we can solve the climate crisis. But we have to engage the public and shift public opinion.

The Alliance attempts to depoliticize the climate crisis. It is led by a board of directors evenly balanced between Democrats and Republicans and works to build a broad base of support in finding solutions to climate change.

Based on my experience, I strongly believe that political leaders in both parties will continue to be timid in their approach to reducing global warming pollution until there is a genuine base of strong support for transforming the way we produce energy and taking the other steps that are necessary. All of the work of the Alliance is aimed at spreading the truth about the choice we now have to make.

The Alliance has organized a nationwide, grassroots, community-based, nonpartisan effort to mobilize support for these changes. Through different media, including television, radio, newspaper and magazine advertisements, emails, Internet campaigns, and live concerts, the Alliance provides information to the public to support solutions. Its major projects include Repower America (repoweramerica.org), We Can Solve It (wecansolveit.org), and This Is Reality (thisisreality.org). The Alliance also sponsored the Solutions Summits that brought together an extraordinary number of experts on 32 topics related to solutions to the climate crisis.

I have donated 100 percent of my earnings from *An Inconvenient Truth*—the movie and the book—to the Alliance. All of my earnings from this book are also going to the Alliance.

The Alliance's website is climateprotect.org.

Protection—which has presented television, radio, Internet, newspaper, and magazine advertisements aimed at building a critical mass of support for action against global warming. It has also mobilized more than two million people to work in their communities and within the framework of our national political system to encourage the adoption of speedy solutions to the climate crisis. Along with others, we focused during the 2008 presidential campaign on getting the candidates in both major parties to adopt positions in favor of taking action. Both major-party nominees, Barack Obama and John McCain, endorsed the adoption of new national laws to cap and reduce the emissions of global warming pollution.

The lesson we should take from looking at the way carbon polluters hijacked the political process on global warming is that grassroots activism is essential to building a base of support strong enough to overcome well-funded opposition. That is the political task at hand for anyone who wants to be part of the solution to the climate crisis. It is also important to hold self-interested corporate deniers of global warming accountable for any continuation of their efforts to intentionally undermine the integrity of the scientific process on which the world must be able to depend in order to solve the crisis. In addition, news media executives must adopt higher standards to insulate the integrity of their reporting from this determined and ongoing effort by large polluters to corrupt their mission.

PROTESTERS OUTSIDE AN EXXONMOBIL
SHAREHOLDERS MEETING, MAY 2006

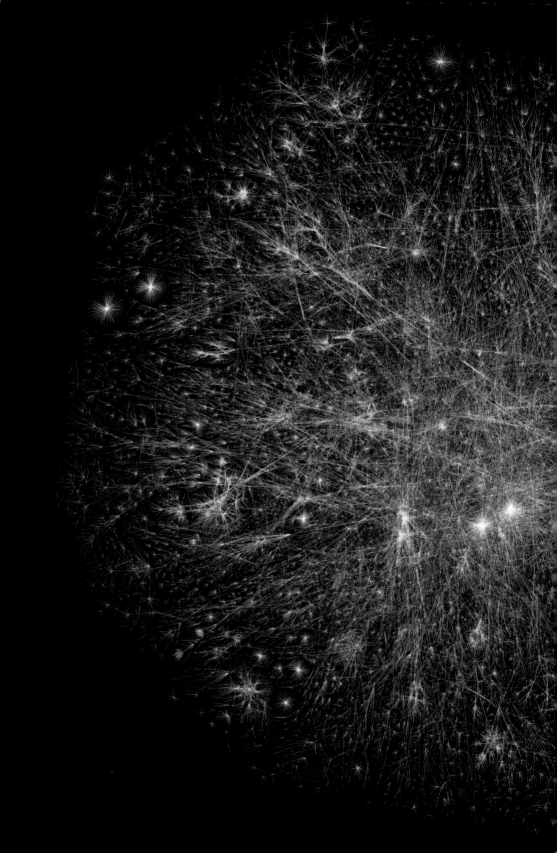

CHAPTER SEVENTEEN

THE POWER OF INFORMATION

THIS VISUALIZATION OF THE INTERNET MAPS
THE LOCATIONS OF AND CONNECTIONS AMONG
THE MILLIONS OF NETWORKS THAT MAKE UP THE
EARTH'S DIGITAL INFRASTRUCTURE.

The invention of modern computers, integrated circuits, and the Internet during the second half of the 20th century set the stage for a profound transformation in the role played by information technology in virtually every aspect of human civilization. This Information Revolution and the continuing rapid development of increasingly powerful information technologies have created new possibilities and new tools for solving the climate crisis.

Our ability as human beings to use information in order to make sophisticated mental models of the world around us is arguably the one capacity that most distinguishes us from all other living creatures. Now that we are faced with the unprecedented challenge of rapidly improving our understanding of the earth's ecological system and our place in it, it is time to focus on how we can make the fullest and most creative use of information technology to help us:

▸ Visualize the true nature of the climate crisis.
▸ Model the impact of current and future economic activity on the climate.
▸ Evaluate the potential solutions.
▸ Redesign our processes, technologies, and systems to reduce and eliminate global warming pollution.
▸ Mobilize widespread support for the transformation of civilization.
▸ Assist and support decision makers in their choice of new policies, laws, and treaties.
▸ Monitor our progress toward a solution.

First and foremost, the ability to visualize the true nature of the climate crisis is essential to developing a widely shared understanding of the task we now face.

Because of the way the human brain works, we have a limited ability to absorb data sequentially, bit by bit. In computer terms, we could be said to have a "low bit rate." In the 1940s, after much research, the U.S. telephone industry determined that seven numbers were the most that could be remembered easily by the average person. (And then they added four.) But an infant at the age of only a few weeks can recognize faces more accurately than the most powerful computers—until the last few years. Again, in computer terms, we have "high resolution." Luckily, advanced computers have an unparalleled capacity for integrating very large amounts of data into recognizable visual patterns that allow the human brain to comprehend the meaning of billions of bits of data simultaneously.

We recognized the face of our planet for the first time when the first picture of Earth taken by a person was snapped by astronaut Bill Anders on December 24, 1968, during the Apollo 8 mission—the first to leave Earth's orbit and travel around the Moon. That famous image of our world rising above the Moon's horizon, known as "Earthrise," brought about a powerful change in our shared understanding that we live on a beautiful, vulnerable

"EARTHRISE," TAKEN DECEMBER 24, 1968,
DURING THE APOLLO 8 MISSION.

blue sphere surrounded by the black vastness of space. The power of that image led to the first Earth Day, the passage of major environmental laws, the first global conference on the ecosphere, and the modern environmental movement. It has now been almost 40 years since the last picture of the earth was taken by a person far enough away to see the entire planet, during the last of the Apollo missions, Apollo 17.

Imagine what it would be like to have a live, high-quality color television image of Earth rotating in space, 24 hours a day. Imagine that the satellite carrying that television camera could somehow hover a million miles from the earth because so much of the incoming solar energy is absorbed by the oceans and is only slowly released into the atmosphere. Scientists studying global warming have long felt that the single most important stream of information needed to improve our understanding of the climate crisis was information about the difference between the energy coming into the earth's atmosphere compared with the energy going out again.

Imagine if the same satellite could calibrate and coordinate many of the other measurements made by satellites moving quickly in low earth orbit around our planet, and help us integrate all of that data in new ways.

Imagine what it would be like to have a live, high-quality color television image of the earth rotating in space, 24 hours a day.

directly between our planet and the sun, so that the full face of the earth was always illuminated.

Imagine if scientists could put special instruments on the same satellite that would measure, for the first time, the exact amount of energy coming to the earth from the sun and compare it in real time to the amount of energy radiated back into space from the earth itself.

The reason those two measurements are crucial is that the difference equals a precise calculation of global warming. The increasing temperature in the atmosphere of the earth is only an indirect measurement of the underlying problem,

The U.S. National Academy of Sciences (NAS) concluded a decade ago that we should build and launch such a satellite to a special orbit around the sun at a point in space known as the Lagrangian 1 (L1) point, where the gravity of the earth and the gravity of the sun are precisely balanced so that a satellite put there always stays exactly between the earth and the sun, providing a stable platform for continuous Earth observations of the planet as a whole. As a result of the NAS study, the U.S. Congress agreed and approved $250 million to build the satellite and launch it in 2001.

During the time when it was built, experts at the National Oceanic and Atmospheric Administration (NOAA) were deciding how to replace an older satellite that was already at the L1 point warning engineers about large solar storms that can disrupt cellular telephone communications, electricity distribution equipment, and other electronic equipment sensitive to large solar flares. From the L1 point, the light from these solar flares is visible 90 minutes before the plasma from the storm hits the planet. That's enough warning time to harden the sensitive electronic equipment and avoid expensive outages and repairs.

Since the older early warning satellite (called the Advanced Composition Explorer) was about to wear out, NOAA decided to put the replacement for it onto the satellite intended to measure global warming and provide a constant full color picture of the earth.

We still haven't seen that live TV image of the earth. The old satellite has not been replaced, because the Bush-Cheney administration canceled the launch within days of taking office after the inauguration on January 20, 2001, and forced NASA to put the satellite in storage. It is still there, nine years later, waiting to be launched. As a result, the older warning satellite could stop working at any moment; it is already two years past its predicted lifetime.

One of its key instruments is already dead;

DSCOVR'S VIEW OF THE EARTH

Current Earth-observing satellites are limited spatially and temporally. For example, low earth orbit (LEO) satellites are restricted by time of day and can take snapshots of only a sliver of the earth at high resolution (below left). The DSCOVR satellite, using three types of measuring instruments, can integrate its data with that of other satellites to create the only high-resolution, full-sphere view of the earth, 24 hours a day, providing the energy balance of the earth—for the first time. The composite at right shows the whole-Earth view of ozone levels that would be made possible by DSCOVR's instruments.

LEO SATELLITE DATA OF A FRACTION OF THE EARTH AT NOON

DSCOVR HIGH-RESOLUTION DATA FROM SUNRISE TO SUNSET

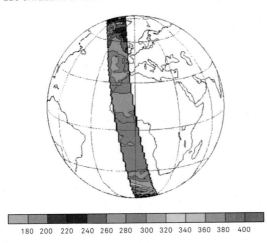

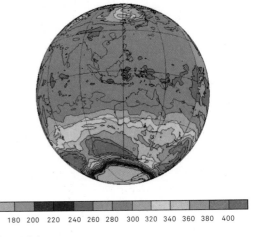

180 200 220 240 260 280 300 320 340 360 380 400

180 200 220 240 260 280 300 320 340 360 380 400

OZONE LEVEL (Dobson units)

TRIANA, OR THE DEEP SPACE CLIMATE OBSERVATORY (DSCOVR)

In 1998, NASA proposed launching the satellite Triana into space to deliver high-resolution images of the entire earth 24 hours a day. In 2000, the spacecraft was approved by Congress and scheduled for launch in 2001. Political resistance from the new administration and delaying tactics have kept the satellite—renamed DSCOVR (Deep Space Climate Observatory) in 2003—grounded and in storage, even though it has already been paid for by taxpayers.

Positioned at a unique point between the earth's orbit and the sun—the Lagrangian 1 point (L1), where the satellite remains directly between the earth and the sun at all times—the satellite will continually provide pole-to-pole images of the sunlit side of the earth, a viewpoint not possible from low earth orbit (LEO) or geosynchronous earth orbit (GEO) satellites.

DSCOVR is outfitted with a television camera and three instruments: EPIC (Earth Polychromatic Imaging Camera), a 10-channel spectroradiometer that provides color pictures; a three-channel radiometer that measures, among other things, albedo and ozone; and a plasma-magnetometer that measures magnetic fields and solar wind. Information from these three devices would be integrated with data from other satellites to provide a synoptic (simultaneous over-the-globe) view of the earth. These images would become an invaluable resource for remote sensing and climate modeling.

SUN

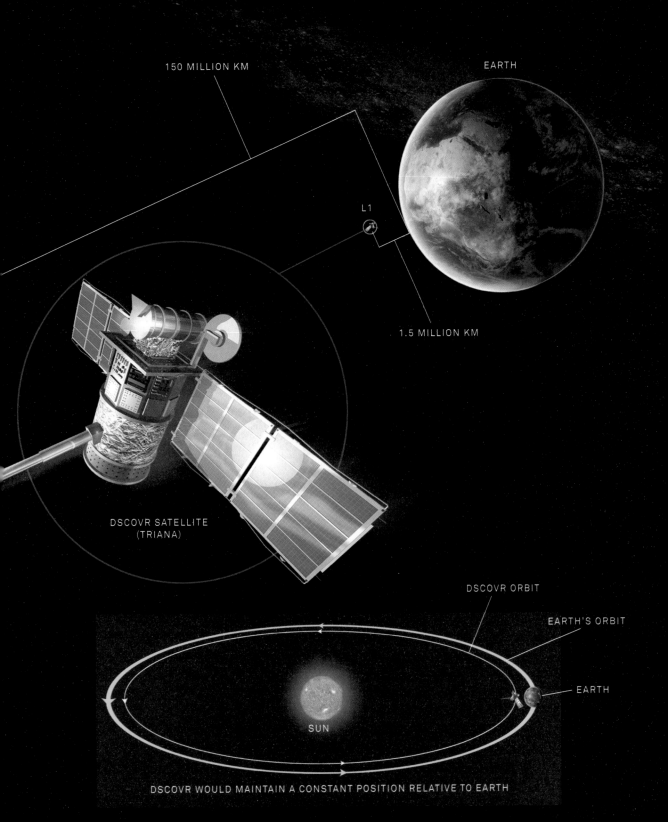

150 MILLION KM

EARTH

L1

1.5 MILLION KM

DSCOVR SATELLITE
(TRIANA)

DSCOVR ORBIT

EARTH'S ORBIT

EARTH

SUN

DSCOVR WOULD MAINTAIN A CONSTANT POSITION RELATIVE TO EARTH

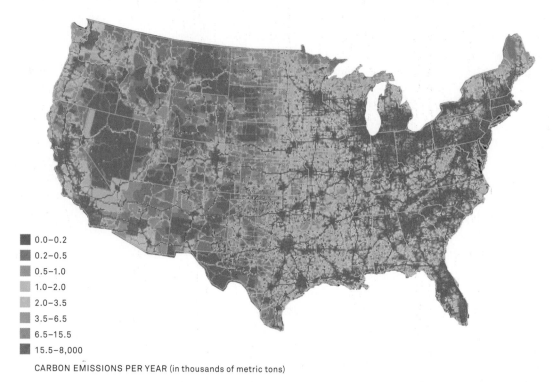

	CARBON EMISSIONS PER YEAR (in thousands of metric tons)
■	0.0–0.2
■	0.2–0.5
■	0.5–1.0
■	1.0–2.0
■	2.0–3.5
■	3.5–6.5
■	6.5–15.5
■	15.5–8,000

A CHANGING VIEW OF OUR CO$_2$ EMISSIONS

The Vulcan Project offers a newly detailed view of U.S. CO$_2$ emissions, giving researchers high-resolution visualizations of exactly how much CO$_2$ is coming from each location. The project is being expanded to cover all nations.

another now routinely fails during peaks of solar flares, when it is needed most. Several important global industries are at risk of being exposed to heavy losses by damage from solar flares.

President Obama and the leadership of the Congress, particularly senators Barbara Mikulski and Bill Nelson, have announced that they are in favor of launching this satellite, which used to be named Triana (after Rodrigo de Triana, the lookout on Columbus's flagship who first saw the New World on October 12, 1492). It was changed under the Bush administration to DSCOVR (Deep Space Climate Observatory) by advocates who hoped that a new name would lead the Bush administration to

feel some ownership of the project. Congressional opponents and bureaucrats inside NASA who want to use the money for other purposes have thus far paralyzed the efforts of the president and the Congressional leadership to follow through.

It's already built and paid for by the taxpayers. All the instruments work. The scientific team assembled a decade ago, before President Bush canceled its launch, has remained in place, volunteering their time under the able leadership of Dr. Francisco Valero of Scripps Institution of Oceanography. As NASA would say, all systems are go—except for the political system. Opponents of any action to solve the climate crisis have helped to block the launch, partly because they know how powerful the constant image of a beautiful rotating Earth on television and computer screens all over the world, in real time, would be in building support for the urgent solutions to the climate crisis.

Of course, computers can help us to visualize some aspects of the climate crisis even without the advantage of a sophisticated satellite at the L1 point. Google Earth, for example, organizes vast amounts of data geospatially in ways that make it easy to find detailed information about geography, buildings are located. There are literally thousands of similar examples.

The ability of advanced computers to integrate, process, and display complex data sets is bringing about a dramatic change in our capacity to understand phenomena that we could never have hoped to grasp in the past. The fastest and most powerful computers can sift through vast quantities of data, searching for the needles in the haystacks that are directly relevant to the questions of interest to us. They can form these data into patterns that are far

Far more information about the earth is collected in secret compared with the amount collected in the open by scientists.

botany, zoology, road networks, population, industry, agriculture, and many other facts that are specifically relevant to every location on the surface of the earth.

A new project developed by a team led by Dr. Kevin Gurney at Purdue University, named Vulcan (after the Roman god of fire), now makes it possible to visualize the amount of CO_2 emissions from any location in North America—and soon from everywhere in the world. Another new computer tool developed by the same Purdue team, Hestia (named after the Greek goddess of the hearth), makes it possible to see a thermal map of buildings, from which communities can gain a clear understanding of where the most inefficient

more accessible to our brains than endless bits of information strung together sequentially. They can artificially alter the scale and speed of the world to make images too large or too small for our comprehension just the right size for us to understand. Processes that are extremely slow can be sped up for our inspection, and processes that occur naturally in the blink of an eye can be slowed down for our convenient analysis.

Supercomputers are now used as tools for developing new designs for renewable energy technologies and advanced efficiency devices. For example, computational biology is now central to the exploration of new enzymes useful in processing cellulose; new, more efficient light-emitting

 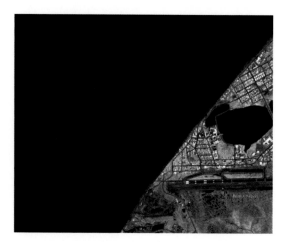

In July 2009, the U.S. released more than 1,000 previously secret satellite photos of the Arctic through the MEDEA program. The new images have a resolution of about one meter—at least 15 times more detailed than earlier images—and provide invaluable new information to climate researchers. This pair of satellite images reveals the dramatic variability of sea-ice breakup near Point Barrow, Alaska, in July 2006 (left) and July 2007 (right).

diodes; exotic solid-state cooling devices; new, more effective algae and other organisms for the production of biofuels; and new generations of photovoltaic cells and associated optics.

The largest and most powerful of these machines have led to the emergence of a completely new form of knowledge creation. In addition to deductive reasoning (formulate a theory and test it against the real world to see how it fits) and inductive reasoning (collect empirical facts and then try to integrate them into an overarching explanation), we now have a new approach that blends aspects of the first two. Computational science can create simulated realities—or "models"—within which experiments can be conducted. Although some scientific purists point out that computational science is conducted in ways that still sometimes fall short of the rigorous requirements of the traditional scientific method, the sheer power of this new

knowledge creation tool is extremely impressive.

The Central Intelligence Agency, when it was headed by Robert Gates (now the U.S. Secretary of Defense) under the first President Bush, approved a plan called MEDEA to enable environmental scientists to gain carefully controlled access to top-secret information relevant to the environment, collected by spy satellites and other secret information-gathering systems managed by the intelligence community for use in better understanding the climate crisis. This information is especially valuable because there is so much of it. Far more information about the earth is collected in secret than the amount collected in the open by scientists.

There are many examples of how the MEDEA program has revolutionized scientific understandings in a number of fields. When environmental scientists first gained access to this formerly secret trove of information, they were overwhelmed by the sheer volume of data. For example, the first measurements of ice cover on the North Polar ice cap came from MEDEA. And when scientists studying whales first gained access to the previously secret arrays of microphones placed on the bottom of the Atlantic Ocean to monitor submarines of the former Soviet Union, they collected more acoustic data on blue whales in one day than were contained in the entire previously published scientific literature. One scientist, Chris Clark, refers to this system as "the acoustic Hubble Telescope."

The Bush-Cheney administration also canceled this program, but the Obama administration has brought it back to life under the leadership of CIA Director Leon Panetta and Senate Intelligence Committee Chairman Dianne Feinstein.

The extraordinary information collected and processed by the MEDEA program will be invaluable to the intelligence community in monitoring and verifying a global climate agreement. Information that was once considered impossible to collect—such as data on tree cover and the carbon content of soils around the world, and

whether they are increasing or decreasing in particular locations—can now be collected through a combination of new information sensors and automatic data transmitters linked to satellite systems.

Computers can also help us make sense of more mundane but still relevant streams of data that are now often lumped together in ways that obscure our understanding. For example, the Google PowerMeter allows homeowners and business owners to monitor their electricity usage in real time. As smart meters are added to the distribution network, it will be possible to monitor the electricity usage of each appliance, television set, hot water heater, and light fixture. There are numerous similar applications and projects in development. All of them hold the promise of validating the old dictum "You manage what you measure."

Perhaps the most powerful application of information technology in actually reducing emissions of global warming pollution will come with the use of cheap semiconductors and "embedded systems" in machinery and in every aspect of industrial processes to eliminate wasteful energy use by

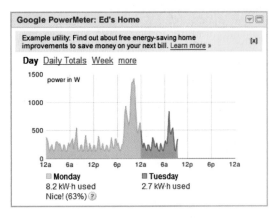

The new generation of smart meters being installed in the U.S. and abroad can communicate power usage in real time. Tools like Google PowerMeter (right) will let customers see that usage; research has shown that when people see their usage, they reduce it.

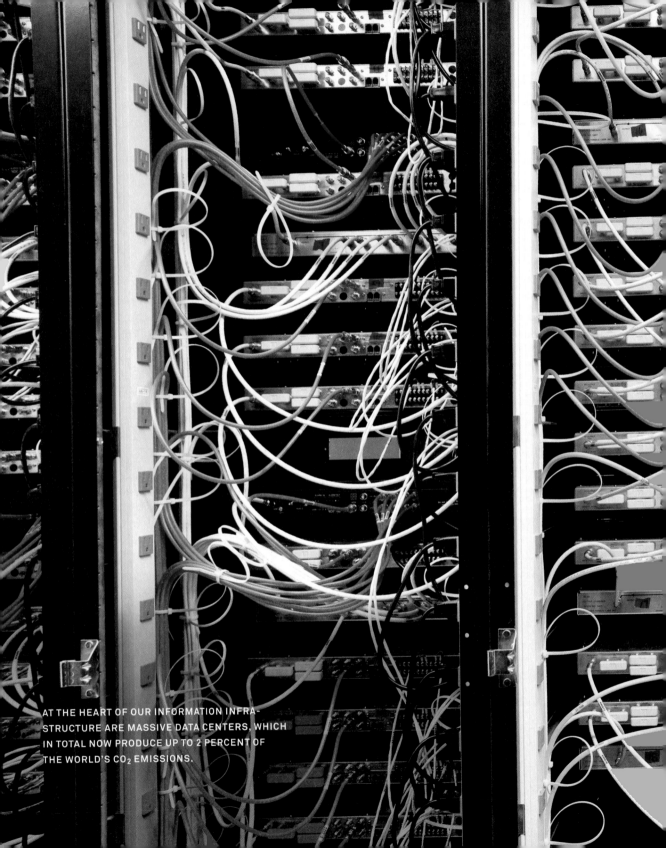

AT THE HEART OF OUR INFORMATION INFRA-
STRUCTURE ARE MASSIVE DATA CENTERS, WHICH
IN TOTAL NOW PRODUCE UP TO 2 PERCENT OF
THE WORLD'S CO_2 EMISSIONS.

optimizing energy efficiency. For example, industrial motors of all sizes often run at a constant rate even if the workload fluctuates; by constantly adjusting the engine rpms to the actual workload in real time, these devices can save large amounts of energy that would otherwise be wasted. The interactions between industrial pumps and their associated piping systems can also be automatically managed with such sensors to optimize fluid handling and minimize the energy needed.

Industrial process automation is not new, of course. Numerically controlled machines were first used during the 1950s when they were programmed with narrow paper tapes punched with regularly spaced holes. Then, in 1962, M.I.T. announced a dramatic improvement in the technology known as Automatically Programmed Tools. The APT— a "universal Numerical Control programming language"—allowed much more flexibility in programming and much greater linkage between different machines and stages in the manufacturing process. The development of CAD/CAM software (computer-aided design/computer-aided manufacture) led to the efficient connection between programs used for the design of products and the computers controlling the machinery that manufacture the same products from the output of the design computers. Now, the vast majority of new machine tools incorporate some version of a seamless integration between design and manufacturing.

As successive generations of software become more sophisticated, market leaders are integrating energy efficiency into the information technology that they use. By some measures, the volume of information flow on the Internet between machines and embedded systems now far exceeds the flow of information between human beings.

Gradually, the use of information technology to eliminate inefficient energy use and unnecessary global warming pollution is extending throughout industrial and business processes to encompass management of the supply chain and product delivery as well. For example, in the United States in 2007, approximately 25 percent of all business truck trips were made while the trucks were empty. By using information technology, some businesses are coordinating the movement of their truck fleets to take advantage of this unused delivery capacity in order to maximize efficiency, cut costs, and reduce CO_2 emissions in a cooperative manner. UPS has saved on fuel costs for its delivery trucks by remapping routes to eliminate as many left turns as possible—thus cutting down on wasteful idling at traffic lights while drivers wait for an opportunity to cut in front of the traffic flowing in the other direction.

Similarly, automatic light sensors measuring the real-time illumination from natural lighting can adjust the output of electric lights to save electricity during the hours when they need not be at full brightness, or need not be used at all. Heating, ventilation, and air-conditioning systems can be linked through inexpensive sensors to maximize the use of outside air coming through windows that actually open when outdoor temperatures make natural ventilation more appropriate and efficient during some hours of each day. Sensors can also alert building owners to gaps in insulation and leaks in ductwork hidden behind walls.

Much larger savings are being made by companies that utilize computer-aided analyses to redesign larger integrated processes in their entirety. "Whole system" redesign often leads to significant breakthroughs that eliminate unnecessary pollution and wasted energy and time. My favorite example is what happened when the leadership of Northern Telecom made a commitment in the late

MANY TOLL ROADS HAVE ADOPTED RADIO-FREQUENCY IDENTIFICATION SYSTEMS THAT LET DRIVERS PAY AUTOMATICALLY. REDUCING TRAFFIC AND IDLING HAS REDUCED TRAVEL TIME, GASOLINE USE, AND EMISSIONS.

1980s to be the first company to completely eliminate the use of harmful chlorofluorocarbons. As a Canadian company, Northern Telecom was alert to the significance of the 1987 Montreal Protocol, which mandated the phaseout of these chemicals. Since Northern Telecom used CFC-113 solvents to clean circuit boards, the company's engineers and scientists searched for appropriate substitutes. When they found none that fit their requirements, one of their engineers finally reframed the question: "How do these circuit boards get dirty in the first place?"

This conceptual breakthrough led to a redesign of the entire process to eliminate the exposure of the newly manufactured circuit boards to the contaminants that had to be removed at the end of the process. The resulting "no clean" process led to better and less costly circuit boards, while eliminating the chemicals that were simultaneously among the culprits that damaged the earth's stratospheric ozone layer and contributed powerfully to global warming. Then, Northern Telecom went one step further and shared its breakthrough with the rest of the industry, thus accelerating the phaseout of these chemicals by its competitors as well.

The company beat the phaseout deadline by an astonishing nine years and made money by doing so. The $1 million it invested in developing the new process earned it almost $4 million within the first three years. And the extra profits have continued to roll in every year since in the form of reduced manufacturing costs due to the elimination of a costly step in the process they used to use. Moreover, I saw firsthand the pride felt by Northern Telecom employees that they had joined together in a collaborative process aimed at a goal larger than simply adding to corporate profits—and the joy they felt when their commitment to doing the right thing and providing leadership for the entire industry ended up boosting their profits as well.

New service providers are now emerging to offer businesses software solutions for the task of redesigning all their processes to save money and emissions. One such company, Hara, uses the metaphor of "organizational metabolism," which encompasses a whole-system analysis of how energy and emissions can be minimized by making every function of the business more efficient as raw materials, energy, and labor are "metabolized" inside the company into products or services, waste, wages, and profits.

In many cases, the redesign of systems, processes, and products leads to the reduction of raw materials consumption through the substitution of innovation for matter. In the U.S. economy as a whole, the last half century has seen a tripling of total output in terms of the value of products manufactured and sold—without any increase in the total tonnage of that output. This effect, known as "dematerialization," is partly due to the increased prominence of information as a percentage of what is sold, but it is due also in significant measure to the clever and efficient redesign of products in ways that improve quality while reducing the physical matter used in the products.

Even as they begin to empower other industries and organizations with new tools for solving the climate crisis, information technology companies are faced with reducing their own global warming pollution. Today, information technology emissions of global warming pollution, principally CO_2, have risen to approximately 2 percent of global emissions. And over the next 10 years, IT emissions are expected to almost double. As a result, industry leaders are taking steps to redesign and reengineer their systems in an effort to become much more efficient and sharply reduce pollution.

Data centers, for example, consume large amounts of electricity. The growth of servers connected to the Internet occurred so quickly that it led to serious inefficiencies that are now being systematically addressed by many companies. Computers, printers, and other IT devices and equipment are all coming under closer scrutiny in the search for ways to cut energy costs and reduce emissions. The growth in online traffic, particularly Internet video, along with data storage requirements and disaster recovery needs, is combining with other factors to further accelerate the growth in data centers and the equipment that fills them. For example, the number of servers in the United States has tripled during the past 10 years. By installing more energy-efficient equipment, consolidating and rationalizing assets, making more use of "virtual servers," and optimizing power usage and cooling requirements, the IT industry is beginning to take responsibility for reducing its contribution to global warming.

The use of increasingly sophisticated information technology to cut global warming pollution in business and industry is also leading to increased awareness of inefficiencies that result from outdated laws and regulations. For example, the federal regulatory framework governing milk production and distribution makes it more profitable for dairies to transport milk and milk products thousands of miles to distributors on the other side of the country than to sell the same products in the regions where they are produced. This absurd, costly, and wasteful pattern continues only because the federal regulatory framework governing milk production and distribution mandates wastefulness.

In other forms of agricultural production, the combination of ground-based sensors and satellite systems are empowering farmers to adopt much more efficient approaches known as "precision farming" that optimize fertilizer mixes and applications to variations in soil types in different parts of the same fields.

Many businesses that begin the process of analyzing and reducing their CO_2 emissions discover early on that employee travel to and from meetings in other cities is one of their largest sources of avoidable CO_2 emissions. As a result, there has been a dramatic increase in telecommuting and the developing of more sophisticated tools, like Cisco's TelePresence, that simulate face-to-face conversations so well that much travel becomes unnecessary.

Transportation costs in the wider society are also being reduced in some cities throughout the world by the use of electronic toll-collection systems, which eliminate wasteful queuing at tollbooths; dynamic road signaling, which reduces traffic congestion and idling time; and congestion charges that more accurately allocate road use. Similarly, the growing use of radio-frequency identification (RFID) on products that travel through the wholesale and retail distribution chain is making inventory management far more efficient.

Privacy advocates have raised alarms about the inappropriate use, and inadvertent consequences, of ubiquitous object tracking to identify the real-time location of individuals who may not want their every move followed. Other privacy consequences of the new intensity of information technology throughout the economy deserve continuing scrutiny.

In addition, the growing importance of IT in the economy further underscores the need for equal access to computers and other important information tools for lower-income individuals and families. Just as telephones were once considered optional but then became essential for full and equal participation in modern society, computers are now approaching the threshold of indispensability.

Inevitably, the growing use of information technology in all sectors of the economy is producing losers as well as winners. Older business models created in a different information environment may no longer be competitive. Newspapers and magazines, for example, are now struggling in most parts of the world as digital electronic forms of communication become far more efficient. As the information revolution continues to gain momentum, similarly revolutionary transformations are taking place in many sectors of business and industry.

The newspaper example illustrates some of the risks that accompany the benefits of these transformations. Prior to facing such daunting electronic competition, newspapers earned revenue streams that allowed the employment of experienced journalists who could take a lot of time to investigate, analyze, and thoroughly report complex stories that regularly connected the public to valuable information about the operations of government and the functioning of societal institutions.

Indeed, during these early stages of the development of Internet-based journalism, the bulk of higher-quality stories still originates with newspapers. With newspaper revenues hemorrhaging, it is still not clear that electronic journalism will, on its own, discover a new standard model that generates sufficient revenue to rebuild a comparable

Hurricane Katrina
August 29, 2005

Photo: NOAA

DIGITAL PRESENTATION SOFTWARE MADE MY
GLOBAL WARMING PRESENTATION BETTER,
EASIER TO UPDATE, AND MORE EFFECTIVE.

cadre of experienced investigative journalists with enough time and resources to do the job formerly done by newspapers. Current TV, a cable and satellite news and information network that I co-founded with Joel Hyatt in 2002, along with its companion online network, Current.com, is devoting considerable resources to support a growing team of investigative journalists. This team, called Vanguard, travels the world reporting in-depth investigative stories that are then distributed electronically on Current TV and Current.com.

Even though investigative journalism is still rare in the new media environment, the new information technologies based on the Internet are connecting people to one another and to vast pools of information relevant to any societal challenge with which they wish to engage. Ultimately, the solutions to the climate crisis will necessitate much broader public engagement in the political process, of a kind that social networks and other Internet tools can make possible.

This book, for example, is being published with a companion website, ourchoicethebook.com, that includes a Solutions Wiki—a moderated forum for the constant improvement and elaboration of the suggested solutions for the climate crisis contained in the book. Many of the expert reviewers who have participated in the Solutions Summits that were so helpful in my efforts to identify the most effective ways to solve the climate crisis have agreed to help moderate ongoing discussions of new ideas, technologies, processes, and innovations that can be used to speed up the needed reductions in global warming pollution worldwide, the more rapid sequestration of pollutants already in the atmosphere, and the redesign of systems and processes that can accelerate the emergence of a low-carbon global civilization.

Internet-based tools also hold great promise for the reinvigoration of democratic self-governance and the mobilization of people at the grassroots level who want to be a part of the urgent task of solving the climate crisis. In my own experience, I remember what a difference it made when I transferred my slideshow about the crisis from film-based slides in Kodak carousels to computer graphics. The ease with which I could integrate new images and new scientific findings was dramatically improved, and soon after that transition, I began to notice a big change in the overall quality and effectiveness of the presentation.

I have now trained more than 3,000 people in dozens of countries to give updated slideshows on a continuing basis in the areas where they live. The Climate Project, managed by Jenny Clad, maintains communication with these presenters around the world primarily by means of the Internet and is able to share new slides with the appropriate explanations and caveats on a regular basis.

The Alliance for Climate Protection, managed by Maggie Fox, maintains contact weekly on the Internet with more than 1.2 million members in an effort to distribute high-quality information about the unfolding of this crisis and the political steps necessary to motivate policy makers to adopt the needed solutions.

Paul Hawken, author of *Blessed Unrest*, has found that more than "one—and maybe even two—million organizations working toward ecological sustainability and social justice" devoted to addressing the multiple challenges facing the earth's ecosystem have already been formed around the world—representing what he calls, "the largest social movement in all of human history." It is hard to imagine that this would have been possible without the new Internet-based tools that most of these groups rely on.

Sometimes, information can bring about change by itself, even in the absence of laws and regulations. For example, the relatively new legal requirement for the display of nutritional information on food labels (in the United States and some other countries) has put pressure on food manufacturers to improve their nutritional content and eliminate unhealthy ingredients like trans fat.

Similarly, after U.S. laws were modified to require the public disclosure of toxic air pollutants being released by industrial facilities, newspapers and electronic media in every city started listing the worst polluters. The public pressure that resulted from the disclosure of that information caused many companies to begin making changes in order to get off the worst-polluter list. Toxic emissions actually declined significantly, even in the absence of a new legal mandate. The information itself, once known to the public, forced the reduction. But the information has to be displayed prominently enough to raise public awareness.

This same principle applies to the display of information about wasteful energy use in homes and businesses. After California changed its laws to give utilities an incentive to reduce energy consumption (by allowing the utilities to share the savings along with their customers), Southern California Edison introduced to their customers a simple but compelling information display that sits inside the home: a glowing glass orb developed by Ambient Devices that changes color depending upon the amount of electricity consumption at any given time; when it turns bright red, that is

SEEING ENERGY USE IN REAL TIME

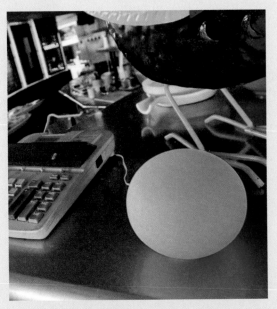 

Real-time energy-use monitors come in many shapes and types, including the glowing Ambient Orb (left) and more information-rich "smart meters" that display detailed cost and usage data. Both of these monitors change color to identify opportunities to save on energy costs.

A number of experts have concluded that when it comes to energy, if we can see it, we will try to save it. That concept may soon apply to our homes in the form of real-time energy monitoring.

With Google's PowerMeter, a project in beta testing at this writing, customers with smart power meters are able to see the peaks and valleys of their own power use, letting them adjust their habits and homes accordingly. Real-time power usage is displayed via the Web; turn off an appliance and the drop in energy use is quickly visible. Microsoft and other companies are testing similar online programs.

Smart meters visible inside the home—from portable models (seen above) to wall-mounted units to Web-based tools—are also under rapid development. Displaying electricity and gas usage, such meters are set to be required equipment in the United Kingdom by 2020.

The Ambient Orb offers similar feedback, though in a more physical design. The desktop sphere changes color based on changes in what it is monitoring.

Originally designed to react to input such as stock market indexes, the device was modified by a manager at Southern California Edison (SCE) to react to input from his utility. The manager then distributed 120 of the orbs to Edison customers, who were told how its changing colors would alert them when peak-period energy rates went into or out of effect. With the orbs as prompts, the utility's customers quickly trimmed their peak-period energy use by a large amount: 40 percent. SCE estimates that by installing smart metering, its customers will reduce greenhouse gas emissions by at least 365,000 metric tons per year.

Like the "Prius effect"—the easing off the gas pedal by drivers who see their fuel dip on the dashboard—smart meters link behavior to use. However, smart meter trials to date show that the real trigger is linking to a price cue. Customers in a smart meter pilot in California reduced their overall use 5 percent. Similar tests in Ireland and Illinois show energy savings of up to 12 percent.

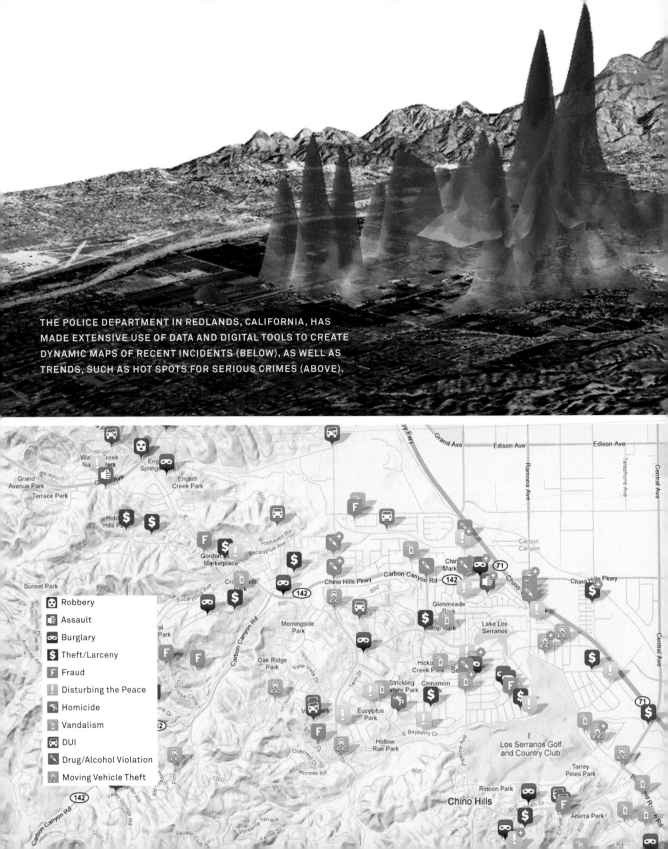

THE POLICE DEPARTMENT IN REDLANDS, CALIFORNIA, HAS MADE EXTENSIVE USE OF DATA AND DIGITAL TOOLS TO CREATE DYNAMIC MAPS OF RECENT INCIDENTS (BELOW), AS WELL AS TRENDS, SUCH AS HOT SPOTS FOR SERIOUS CRIMES (ABOVE).

Robbery
Assault
Burglary
Theft/Larceny
Fraud
Disturbing the Peace
Homicide
Vandalism
DUI
Drug/Alcohol Violation
Moving Vehicle Theft

the signal that electricity consumption is at very high levels. This is a very effective way to alert homeowners and business owners to the level of energy consumption at any given time, and signals them that there is an opportunity in the present moment for them to save money by modifying consumption levels.

Finally, information technology can be used in creative ways to assist and empower decision makers in governments, businesses, civic organizations, and other groups in their efforts to implement these solutions. When serving as vice president, I undertook a challenge called Reinventing Government, which was aimed at redesigning the departments, agencies, and processes of the U.S. Government to make them far more efficient. I learned a great deal from innovators in private business and in state and local governments that had succeeded in similar tasks.

One of the local government projects that most impressed me was a police strategy in New York City innovated by Jack Maple, a former transit policeman. Maple discovered the value of computerized statistics utilized in a group setting where all relevant decision makers are able to easily visualize the patterns revealed by the statistics. The data is organized geospatially, precinct by precinct, and displayed on a large screen visible at the same time to all participants—who share responsibility for quickly implementing solutions to the problems that are identified. When William Bratton was named to head the New York Police Department, he institutionalized this approach and sharply reduced crime rates in almost every category. Since then, this approach—colloquially known as CompStat—has been adopted in many other cities.

One of the most advanced systems has been developed by the police chief of Redlands, California, Jim Bueermann, who has now applied the CompStat technique to consolidate housing, recreation, and senior services into his police department and has enriched the process with ongoing social research in an effort to make his city a safer environment for children, seniors, and families. "We need to understand the nature and location of risk factors—in families, communities, schools, peer groups—and develop strategies to solve and prevent community problems. We are paid to get criminals, but our added value is found in the other, long-term approaches we are taking to make the community safer," Bueermann said, adding, "Mapping risk and protective factors lets us put tax dollars, and the resources of our community partners, where there is a high concentration of risk factors and strategically leverage the community's investment in public safety and problem prevention."

I believe this is one of the best examples of how information technology, properly used, can assist decision makers in their efforts to solve the climate crisis.

Heads of state, governors, other regional leaders, and mayors of cities and towns could benefit by developing computerized statistics on each of the major challenges they face and integrate them and display them visually for groups that include department heads and other stakeholders in a shared effort to discover what really works and what does not. The task confronting policy makers in the historic effort to solve the climate crisis will require the innovative use of every new tool available.

GOING FAR QUICKLY

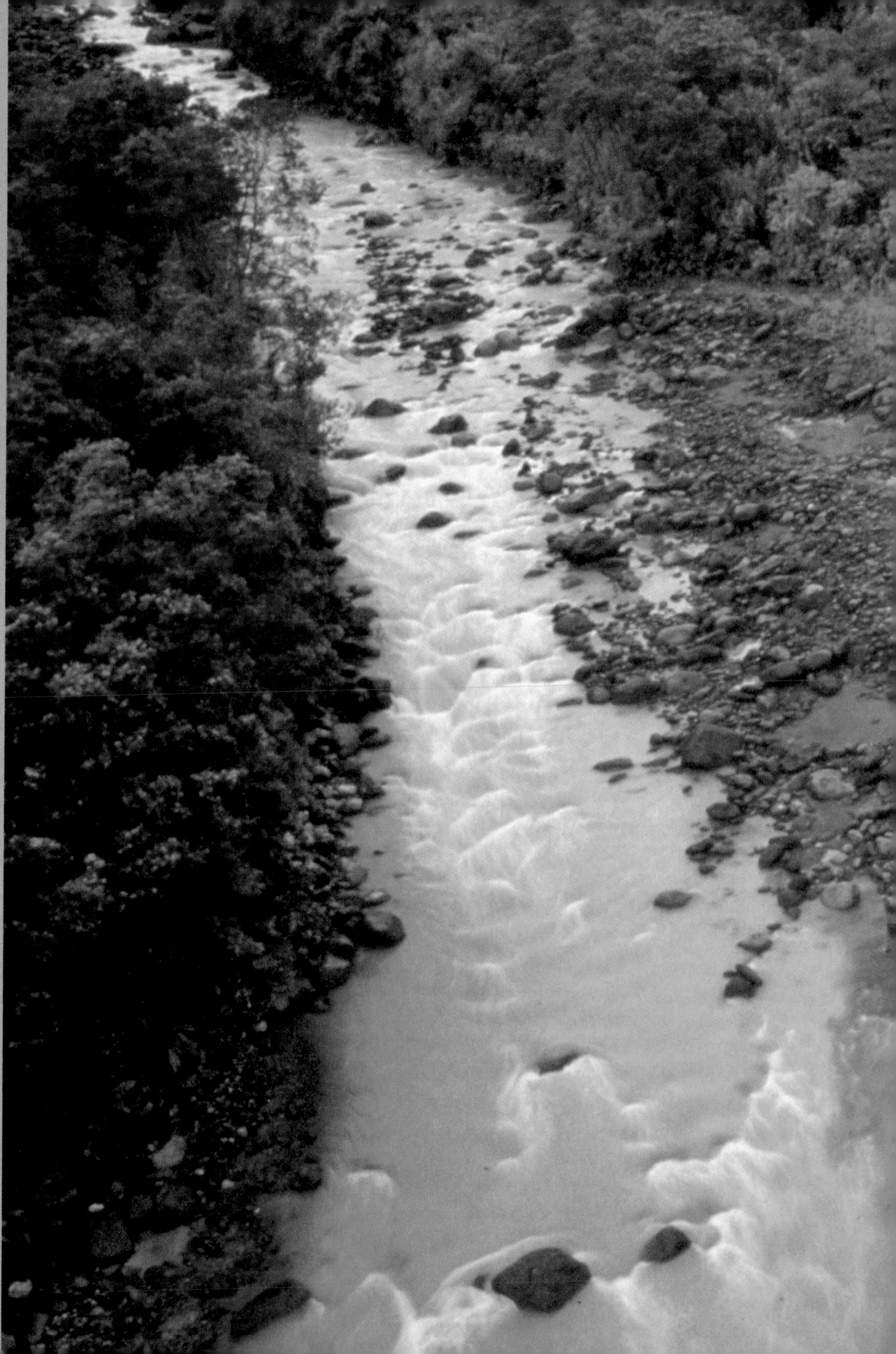

CHAPTER EIGHTEEN

OUR CHOICE

TWO STREAMS CONVERGE IN THE
COSTA RICAN JUNGLE.

Like anyone, I sometimes wish that I could go back and change some of the mistakes I made when I was younger. But none of us can travel backward in time to undo errors, no matter how clear their consequences make them over time.

Yet all of us, by virtue of the moral imagination we possess, can often glimpse the future conceived in the choices we make together today, even before that future is born into the lives of those who will live with the consequences of what we do or fail to do in the present.

Not too many years from now, a new generation will look back at us in this hour of choosing and ask one of two questions. Either they will ask, "What were you thinking? Didn't you see the entire North Polar ice cap melting before your eyes? Didn't you hear the warnings from the scientists? Were you distracted? Did you not care?"

Or they will ask instead, "How did you find the moral courage to rise up and solve a crisis so many said was impossible to solve?"

We must choose which of these questions we want to answer, and we must give our answer now—not in words but in actions.

The answer to the first question—what were you thinking?—is almost too painful to write:

"We argued among ourselves. We didn't want to believe that it really was happening. We waited too long. We couldn't imagine that it was even possible for human beings to cause such profound changes on a planetary scale. We didn't understand that so much could go so wrong so quickly.

"Somehow, we lost confidence in our own ability to reason together on the basis of the best evidence provided to us by our finest scientists. Even when the facts were already clear, we found it impossible to break free of the political paralysis induced in part by those who felt with passionate intensity that we should do nothing.

"Change, after all, is difficult. Please try to understand that it is nearly impossible to make big changes quickly on a global scale.

"We had so many other problems crying out for attention. We didn't see how the solutions to those problems were connected to the very same changes we should have made to save the integrity of the earth's ecosystem. I know this is of little comfort, but we did try. I'm sorry."

The second question—how did you solve it?—is the one I much prefer that we answer, and here is the answer I hope we can give:

"The turning point came in 2009. The year began well, with the inauguration of a new president in the United States, who immediately shifted priorities to focus on building the foundation for a new low-carbon economy. The resistance to these

PRESIDENT OBAMA SPOKE OF NEW ENERGY
TECHNOLOGY AS A PILLAR OF OUR FUTURE
ECONOMY AT NELLIS AIR FORCE BASE IN MAY 2009.

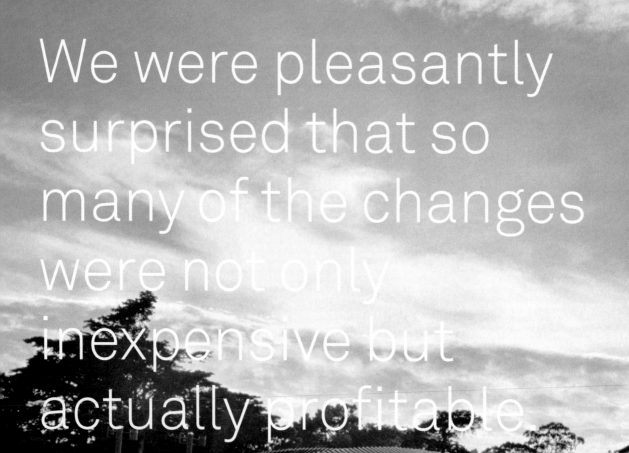

We were pleasantly surprised that so many of the changes were not only inexpensive but actually profitable.

HILLOCKS AND SKYLIGHTS ARE PART OF THE GREEN ROOF ATOP THE CALIFORNIA ACADEMY OF SCIENCES IN SAN FRANCISCO, THE FIRST LEED PLATINUM–CERTIFIED MUSEUM.

changes—especially by corporations that were making a lot of money from producing, selling, and burning coal, oil, and gas—was ferocious. There were times when I feared that we might not be capable of doing so much so quickly.

"Our public debates were confusing to most for a while. Our political culture was distorted by the fact that only those with large sums of money could present their points of view over and over again on television—which, in the early days of the Internet, was the dominant medium of communication. Advocates of the public interest—and of your future—were at a disadvantage.

"But the truth about the global emergency gained ground. The evidence presented by the scientists accumulated, slowly at first, but then something happened that is hard to describe. A few of the opponents of change changed themselves. One of them told me that his daughter asked him some questions he found difficult to answer in the same old way.

"Whatever happened, it made a powerful difference when these former opponents became passionate advocates for taking a new direction. The momentum shifted. One by one, others joined in what became a powerful consensus that we had to act, boldly and quickly. At the end of 2009, two things happened to decisively turn the tide. First, the United States of America passed legislation that changed the way business and civic leaders made plans for the future.

"By putting a price on the pollution that had been previously ignored, the United States established powerful incentives to begin the historic shift away from burning coal without trapping and sequestering the CO_2 it contains. The new incentives to shift our energy production from fossil fuels to solar, wind, and geothermal sources unleashed a wave of improvements in those technologies and others that avoided pollution.

"We were pleasantly surprised that so many of the changes were not only inexpensive but actually profitable. Many of our industries found ways to change wasteful practices and become more efficient. Farmers, ranchers, and large landowners began planting trees by the millions and changing the way crops were grown and animals were raised.

"The new policies caused home owners and business owners to insulate their buildings and change their roofs, lights, and windows. Architects, developers, and construction companies began to design and build zero-carbon structures. It became a point of pride. We thought of you, but we took pride in what we were doing for ourselves. Slowly at first, then powerfully, a sense of shared purpose arose among us. It lifted us and encouraged us to make bigger changes in industry, agriculture, transportation, even the design of our cities.

"Soon after the U.S. began to change, in December, all the nations of the world gathered in Copenhagen, Denmark, to negotiate a global treaty that many, even then, thought was impossible. But something happened there, too. It was only the beginning of the global change, and at the time, many felt that too little was done—just as many in the U.S. had felt the legislation passed on the eve of the treaty talks accomplished too little. But the new ground rules established in the treaty turned out to have had more power than we realized at the time. They shifted expectations, planning, thinking, and then—again, slowly at first—behavior.

"China, it turned out, had been quietly changing on its own. India was slower to begin the change, but in 2009, the combination of the United States and China—then the two largest global warming polluters on the planet—made all the difference. Europe, which was then in the early stages of its unification process, joined with Japan in supporting the U.S.-China proposal to place broad limits on emissions of CO_2 and the other five heat-trapping pollutants that were causing the crisis. Indeed, both Japan and Europe had provided crucial leadership for the world during the early years of the 21st century when the United States had abdicated its responsibility.

"Brazil and Indonesia—the two biggest deforesting nations—rallied the developing countries to join in an agreement that linked, for the first time, the reversal of deforestation in poor countries to the sharp reduction of industrial emissions in the wealthier countries.

"All over the world, as awareness of the climate crisis grew, people concerned about you found ways to put pressure on their leaders. Hundreds of thousands, then millions of grassroots networks emerged. Connecting with one another, largely online, these groups formed a powerful 'Grand Alliance' of non-governmental organizations (NGOs) that agreed to support a common agenda for systemic transformation of agriculture, manufacturing, business, and commerce. The new incentives for reducing carbon unleashed flows of resources to finance tree planting, organic farming, restoration of soil fertility, education reform—focusing on girls as well as boys—and new health care initiatives, with particular emphasis on child and maternal health care that lifted child survival rates and accelerated a shift worldwide toward smaller families.

"The agreement in Copenhagen, though criticized at the time for the timidity of what was approved there, turned out to be only the first step. Soon afterward, the treaty was strengthened. And then it was strengthened again. We should have known, I suppose, that this is the way it would happen. Back in 1987, when the sudden appearance of a large hole in the stratospheric ozone layer above

Antarctica alerted the world to the onset of the first crisis of the global atmosphere, the same thing happened. The first treaty, in Montreal, was also criticized as too weak. But then it was revisited in London shortly afterward and strengthened. Two years after that, in Copenhagen, ironically, the big changes were made. And finally, it worked! The stratospheric ozone layer, as you know, has almost fully recovered. As a result of what we did then, you no longer have to worry about that any more than you have to worry about global warming.

"None of this was clear at the end of 2009. Even had suffered so much from war and division for thousands of years.

"We should have also known that once we put our minds to it, we would succeed brilliantly at the development of new technologies that allowed much faster progress than anyone believed was possible in 2009. After all, it took the United States only eight years and two months to land men on the moon and return them safely—once we made up our minds to do it in 1961.

"We in the United States had forgotten how good we were at policies supporting sustained inno-

Although leadership came from many countries, once the United States finally awakened to its responsibilities, it reestablished the moral authority the world had come to expect from the U.S.

as the change had already begun, the arguments continued.

"Although leadership came from many countries, once the United States finally awakened to its responsibilities, it reestablished the moral authority the world had come to expect from the U.S. during the 40 years after World War II. It's worth remembering that no one could have imagined the success of the Marshall Plan, begun in 1947, to establish the foundation for such amazing prosperity and lasting peace on a continent that

vation in technology. We had done it before—lots of times. Our experience in building railroads coast to coast, a nationwide electric grid, the Interstate Highway System, the Manhattan Project, and the Internet turned out to be invaluable as we went to work to build a nationwide super grid that allowed the use of nearly unlimited amounts of electricity from solar, wind, and geothermal generators.

"The automobile industry—which had appeared to be dying in the U.S. in early 2009—retooled with government help, refocused, and led the

WORKERS MOVE SOLAR TROUGHS INTO
PLACE AT A CST PLANT IN SPAIN.

historic conversion to electric cars powered by the newly available electricity from renewable, no-carbon sources.

"Once the new direction was established, nations competed to develop better and cheaper technologies that accelerated the shift toward lower carbon emissions. Indeed, many of the new technologies you take for granted had their beginnings in this powerful new surge of innovation that was unleashed by the Great Transformation that began in 2009.

"And the entrepreneurial energy and innovative thinking that originated in parts of the developing world that were once thought of as hopeless made a bigger difference than any of us would have thought when the transformation began.

"The most important change that made this transformation possible is something that is hard to describe in words. Our way of thinking changed. The earth itself began to occupy our thoughts. Somehow, it became no longer acceptable to participate in activities that harmed the integrity of the global environment.

"Young people throughout the world led the way in this change in our thinking. Businesses that lagged behind lost customers and employees and then they, too, began to change.

"The ability we gained in 2010 to see Earth from space all the time also played a role in subtly but powerfully raising our constant awareness that we all share the same home. Once the argument about whether the crisis was real was over, we shifted into high gear, and the change became unstoppable.

"Once again, we should have known that we were capable of such changes in consciousness. But we had forgotten that earlier generations had ended slavery by first of all changing the way human beings thought about slavery. We had forgotten the days when the United States and the former Soviet Union had tens of thousands of nuclear warheads poised on the top of intercontinental ballistic missiles, ready to be launched at one another on a moment's notice. The dismantling of those horrific arsenals was preceded by a similarly powerful change in our thinking about the prospect of nuclear war.

"I know that we waited too long. I wish we had acted sooner than we did. But the problems we left you, as difficult as they are, are nothing compared with what would have happened if the Great Transformation had not succeeded. The outlook for your future is now bright. The wounds we inflicted on the atmosphere and the earth's ecological system are healing. In a few centuries, your descendants will thank you for staying this course when the healthy climate balance of our planet is fully restored. It might have been otherwise.

"The establishment of an ongoing global dialogue on the Internet focused on ever-more effective solutions to each of the problems that needed to be solved in order to make the transformation was one of the keys to success. The new tools for evaluating the best solutions and then developing and perfecting them cooperatively accelerated the progress far beyond what anyone believed was possible. Everyone, it seemed, was suddenly doing his or her part. And here, too, we should have known we were capable of coming together in support of such an urgent cause. During World War II, our parents did much the same with victory gardens and recycling and a thousand other changes in routine behavior that they themselves would not have imagined possible before that war began. That cohesion and common purpose was what gave them the confidence after World War II to establish the United Nations, the global trading system—the global institutions that, for all their faults, avoided depressions of the kind that were common before the latter half of the 20th century.

"We actually feared, in the early months of 2009, that another global depression might be in store for us after all. We also worried about war in the Middle East, partly because we were so dependent on oil—and most of the easily recoverable oil was concentrated in the Persian Gulf. The competition to ensure access to what were already dwindling supplies was leading to tensions that might well have triggered yet another, larger war.

"It seems ironic now that our commitment during the Great Transformation to a low-carbon economy was what restored economic prosperity and defused the tensions that gave rise to the fears of war. Once the world embarked on the journey to heal our world and save your future, tens of millions of new jobs—including whole new professions—began to emerge. The shift away from oil and coal lessened tensions in the Middle East. The new focus on reforestation and re-carbonization of the soil created millions more new jobs in developing countries. The spread of photovoltaic panels and small windmills triggered the transformation of economic activity in poor countries.

"The single biggest part of the solution to the crisis turned out to be the incredible changes in the efficiency with which we used energy. As soon as we secured a global agreement to put a price on the emissions of CO_2 and the other global warming pollutants, all of the business analyses of the future began to change.

"When the Copenhagen agreement was toughened in subsequent agreements, everyone understood the direction in which we were moving and saw the advantages of moving more quickly than competitors. Since efficiency gains were the easiest, most cost-effective, and most readily available options, there was a wholesale reexamination of business practices across the board.

"The global citizens' movement against corruption played an important role. We had not focused beforehand on how many of the mistakes we had made were actually driven by corrupt decision-making. Once that corruption began to be routinely exposed by citizen activists, a new ethic of public service emerged.

"It turned out that the political system shared one thing in common with the climate system: it too is what the scientists describe as 'nonlinear.' It can appear to move only at a snail's pace, but then it can cross a tipping point beyond which it suddenly moves with lightning speed. That's what happened in the decade beginning in 2010. Once the change began, it picked up speed. As the momentum grew, it became unstoppable. Then, once we began to think as a global civilization, we started solving other problems far more effectively.

"Young people brought incredible passion and commitment to meeting the challenge. Many who were in school when we decided to act were so inspired by the undertaking that they changed their courses of study to prepare for careers that allowed them to play meaningful roles in ensuring our success. Their idealism and seemingly inexhaustible energy was a renewable resource we had forgotten about.

"It recalled for me what happened the day the first man landed on the moon in 1969, when the systems engineers at Mission Control in Houston raised their arms and cheered with joy. Their average age that day was 26—which meant, of course, that when they first heard President John F. Kennedy's challenge in 1961, their average age was 18.

"In 2010, all over the world, an idealistic new generation took the initiative. They brought new ideas along with their passion, sharing both with one another. In nation after nation, they changed the political and cultural tone. They saw the world

ACTIVISTS IN BRITAIN WORK TO GET THEIR MESSAGE HEARD OUTSIDE THE EUROPEAN CLIMATE EXCHANGE IN LONDON (TOP) AND AT THE CAMP FOR CLIMATE ACTION IN BLACKHEATH, ENGLAND.

with fresh eyes. For them, it was simply unthink-able that we would fail. Because of them, we didn't.

"With God as our witness, we made mistakes. But then, when hope seemed to fade, we lifted our eyes to the Heavens and saw what we had to do.

"I ask only one thing of you in return for what we have done on your behalf: pass on to your chil-dren the courage and resolve to act boldly and wisely whenever the future is at risk. We found strength from the courage and heroism of those who came before us. Pass on to those who depend on your decisions now the unshakable confidence that collective will and vision are forces too potent to fail. You will be challenged, as we were. But I know that you will not fail those who come after you, as we did not fail you.

"Take care, because some of the new knowl-edge we gained in solving this crisis has led to pow-erful new tools that you must use with discretion, restraint, and wisdom.

"The account in the Bible, the Torah, and the Koran of Adam and Eve's temptation to eat fruit from the forbidden tree is a warning of the power of knowledge to destroy.

"Seen from the vantage point of space, our beautiful planet is the Garden of Eden for all of humankind, both living and yet to be born. In our time, without realizing it at first, we attained the knowledge and the power to destroy it. For us then and for you now—once again as in the ancient scriptures—the issue is whether we have the wis-dom and self-restraint needed to avoid that out-come, and whether we would use them.

"The choice is awesome and potentially eter-nal. It is in the hands of the present generation: a decision we cannot escape, and a choice to be mourned or celebrated through all the generations that follow."

The choice is awesome and potentially eternal. It is in the hands of the present generation.

SOURCES

For a complete list of sources for *Our Choice*, please visit www.ourchoicethebook.com.

INDEX

Note: Page numbers in *italics* refer to illustrations.

A

Abdullah, King, 348–49
abortion, 236, 240, 241
accountability, 192, 194
acid rain, 137, 344
activism, shared commitment to, 18, 327
Adorno, Theodor, 24
advertising:
  consumer, 310–11
  political, 352
  propaganda, 355, 358, *361*, 362–65, 366, 368
AEP (American Electric Power), 287–88
AES Corporation, 287
Africa:
  cities in, 224–25
  deforestation in, 41, 173–74
  genocide in, 240
  population growth in, 206, 232, 239
  soil quality in, 198, 203, 204, 206, 207, 215, 222
  super grid in, 278, 279
  tree planting in, 194, *195*
  water shortages in, 26, 232–33, 240
  wind energy in, 87
Agricultural Revolution, 200, 203, 204
agriculture:
  biomass energy from, 116, 122, 123
  burning, 40, 44
  clearing land for, 41, 172, 174
  crop rotation, 49, 123, 211, 220
  crop yields, 49, 198, 206, 209
  fertilizers, 48–49, 146, 204, 211–12, 215, 386
  genetically modified crops, 124–25, 209
  and global warming, 206, 208, 209
  in history, 200–204
  insect pests, 209
  methane released in, 39
  mixed cropping, 123
  nitrogen-intensive, 46, 48, 49
  no-till farming, 117, 203, 215
  organic farms, 211, 216, 217
  petroleum-intensive, 118, 120
  plowing, 200, 202–3, 221
  precision farming, 117, 386
  and soil, *see* soil
  taxpayer subsidies to, 212
air pollution:
  black carbon, 41–42, 44, 46, 47
  from burning coal, 30–31, 32, 41, 54, 55, 137–38, 261, 304, 318–19, 337
  carbon dioxide, 32, 36, 39, 44, 46, 147
  carbon monoxide, 46, 47
  costs of, 320, 327
  as externality, 320, 325
  and greenhouse gases, 34–35
  halocarbons, 44, 46–47
  methane, 36, 39, 41
  nitrous oxide, 49
  sources of, 32, 34–35, 46, 47
  VOCs in, 47, 49
albedo, 44, 45
algae blooms, 49, 212
algal fuels, 129–30, *131*
Alliance for Climate Protection, 366–68, 388
Allison, Graham, 164
Amazon rain forest, *170–71*, 174, 184, 190–91
Ambient Devices, 388, *389*, 391
American Council for an Energy-Efficient Economy (ACEEE), 254–55
ammonia, 49, 211–12
Anders, Bill, 4, 372
Anderson, Ray, 128
Antarctica:
  melting ice in, *13*, 232
  ozone hole over, 47
APT technology, 383
Arctic:
  melting ice in, 42, 44, 206, *380*
  soot blown into, 41
asteroids, 98
Atomic Energy Commission, 152, 161
Australia:
  drought in, 120, 208
  geothermal energy in, *104*, 109–10
  gold mining in, *328–29*
automobiles:
  electric, 71, 91, 129, 285, 286, 287–88, *287*
  and energy use, 260
  hybrids, 71, 118, 120, 284, 286, 290
  traffic jams, *22–23*, 266–67
  value of, 336
Ayres, Robert, 244

B

bagasse (sugarcane waste), 44, 118
Bank of America Tower, New York, 292, *293*
Ban Ki-moon, 240

Begley, Sharon, 365
Berns, Greg, 305
Bill and Melinda Gates Foundation, 222
Billion Tree Campaign, 194
biobutanol, 129
biochar, 132, 216, *218*, 219–20
biodiesel fuels, 114, 122, 175
biomass energy, 112–33
  as baseload power, 130
  cellulosic, 123–24, 132
  CHP generators, 130
  cost of, 130
  fermentation, 124, 129, 284
  first generation, 114, 116–18, 120–23, 125
  genetic modification in, 124–25
  second generation, 123–24, *125*, 127–29
  sources of, 114, 116–17, 122
  sustainability, 116
  thermochemical process in, 125
  third generation, 129–31
bird deaths, 84
black carbon, 40, 41–42, *42*, 44, 46, 47
Boeing Aircraft, 269
Bonneville Power Administration, 280
Borneo, orangutans in, 175, *180*, 184
Boykoff, Jules, 363
Boykoff, Maxwell, 363
brain:
  data patterns in, 379
  in decision-making, 304–5, 307
  information overload in, 314
  "low bit rate" of, 372
  prefrontal cortex, 305
  self-control center of, 310
Bratton, William, 391
Brazil:
  deforestation in, 41, 174, 194
  and global warming, 173
  sugarcane in, 40, 113, 117–18, 118–19
Brookhaven National Laboratory, 222
Bueermann, Jim, 391
Buffett, Warren, 331
Bunn, Matthew, 164
Bush, George H. W., 240, 344
Bush, George W., 348, 365, 378
butane, 46

C

CAD/CAM software, 383
California:
  energy efficiency in, 247, 249, 257, 265, 271, 388
  geothermal energy in, 99, 100
  solar energy in, 74
  wind energy in, 88, 89, 90, 346
Calvin cycle, 185
Canadian Forest Service, 188

cap and trade system, 342–45
carbon:
  in ecosystem, *210*
  price on, 148, 152, 172, 186–87, 191, 192, 222, 320, 327, 330–31, 336, 340
  tax on, 343, 345
carbon-based fuels, 21, 36
carbon capture and sequestration (CCS), 134–49
  obstacles to, 136, 145
  process of, *141*, 143
  research, 138, 146, 148
  retrofit for, 140
  safety issues, 136–38, 143–46, 147–48
  sites for, 134–35, 137, 143, 145, 148
carbon cycle:
  in forests, 185, 187, 188–89, *191*, 315
  in soil, 199, 203, 215, 222
carbon dioxide (CO_2):
  and air pollution, 32, 36, 39, 44, 46, 147
  from carbon-based fuels, 54, 55, 136
  fertilization effect, 189
  and global warming, 30–31, 32, 36, 39, 46, 172
  invisibility of, 320, 327
  and population growth, 227
  supercritical, 102–3, 140, 143
  transporting, 140–43
  visualization of, *378*, 379
carbon monoxide, 46, 47, *47*, 49
carbon tetrafluoride, 47
Carey, Al, 271
cars, *see* automobiles
Carter, Jimmy, 88, 156, 247, 340
Casten, Tom, 253, 255
cattle ranching, 38–39, 172, 174
cellulosic biomass, 123–24, 132
Central Intelligence Agency (CIA), 380
change:
  opponents of, 27
  resistance to, 394
  transformative, 312
Chartres Cathedral, 306–7
Chernobyl, 155, *155*
Chevron, 129
Chicago Climate Exchange (CCX), 222
China:
  African land owned by, 174, 194
  black carbon in, 42, *42*, 44
  coal in, 140, 147
  desertification in, 20
  economic growth of, 21, 346
  garbage in, 127
  and global warming, 173
  solar energy in, 74

super grid in, 21, 279
tree planting in, 189, 193, 194, 316–17
urbanization, 231–32
U.S. loans from, 21
wind energy in, 78, 86
Chinese National Greening Committee, 194
chlorofluorocarbons (CFCs), 44, 46, 384
choice:
    collective, 16, 315–16
    consequences of, 320, 324
    see also decision-making
CHP (combined heat and power), 130, 244, 253–55, 258, 261
Christner, Brent, 190
Chu, Stephen, 94
cities, 10–11, 224–25
    new designs for, 270
    urbanization, 231–32, 234–35, 236
Clad, Jenny, 388
Clark, Chris, 381
Clean Air Act, 262, 321, 343, 344, 345
climate crisis, 30–49
    causes of, 32, 46, 49, 172–73, 304
    communicating the urgency of, 315
    deniers of, 24, 27, 351, 354, 355, 358–66
    global scale of, 304
    global warming, *see* global warming
    impact of, 21, 27, 32, 36, 304
    media propaganda on, 354, 355, 358, 361, 362–65, 366, 368
    onset of, 304
    policy intervention required for, 325
    political obstacles to solution of, 350, 352, 355, 358–66
    rich-poor divide in, 350
Climate Project, 388
Clinton, Bill, 345
clouds:
    brown, 44
    seeding, 190
    sunlight blocked by, 70, 280
coal:
    and CCS, 136–38, 140, 141
    as energy source, 52, 54, 55, 57, 130
    fluidized bed combustion, 140
    mountaintop mining, 137, 139, 337
    and natural gas, 55
    pollution from burning of, 30–31, 32, 41, 54, 55, 137–38,

261, 304, 318–19, 337
    price of, 156, 336–37, 339
    scrubbers, 321, 344
    shifting away from, 21
    supplies of, 147
    and toxic waste, 166–67, 333, 339
    transport of, 53
cogeneration, *see* CHP
communication revolution, 236, 289, 366
compact fluorescent lightbulbs (CFLs), 258, 258–59, 264
compressed air energy storage (CAES), 283
CompStat, 391
computational biology, 379–80
concentrated solar thermal (CST), 62–63, 64, 65, 66–67, 68, 74, 91
Constitution, U.S., 301
consumer goods, 312–13
consumer spending, 310–12
Container Recycling Institute, 252
Cooney, Philip A., 364, 365
Copenhagen negotiations, 192, 222, 345
corn:
    burning waste of, 44
    ethanol from, 114, 116, 117, 118, 120–23
    and soil quality, 211
Current TV, 387

D
Daly, Herman, 331
dead zones, 49, 212
decision-making:
    brain system for, 304–5, 307
    collective, 16, 315–16
    in democracy, 327
    future impact of, 342
    information about, 312
    learned responses in, 303–4
    long-term, 304–7, 310, 331, 332, 335
    market incentives for, 342
    rational, 300–301, 355
    short-term, 331–32, 335, 346
Declaration of Independence, 300, 327
Deere, John, 200
deforestation, 41, 122, 172–84
    causes of, 172, 181, 188
    clear-cutting, 326
    CO_2 emissions from, 36, 116, 172
    economic impact of, 186–87
    global, 173, 174
    measurement of, 192, 194
    slash-and-burn, 172, 173
dematerialization, 385
democracy:

decision-making in, 327
    grassroots, 388
    and market capitalism, 301, 303, 327, 330
    propaganda vs., 27, 355, 358
    rule of reason in, 300, 303, 304
Demographic Transition, 229, 231, 236, 239
Denmark, wind energy in, 78, 80, 86
DESERTEC, 278, 279
direct exchange (DX) systems, 111
district heating, 260–61
drought, 120, 168, 188, 189, 208, 209, 231
DSCOVR, 375, 376–77, 378
Dust Bowl, 198, 203

E
earth, 5, 373, 375
    core of, 105
    crust of, 105, 143–44
    hot spots, 95, 95, 98
    human civilization's impact on, 32, 226
    reflectivity of, 44, 45, 374
    resiliency of, 325
    temperatures of, 359
Earth Day, 374
"Earthrise," 372, 373
Earth Summit (1992), 358
Easter Island, 298–99
economic crisis, 18, 21, 320, 330, 346, 350, 365
ecosystem, value of, 335, 339
Edison, Thomas, 55, 57, 258, 259, 261, 277
EGS (enhanced geothermal systems), 100–103, 105, 107–11
Eisenhower, Dwight D., 240, 258
electricity:
    AC and DC, 261, 277
    batteries, 280, 284–87, 295
    blackouts, 276, 277
    carbon footprint, 165
    and coal, 52, 55
    costs of, 156–57
    disadvantages of, 55
    growing demand for, 55, 157, 272–73
    intermittency problem of, 70–71, 91, 280
    micro-power, 288, 297
    net metering, 68–69, 288
    peak use of, 281, 285
    production of, 55, 59–60, 61, 69, 242–43
    pumped hydro, 282–83, 283
    renewable, 279, 288
    smart grids, 70, 91, 129, 273–97
    smart meters of, 381, 389, 391
    sources of, 55, 129, 165
    storage cost of, 55, 91, 283
    undersea transmission of, 86

Electric Power Research Institute, 279
electromagnetic spectrum, 325
Eliot, T. S., 136
energy efficiency, 244–71, 292
    CHP, 130, 244, 253–55, 258, 261
    district heating, 260–61
    fluid handling for, 240, 252
    homes retrofitted for, 262–65, 269
    leadership for, 271
    in lighting, 258–59, 383
    new processes for, 248, 250, 252, 270, 381, 383
    opportunities for, 244, 247, 249
    political obstacles to, 255, 257, 295, 342
    recycling for, 250–51, 252
    replacement of older systems, 249–50, 262, 269, 270
    waste heat loss, 257–58, 259, 261–62
    waste heat recaptured, 252–55, 259–61
    whole system redesign for, 383, 385
energy sources, 50–61
    electricity, 52, 55, 59–60, 61
    fossil fuels, 52–55, 57
    renewable, 21, 57, 58
    solar, 55, 56–57, 57
    wind, 55, 58, 78, 88
    wood, 52, 54
Enlightenment, 300
environmental movement, birth of, 374
ethanol, 114, 116, 117–18, 120–23, 124, 125, 129, 175
Europe:
    biomass energy in, 131–32
    geothermal energy in, 109
    super grid, 277–79
European Union Greenhouse Gas Emission Trading System (EU ETS), 222
extinction, 184, 186, 186, 304
ExxonMobil, 129, 145, 358, 360, 366, 368–69

F
family planning, 229, 236, 239, 240, 241
Federal Energy Regulatory Commission, 295, 296
Feinstein, Dianne, 381
fertilizers, 48–49, 146, 204, 211–12, 215, 386
Flannery, Tim, 219
Florida Power and Light, 90
food crops:
    and biodiversity, 184

biomass energy from, 116, 122, 123
food prices, 116, 120, 212
forest fires, 41, 42, 188, 189, 210
forests, 170–95
  beetle infestations, 188, 189, 189
  biodiversity of, 184, 186, 189, 190
  in carbon cycle, 185, 187, 188–89, 191, 315
  carbon density of, 184
  deforestation, see deforestation
  in hydrological cycle, 190–91
  primary, 181
  replanting, 189–90, 193, 194
  secondary growth, 189
  sustainability, 124, 192
  value of, 172, 186–87, 190
Forests Dialogue, 192
forest waste, biomass energy from, 114, 123–24, 130
fossil fuels:
  burning of, 32, 36, 36–37, 44, 49, 203–4, 216
  energy source, 52–55, 57
  materials recycled from, 114
Fox, Maggie, 388
France, nuclear power in, 156
Franklin, Benjamin, 301
Frito-Lay, 271
fungi, 199, 220–21
fusion, 166
FutureGen, 146

G
Galbraith, John Kenneth, 350
Galvin Electricity Initiative, 277
gasohol, 117
Gates, Robert, 380
GDP (gross domestic product), 324
General Electric, 88, 285
generators, 55, 61, 280, 281
genetic modification, 124–25, 209
geoengineering, 315
geothermal energy, 55, 92–111
  advantages of, 95
  characteristics of, 100
  and coproduction, 110
  in earth's core, 105
  EGS, 100–103, 105, 107–11
  geysers and hot springs, 95, 96–97, 98, 99, 100
  heat pumps, 110–11
  in the home, 111
  hydrothermal sites, 94, 98, 100
  methane reclaimed in, 110
  origin of, 98
  potential of, 94–95
  price of, 100, 107
  Ring of Fire, 95, 95
  seismic risk, 105, 107
Germany:

biomass energy in, 133
solar energy in, 74, 289
wind energy in, 78, 80
Gibbon, Edward, The History of the Decline and Fall of the Roman Empire, 300
Ginkel, Hans van, 204
Global Climate Coalition (GCC), 363, 365
GlobalSoilMap.net, 222
global warming:
  and agriculture, 206, 208, 209
  and black carbon, 40, 41–42, 44
  and carbon cycle, 188
  carbon dioxide, 32, 36, 39, 172
  carbon monoxide, 47
  causes of, 32, 46, 47, 49, 57
  consequences of, 304, 330–31
  deniers of, 24, 27, 354, 355, 358–66
  and drought, 168, 188, 208, 209
  and greenhouse gases, 34–35, 38, 47, 49
  halocarbons, 44, 46–47
  as market failure, 303
  measurement of, 374
  melting ice and permafrost, 13, 16–17, 39, 41, 42, 44, 206, 232, 359, 380
  methane, 36, 39, 41, 127
  nitrous oxide, 49
  science of, 54
  slowing, 41
  sunlight absorption in, 44
  and tree deaths, 188
  VOCs in, 47, 49
GNP (Gross National Product), 327
gold mining, 328–29
Google Earth, 379
Google PowerMeter, 381, 389
Gore, Al:
  and Chernobyl, 155
  and ethanol, 117, 120
  and geothermal energy at home, 111
  media presentations, 387, 388
  and nuclear weapons proliferation, 161
GPI (Genuine Progress Indicator), 324
grasslands, 223
  biomass energy from, 115, 123
  burning of, 41
  restoration of, 204
Great Depression, 301, 324
Great Transformation, 394–404
Green, Bruce, 94
greenhouse gases, 34–35
  halocarbons, 44, 46–47
  reduction of, 41, 129, 130, 246
Greenpeace, 174
ground source heat pumps, 265

Gurney, Kevin, 379

H
halocarbons, 44, 46–47
Hande, Harish, 289
happiness, 311
Hara, 385
Hartemink, Alfred, 222
Hatfield, Jerry L., 208
Hawken, Paul, 128, 388
Heisenberg, Werner, 325
herbicides, 203, 204
Herzog, Howard, 137
Himalayas:
  melting ice and snow in, 44
  pollution in, 42, 43
homes:
  built on wetlands, 340–41
  energy efficiency in, 262–65, 269, 292
  geothermal energy in, 111
  net metering in, 68–69
  on-site electricity generation in, 289, 290
  passive solar energy for, 75, 269, 292
  poorly constructed, 342
  television sets in, 269
  windmills for, 91
hot spots, 95, 95, 98
Hugo, Victor, 330
Hu Jintao, 194
Hurricane Katrina, 19, 350, 387
Hyatt, Joel, 387
hydroelectric dams, 55, 283
hydroelectric generators, 55
hydrofluorocarbons, 46
hydrological cycle, 190–91

I
Iceland, geothermals in, 92–93, 102
India:
  black carbon in, 42, 42, 43, 44
  coal in, 147
  economic growth, 346
  wind energy in, 78
Indonesia:
  black carbon in, 44
  deforestation in, 41, 122, 174–75, 176–79, 192
  geothermal energy in, 100
  and global warming, 173, 174–75
industrial process automation, 383
Industrial Revolution, 52
inelastic neutron scattering, 222
information:
  data centers, 382–83, 385
  flawed, 320, 325, 358, 363–65
  freedom of, 301
  Internet flow of, 383
  in mental models, 372
  political opinion vs., 362

from polls, 332
  power of, 372, 379–82, 388
information technology, 372, 379–82, 385–88, 391
Inglis, Bob, 343
Interface Flooring, 128
Intergovernmental Panel on Climate Change, 21, 24, 143, 204, 209, 358, 360, 364
intermittency problem, 70–71, 91, 280
International Energy Agency (IEA), 18, 244, 254
International Union of Forest Research Organizations (IUFRO), 188
Internet, 236, 274, 289, 366, 383, 385

J
Jackson, Lisa, 345
Jefferson, Thomas, 300, 327
jet airplanes, 268
job creation, 265, 320
jobs-loss fears, 350
Johnson Controls, 247

K
Kamkwamba, William, 87
Keeling, Charles David, 190
Keeling curve, 191
Kennedy, Robert F., 327
Kerguelen Islands, 98
Keynes, John Maynard, 324
Kidd, Steve, 156
Kyoto Protocol (1997), 47, 192

L
Laffer, Arthur, 343
Lagrangian 1 (L1) point, 374, 375, 376, 379
Lake Nyos, Cameroon, 144
Lal, Rattan, 203–4, 209, 215, 216
Landfill Rule (1996), 127
landfills, 126, 127–29, 252
land use:
  measurement of, 192
  pollution caused in, 32, 36, 204
  and soil quality, 200, 206
LaSalle, Timothy J., 215–16
laser-induced breakdown spectroscopy, 222
leachate, 127
LEED program, 245, 265, 293, 396–97
Lehmann, Johannes, 219
light emitting diodes (LEDs), 259
lightbulbs, 258, 258–59
lightning strikes, 188
lights, electric, 258–59, 383
Lincoln, Abraham, 332
Linden, Larry, 187
Linden, New Jersey, CCS plant in, 146
London, smog in (1952), 41, 41

long-term value creation, 346
Los Alamos National Laboratory, 222
Lovejoy, Tom, 186, 191, 206
Lovelock, James, 219
Lovins, Amory, 168, 249

M
Maathai, Wangari, 194
MacCracken, Mark, 292
Madoff, Bernie, 365
Mantria Industries, 220
Maple, Jack, 391
market capitalism, 301, 303, 327, 330
market deregulation, 327
market fundamentalism, 327, 330, 343
market system:
    inadequacy of, 303, 320, 325, 327, 330–32, 340, 346
    options for fixing, 342–44
Maslow, Abraham, 325
McCain, John, 368
McKinsey & Company, 246, 247, 257
meat production, 215
MEDEA program, 380–81
membrane separation, 252
mercury, 137, 328, 337
methane:
    and agriculture, 39
    capture and reuse of, 39, 110, 126, 127–28
    as energy source, 54–55
    in flaring excess fuel, 33
    in global warming, 36, 39, 41, 46, 127
    from livestock, 39
    and ozone, 39
methylbromide, 46
Mexico:
    cities in, 10–11
    flooding in, 25
    rainfall in, 231
micro-power, 288, 297
Mikulski, Barbara, 378
milk industry regulation, 385–86
M.I.T.:
    on APT technology, 383
    on CCS, 137, 138, 143, 145, 146, 147
    on geothermal energy, 94–95, 109
    on nuclear power, 152, 165
Montreal Protocol (1987), 44, 46, 384
Moore's Law, 58, 68
Moynihan, Daniel Patrick, 24
mycorrhizal fungi, 220, 221
Myers, Norman, 172, 184

N
National Academy of Sciences (NAS), 374

national accounts, 320, 324
National Oceanic and Atmospheric
    Administration (NOAA), 375
National Renewable Energy
    Laboratory, 124, 292
National Research Council, 147
natural gas:
    and coal, 55
    as energy source, 52, 54–55, 57
    flaring of, 33, 127
    methane as, 39, 54, 127–28
    pollution from burning of, 32
    production of, 55
    shifting away from, 21
Natural Resources Defense Council, 130
Nelson, Bill, 378
NETL (National Energy Technology
    Laboratory), 274
net metering, 68–69, 288
news, as entertainment, 21, 24, 27
newspapers, decline of, 362, 366, 386–87
Newton Falls Fine Paper, 252
New Zealand, geothermal energy
    in, 106–7
NGK, 285
nitrous oxide, 46, 49, 137
Nix, Gerald, 94
Nixon, Richard M., 156
Northern Telecom, 384–85
Nourai, Ali, 288
nuclear power, 55, 150–69
    advantages of, 152
    capacity factor, 152
    chain reaction, 152, 154–55, 155
    costs of, 155–56, 158, 161, 168
    decline of, 152, 155, 157–58, 168
    fuel available for, 164
    fusion, 166
    global, 159
    government subsidy of, 156, 168
    management challenges of, 158, 161
    pebble bed reactor, 160
    pressurized water reactors, 155
    radioactive waste, 153, 155, 156, 165–68
    reprocessing, 156, 164–65
    safety issues, 155–56, 157, 164–68
    and weapons, 161, 164
Nuclear Regulatory Commission, 166, 168
Nyerere, Julius K., 236

O
Oak Ridge National Laboratory
    (ORNL), 253, 254, 255
Obama, Barack, 21, 24, 241, 274, 345, 355, 368, 378, 395

oceans:
    solar energy absorbed by, 374
    transmission cables
        underneath, 86
oil:
    dependence on, 18, 21, 54, 116, 117, 129, 247, 249
    as energy source, 52, 54, 57, 130
    enhanced oil recovery, 145
    materials recycled from, 114
    Mideast embargoes (1970s), 88, 107, 109, 117, 156, 247, 283, 339–40
    offshore production, 19
    peak production, 18
    political control of, 339
    pollution from burning of, 32, 54
    prices of, 21, 54, 74, 88, 120, 156, 249, 270, 336–37, 339, 357
    shifting away from, 21
    transport of, 50–51
OPEC (Organization of Petroleum
    Exporting Countries), 339, 340
Oppenheimer, Michael, 362
Oreskes, Naomi, 360
organizational metabolism, 385
ozone formation, 49, 210
ozone layer, 39
    hole in, 46, 47
    protection of, 47

P
palm oil plantations, 172, 175, 194
Panetta, Leon, 381
parabolic trough mirrors, 62–63, 64
Persia, wind energy in history in, 86
Peterson, Pete, 21
Peterson Institute, 21
photosynthesis, 114, 185, 198, 216
photovoltaic (PV) cells, 55, 65, 67–70, 69, 72–73, 74, 288–89
phthalates, 252
Pickens, T. Boone, 18
plankton blooms, 315
plastic, recycling, 252
plug-in hybrid electric vehicles
    (PHEVs), 71, 91, 286
plutonium, 164
poison ivy, 210
police, digital tools for, 390, 391
Pollan, Michael, *The Omnivore's
    Dilemma*, 212
population growth, 18, 52, 224–41
    and death rates, 239
    impact of, 206, 226, 226–27, 231
    measurement of, 231
    and social unrest, 239–40
    stabilizing, 226, 228–29, 241
    urban, 231–32, 234–35, 236

    women and, 228–29, 236, 240–41
Power Towers, 56–57, 64, 66–67, 70
principal-agent problem, 342
Prius effect, 389
psychological subsidies, 312
pumped hydro, 282–83, 283
pyrolysis, 132

R
Radford, Bruce, 296
radiation:
    infrared, 44, 54
    solar, 44
radioactive waste, 153, 155, 156, 165–68
radio-frequency identification
    (RFID), 386
rainfall:
    and agriculture, 209
    Amazon, 191
    cloud formation, 190
    lack of, *see* drought
    in landfills, 127
    soot washed away by, 44
rain forests, 170–71, 180, 184, 190
Reagan, Ronald, 88, 247, 340, 343
reason, rule of, 300–301, 303, 304
Recycled Energy Development, 253
recycling, 114, 250–51, 252
"Reinventing Government," 391
Renewable Electricity Standard, 132
renewable energy sources, 21, 57, 58, 315
    price of, 346
    shift to, 340, 345
    see also biomass; geothermal;
        solar; wind
Renewable Fuel Standard, 118
Renewable Portfolio Standard, 109, 255
Revelle, Roger, 190
Ring of Fire, 95, 95
Roosevelt, Franklin Delano, 198
Rosenfeld, Art, 249
Royal Dutch Shell, 129, 145
Royal Society of London, 366

S
satellites:
    earth-observing, 375, 376–77
    low earth orbit, 375
    and solar power, 71
    see also DSCOVR; MEDEA
        program; Triana
Schlesinger, William H., 204, 216
Schwarzenegger, Arnold, 271
Schweiger, Larry, *Last Chance*, 187
Science, 132
Scientific Revolution, 52
security threats, 18, 21
selenium, 70
self-powered buildings, 261

sewage, management of, 49, 231
Shearer, David, 216
Shelton, Chris, 281
silicon, 69–70
Silverstein, Alison, 296
single-action bias, 314
Sixth Great Extinction, 184, *186*
smart grids, 70, 91, 129, 272–97
   and China, 21, 279
   DESERTEC, 278, 279
   elements of, 275, 290–91
   incentives for, 296–97
   obstacles to, 295–96, 297
   potential of, 279–81, 284–86,
     288–92
   transmission systems, 277, 297
Smith, Adam, *The Wealth of Nations,*
   300, 327
Smits, Willie, 175, *180*
smoking, 305, 355
soil, 196–223
   biochar in, 132, 216, *218,*
     219–20
   in carbon cycle, 199, 203,
     215, 222
   erosion of, *196*–97, 198, *201,*
     203, 204
   fertility of, 116, 132, 198,
     211–12, 215, 219, 222
   quality of, 184, *184,* 198, 200,
     203–4, 206, 209, 212,
     215, 221–22
   regeneration of, 220, 222
   temperature of, 189
   top layer of, 198
   in wetlands, 206
solar energy, 55, 57, 58, 62–75
   CST, 62–63, 64, 65, 66–67, 68,
     74, 91
   government support for, 74
   intermittency problem of,
     70–71, 280
   opponents of, 74
   passive, 75, 269, 292
   peak output, 70
   photovoltaic (PV) cells, 55,
     65, 67–70, *69,* 72–73, 74,
     288–89
   Power Towers, 56–57, 64,
     66–67, 70
   price of, 68, 71, 74, 80, 130
   smart grid, 70
   space-based, 71
   transmission lines, 91
Solomon, King, 27
"Solutions Summits," 12, 387–88
soot (black carbon), 40, 41–42, 44,
   46, 47
Southern California Edison (SCE),
   388, 389
Spain:
   solar energy in, 56–57, 65,
     66–67, 74

   wind energy in, 78, 80
Stamets, Paul, 220
steam generators, 55
Stern, Sir Nicholas, 360
Stirling engines, 64
stocks, holding time of, 331–32, *331*
stover (corn waste), 44, 123
Strategic Petroleum Reserve, 283
Streck, Charlotte, 204, *206*
sugarcane:
   biomass energy from, 117–18
   burning, *40,* 44
   harvesting, *112*–13, *118*–19
sulfur dioxide, 49, 137, 314, 344
sulfur hexafluoride, 46
sun:
   albedo (reflection) of, 44, *45*
   as energy source, *see* solar
     energy
   and fusion, 166
   solar radiation from, 44
sunlight:
   absorption of, 44
   blocking, 70, 280, 314
   photosynthesis, 114, 185
super grids, *see* smart grids
sustainability, 116, 132, 332, 346,
   388
syngas, 125

T
tar sands, 36–37
tectonic plates, 95
television, 269, 311
Teller, Edward, 314
Tennessee, toxic waste in, 138,
   333, 339
Tennessee Valley Authority (TVA),
   156, 157, 168
terrorism, threats of, 18, 21, 164
Tesla, Nikola, 55, 261, 277
tetrafluoroethane, 46
thermal heat maps, 379
Three Mile Island nuclear plant,
   150–51, 155, 157
tobacco industry, 355, 356
Tokyo Electric Power Company, 285
torrefaction, 132, 284
toxic assets, 330
toxic emissions, public disclosure
   of, 388
transportation:
   cars, *see* automobiles
   diesel engines, 44, 122
   electronic toll collection,
     384, 386
   energy efficiency in, 260,
     269–70
   fuel-based pollution, 32
   inefficient internal combustion
     engines, 129, 286
   jet engines, 268
   landfill gas for, 127–29

   liquid fuels, 128–30
   mass transit, 236, 266–67, 269
trees:
   biomass energy from, 114,
     123, 124
   carbon uptake in, 187
   fast-growing, 123
   planting, *14,* 189, 190, 193, 194,
     315, *316*–17
   see also forests
Triana, 376–77, 378
turbines, 59–60, 242–43
   gas, 280
   hydroelectric, 283
   hydrothermal, 94, 100, 102
   steam-driven, 154–55
   wind, 76–77, 79, 80, 81, 82–83
Tyndall, John, 54

U
United Kingdom:
   solar energy in, 347
   wind energy for, 85, 86
United Nations:
   Billion Tree Program, 194
   Environment Program, 194,
     262
   Food and Agriculture
     Organization, 173, 181
   International Conference
     on Population and
     Development, 228
   Population Fund, 232
United States:
   biomass energy in, 127–32
   budget deficits, 18, 21
   Constitution, 301
   Democratic Capitalism in, 327
   geothermal energy in, 103,
     108–9
   and global warming, 173
   political shifts in, 74
   Social Security, 239
   solar energy in, 74
   as superpower, 327
   wind energy in, 78, 88–91,
     89, 90
uranium, 152, 154, 156, 164, 167
urbanization, 231–32, 234–35, 236

V
Valero, Francisco, 378
Vatican, solar energy for, 72–73
Vattenfall, 146, 147, 246, 282
Venter, Craig, 129
VOCs (volatile organic compounds),
   46, 47, 49
volcanoes, 95
Vonnegut, Kurt, 12
Vulcan Project, 378, 379

W
Wales, wind energy in, 82–83

Ward, Bob, 366
waste, 302–3
   agricultural, 219
   biomass energy from, 114, *117,*
     123–24, 130
   consumption and, 311
   industrial, 137–38
   landfill, 126, 127–28
   radioactive, 153, 155, 156,
     165–68
   untreated, 231
water:
   aquifers, 144
   availability of, 209, 231,
     232–33, 240
   dead zones in, 49, 212
   for energy production, 167
   from ice melting, 339
   pollution of, 137, 212, 392–93
   in pumped hydro systems, 283
water vapor, 39, 49
wealth gap, 350
West, Ford B., 212
Westerling, A. L., 188
Westinghouse, George, 261
wind energy, 55, 58, 76–91, 346
   advantages of, 84
   costs of, 91, 130
   distributed energy approach, 91
   global production, 80
   limitations on, 91, 280
   opponents of, 86, 88
   turbines, 76–77, 79, 80, 81,
     82–83
wind farms, scalability of, 84
windmills:
   bird deaths from, 84
   building, 84, 87
   designs of, 80, 84
   maintenance costs, 84
   siting of, 85, 86
   small, 91
wind patterns, 78
women:
   education of, 228, 229, 236,
     240, 241
   and population growth, 228–
     29, 236, 240–41
wood:
   demand for, 172, 184, 194
   as energy source, 52, 54, *117,*
     130–32
   firewood, 44, 114
World Agroforestry Center, 194
World Resources Institute, 173, 181,
   192, 320, 335

Y
Yap, Arthur, 208
Yellowstone National Park, 96–97

Z
"zero landfill," 271

ACKNOWLEDGMENTS

I am grateful first of all to my wife, Tipper, for her support and encouragement—my children, Karenna, Kristin, Sarah, and Albert, my brother-in-law, Frank Hunger, and my entire family for their encouragement, assistance, and love.

A special thanks at the outset to my two outstanding research assistants, Brad Hall and Jordan Pietzsch, for doing a truly extraordinary job in tracking down, assessing, and verifying thousands of facts, figures, quotations, studies, and analyses that have been essential for this book. Under the direction of Kalee Kreider (who has been of invaluable assistance in all of my climate work), they also organized on my behalf most of the more than 30 "Solutions Summits" over the past three years. In this respect, they took over the excellent work of Elliot Tarloff, who organized the first wave of Solutions Summits and also did voluminous research. One of my two research assistants on *The Assault on Reason*, Elliot generously stayed another year, delaying his entrance to law school to begin the research for me on this project. Roy Neel has ably managed the entire staff here in Nashville to provide sustained support for all aspects of this project, even as the staff continued to handle the rest of their ongoing work. Beth Prichard Alpert has made the trains run on time and has coordinated all of the telephone calls and meetings for this project over the past few years. Her role as my Deputy Chief of Staff is indispensable. Conor Grew has also been indispensable and indefatigable in helping me in multiple ways—especially on the road. Lisa Berg and Patrick Hamilton have also played important supportive roles, as have Elizabeth Spencer, Bill Huskey, Anna Katherine Owen, and every other member of the staff. And a special thanks to Dwayne Kemp for his culinary mastery—including on most Saturdays and Sundays as the work on this project intensified over the past year.

I am especially grateful to the many distinguished scientists and engineers who participated in the Solutions Summits and who, in most cases, continued to remain involved in the project by sending new material, research findings, papers in the final stages of review prior to publication, and answers to questions that came up after the sessions in which they participated. Their insights and explanations really make up the heart of this book. Before listing those who took part, I want to single out those who undertook a complete review of the manuscript after it was finished and pored over all 416 pages to ensure they reflected the best scientific and engineering understandings available. Most of them also participated in one or more of the Solutions Summits. Professor Rosina Bierbaum, Dean of the School of Natural Resources and Environment at the University of Michigan, assembled and led this team, which included Prof. Jim McCarthy, Alexander Agassiz Professor of Biological Oceanography at Harvard University; Dr. Henry Kelly, President of the Federation of American Scientists; Dr. Mike MacCracken, Chief Scientist for Climate Change Programs at the Climate Institute; and Prof. Henry Pollack, Professor of Geophysics at the University of Michigan. Others who conducted detailed reviews of one or more chapters included Prof. Rattan Lal, Professor of Soil Science at The Ohio State University; Prof. V. Ramanathan, Victor Alderson Professor of Applied Ocean Sciences and Distinguished Professor of Climate and Atmospheric Sciences at the Scripps Institution of Oceanography, University of California, San Diego; Dr. Amory Lovins, Cofounder, Chairman, and Chief Scientist of the Rocky Mountain Institute; Prof. Dan Schrag, Professor of Earth & Planetary Sciences at Harvard University and Director of the Harvard University Center for the Environment; Prof. Mike McElroy, Gilbert Butler Professor of Environmental Studies at Harvard University; Prof. Matthew Bunn, Associate Professor of Public Policy at Harvard University's John F. Kennedy School of Government; Dr. Joe Stiglitz, Nobel Laureate and University Professor, Columbia University; Prof. Laura Tyson, S.K. and Angela Chan Chair in Global Management at the Haas School of Business, University of California, Berkeley; Alison Silverstein, Former Senior Energy Policy Advisor at the Federal Energy Regulatory Commission; Ross Gelbspan, author and Pulitzer Prize–winning journalist; Dr. Drew Shindell, ozone specialist and climatologist at the NASA Goddard Institute for Space Studies; Dr. Gavin Schmidt, climate modeler at the NASA Goddard Institute for Space Studies; Dr. Joe Romm, editor of Climate Progress and a Senior Fellow at the Center for American Progress; Prof. Jeff Tester, Croll Professor of Sustainable Energy Systems at Cornell University; Dr. Howard Herzog, Principal Research Engineer, MIT Laboratory for Energy and the Environment; Tom Casten, Chairman of Recycled Energy Development; Prof. Greg Berns, Distinguished Chair of Neuroeconomics and Director of the Center for Neuropolicy at Emory University; Prof. Leon Fuerth, Former National Security Adviser to Vice President Al Gore and Research Professor of International Affairs at George Washington University; Dr. Thomas Lovejoy, Heinz Center Biodiversity Chair; Prof. Hans Rosling, Professor of International Health at Karolinska Institutet and Director of the Gapminder Foundation; and several of my partners and associates at Generation Investment Management: Mark Ferguson, Colin le Duc, Peter Knight, Lila Preston, Duncan Austin, and Nicholas Kukrika.

The scientists and experts who contributed greatly to this project—and to whom I am extremely grateful—include Gail Achterman, Director of the Institute for Natural Resources at Oregon State University; Justin Adams, Business Unit Leader for Venturing at BP Alternative Energy; Brent Alderfer, Executive Vice President, Iberdrola Renewable Energies USA; Dr. Paul Alivisatos, Professor of Chemistry, Materials Science and Nanotechnology at the University of California, Berkeley; J. Norman Allen, President and CEO of CCS Materials, Inc.; Dr. Graham Allison, Director of the Belfer Center for Science and International Affairs and Douglas Dillon Professor of Government at Harvard's John F. Kennedy School of Government; Dr. Alan Andreasen, Professor of Marketing at the McDonough School of Business at Georgetown University; Duncan Austin, Director, Generation Investment Management; Ricardo Bayon, Partner and Cofounder of EKO Asset Management Partners; Dr. Sally Benson, Director of the Global Climate and Energy Project at Stanford University; Dr. Greg Berns; Scott Bernstein, President of the Center for Neighborhood Technology; Stan Bernstein, Senior Policy Adviser at the United Nations Population Fund; Dr. Roger Bezdek, President, Management Information Services Inc; Dr. Rosina Bierbaum; Tom Blees, author; David Blood, Cofounder and Senior Partner, Generation Investment Management; Hon. Sherwood L. Boehlert, former member of the U.S. House of Representatives; Dr. James Boyd, Senior Fellow at Resources for the Future; Wes Boyd, Cofounder of MoveOn; Dr. Peter Brewer, Senior Scientist, Monterey Bay Aquarium Research Institute; Carol Browner, Assistant to the President of the United States for Energy and Climate Change; William Bumpers, Head of the Global Climate Group at Baker Botts LLP; Dr. Matthew Bunn; Brett Caine, Senior Vice President and General Manager of the Online Services Division at Citrix; James Caldwell, Assistant General

Manager for Environmental Affairs at the Los Angeles Department of Water and Power; David Calley, Founder, Chairman and President of Southwest Windpower, Inc.; Andres Carvallo, Chief Information Officer, Austin Energy; Tom Casten; Ed Cazalet, Vice President and Cofounder of MegaWatt Storage Farms, Inc.; Andrew Chang, Senior Energy Analyst at the Alliance for Climate Protection; Robin Chase, Founder and CEO of GoLoco; Amit Chatterjee, Founder and CEO of Hara Software, Inc.; Nathan Cheng, Director at Johnson Controls; Dr. Ellen Chesler, Director of the Eleanor Roosevelt Initiative on Women and Public Policy at Hunter College of the City University of New York; Yet-Ming Chiang, Kyocera Professor in the Department of Materials Science and Engineering at the Massachusetts Institute of Technology; Dr. Robert Cialdini, Regents' Professor of Psychology and Marketing at Arizona State University; Dr. Chris Clark, I.P. Johnson Director of the Bioacoustics Research Program at Cornell University; Craig Collar, Senior Manager, Energy Resource Development, Snohomish County Public Utility District No. 1; Craig Cornelius, Principal at Hudson Clean Energy Partners; Dr. Robert Correll, Vice President of Programs at the Heinz Center; Peter Corsell, President and Chief Executive Officer, GridPoint; Dr. Pedro Moura-Costa, President of EcoSecurities; Peter Darbee, Chairman, CEO, and President of the PG&E Corporation; Dr. Stefano DellaVigna, Associate Professor of Economics, University of California, Berkeley; John Doerr, Partner, Kleiner Perkins Caufield & Byers; Dr. David Eaglesham, Vice President of Technology at First Solar; Carter Eskew, Founding Partner and Managing Director of the Glover Park Group; Mark Ferguson, Chief Investment Officer, Generation Investment Management; Maggie Fox, CEO and President of the Alliance for Climate Protection; Dr. Peter Fox-Penner, Principal at The Brattle Group; Joel Freehling, Manager of Triple Bottom Line Innovations at ShoreBank; Dr. Stephan Freyer, Senior Research Manager in the Chemicals Research and Engineering Division at BASF; Leon Fuerth; Dr. Yang Fuqiang, Vice President of the Energy Foundation; John Gage, Partner, Kleiner Perkins Caufield & Byers; Dr. Kelly Sims Gallagher, Associate Professor of Energy and Environmental Policy at Tufts University; Ross Gelbspan; Julius Genachowski, Chairman of the Federal Communications Commission; Paul Gipe, Wind Energy Author and Analyst; Paul Gorman, Founder and Executive Director

of the National Religious Partnership for the Environment; Kevin Grandia, Operations Manager at DeSmogBlog; Dr. Martin Green, Scientia Professor at the University of New South Wales; Dr. Kevin Gurney, Assistant Professor in the Department of Earth and Atmospheric Sciences and the Department of Agronomy at Purdue University; Dr. James Hansen, Director, NASA Goddard Institute for Space Studies; Dr. Volker Hartkopf, Professor of Architecture and Director of the Center for Building Performance and Diagnostics at Carnegie Mellon University; Professor Syed Hasnain, Chairman of the Glacier and Climate Change Commission, The Government of Sikkim; Dave Hawkins, Lead Industry Relations Representative in the External Affairs Division at the California ISO; David Hayes, Deputy Secretary of the Interior; Dennis Hayes, President and CEO of the Bullitt Foundation; Tim Healy, Cofounder, CEO, and Chairman of EnerNOC, Inc.; Dr. Stefan Heck, Director at McKinsey & Company, head of the McKinsey Global Cleantech Practice and the North American Climate Change Special Initiative; Ben Heineman, Senior Counsel at WilmerHale, Senior Fellow at the Belfer Center for Science and International Affairs at Harvard University's Kennedy School of Government; Dr. Howard Herzog; Michael Heyek, Senior Vice President, Transmission, for American Electric Power; Dr. William Hogan, Raymond Plank Professor of Global Energy Policy at Harvard University's Kennedy School of Government; James Hoggan, President of James Hoggan & Associates and Founder of DeSmogBlog; Dr. John Holdren, Assistant to the President for Science and Technology, Director of the Office of Science and Technology Policy; Dr. Wen Hsieh, Partner, Kleiner Perkins Caufield & Byers; Chi-Hua Chien, Partner, Kleiner Perkins Caufield & Byers; Emily Humphreys, Fellow at the Alliance for Climate Protection; Dr. Meg Jacobs, Associate Professor of History at the Massachusetts Institute of Technology; Dr. Mark Jacobson, Professor of Civil & Environmental Engineering at Stanford University; Michael Jones, Chief Technology Advocate at Google; Van Jones, Founder of Green for All; Dr. Bill Joy, Partner, Kleiner Perkins Caufield & Byers; Thomas Kalil, Special Assistant to the Chancellor for Science and Technology at the University of California, Berkeley, and Deputy Director for Policy with the White House Office of Science and Technology Policy; Dr. Elaine Kamarck, Lecturer in Public Policy at the Belfer Center

for Science and International Affairs, Harvard University's Kennedy School of Government; Erin Kassoy, Director of Solutions Development and Analysis at the Alliance for Climate Protection; Dr. Henry Kelly; Joseph Kerecman, Managing Director, APX, Inc.; Dr. Gerhard Knies, Project Manager of DESERTEC for the Club of Rome; Kevin Knobloch, President of the Union of Concerned Scientists; Ben Kortlang, Kleiner Perkins Caufield & Byers; Orin S. Kramer, General Partner at Boston Provident LP; Dr. Tim LaSalle, Chief Executive Officer of the Rodale Institute; Dr. Rattan Lal; Dr. W. Henry Lambright, Professor of Public Administration and Political Science at the Maxwell School of Citizenship and Public Affairs at Syracuse University; Jonathan Lash, President of the World Resources Insitute; Dr. David Lashmore, Founder of Nanocomp Technologies; Anne Lauvergeon, Chief Executive Officer of AREVA; Dr. Henry Lee, Lecturer in Public Policy and Jassim M. Jaidah Family Director of the Environment and Natural Resources Program, Belfer Center for Science and International Affairs at Harvard University's Kennedy School of Government; Dr. Anthony Leiserowitz, Director of the Yale Project on Climate Change and a Research Scientist at the School of Forestry and Environmental Studies at Yale University; Lawrence Lessig, Professor at Harvard Law School, Director of the Edmond J. Safra Foundation Center for Ethics at Harvard University; Dr. David Lewis, Senior Vice President with HDR Decision Economics and Chief Economist with HDR Consultancy; Dr. Lawrence Linden, Founder and Trustee of the Linden Trust for Conservation; Dr. Thomas Lovejoy; Dr. Amory Lovins; David Lowish, Director, Generation Investment Management; Dr. Lee Lynd, Professor of Engineering and Adjunct Professor of Biology at Dartmouth College; Jim Lyons, Vice President of Policy and Communications at Oxfam America Inc; Wangari Muta Maathai, Nobel Laureate and Founder of the Green Belt Movement; Mark MacCracken, CEO of CALMAC Manufacturing Corporation; Dr. Mike MacCracken; Dr. Allison Macfarlane, Associate Professor of Environmental Science and Policy at George Mason University; Birger Madsen, Director and Partner, BTM Consult ApS; Arni Magnusson, Managing Director of Glitnir Bank's Global Sustainable Energy team; Dr. Charles Maier, Leverett Saltonstall Professor of History at Harvard University; Dr. Arjun Makhijani, President and Senior Engineer at the Institute for Energy and

Environmental Research; Dr. Ulrike Malmendier, Associate Professor of Economics, University of California, Berkeley; Dr. Thomas Mancini, Program Manager for Concentrating Solar Power at Sandia National Laboratories; Dr. Asgeir Margeirsson, CEO of Geysir Green Energy; Dr. James McCarthy; Bill McDonough, Founding Principal of William McDonough + Partners; Dr. Mike McElroy; Des McGinnes, Business Development Manager for the Pacific North West at Pelamis Wave Power Ltd; Kathleen McGinty, Former Secretary of the Pennsylvania Department of Environmental Protection; Dr. Richard McGregor, Assistant Professor of Religious Studies, Vanderbilt University; Bill McKibben, Scholar-in-Residence in Environmental Studies at Middlebury University and Founder of 350.org; Mark Mehos, Program Manager for Concentrating Solar Power at the National Renewable Energy Laboratory; Dr. John Melo, Chief Executive Officer of Amyris Biotechnologies; Dr. Gilbert Metcalf, Professor of Economics at Tufts University; Dr. David Meyer, Professor of Sociology at the University of California, Irvine; Dr. David Mills, Founder, Chairman and Chief Scientific Officer at Ausra; Lesa Mitchell, Vice President, The Kauffman Foundation; David Mohler, Vice President and Chief Technology Officer for Duke Energy; Dr. Mario Molina, Nobel Laureate, Professor of Chemistry and Biochemistry at the University of California, San Diego, and President of the Molina Center for Strategic Studies in Energy and the Environment; Dr. Ernest Moniz, Professor of Physics and Cecil & Ida Green Distinguished Professor at the Massachusetts Institute of Technology; Dr. Read Montague, Professor in the Department of Neuroscience at Baylor College of Medicine, Director of the Human Neuroimaging Lab, and Director of the Center for Theoretical Neuroscience; Dr. Fred Morse, Senior Adviser of U.S. Operations for Solucar Power Inc; Michael Mudd, Chief Executive Officer of the FutureGen Alliance; Dr. Sendhil Mullainathan, Professor of Economics at Harvard University; Walter Musial, Principal Engineer at the National Renewable Energy Laboratory; Steve Nadel, Executive Director of the American Council for an Energy-Efficient Economy; Dr. Michael Nelson, Visiting Professor of Communication, Culture and Technology at Georgetown University; Dr. Ken Newcombe, Chief Executive Officer of C-Quest Capital; Bo Normark, Senior Vice President for ABB Grid Systems; Dr. Ali Nourai, Chairman of the Electricity Storage Association and Manager of Distributed Energy Resources at American Electric Power; Dr. Thoraya Obaid, Executive Director of the United Nations Population Fund; Dr. Eddie O'Connor, Founder and CEO of Mainstream Renewable Power; David Palecek, Partner at McKinsey and Company; Ben Parco, Founder and Chief Executive Officer of Parco Homes; Dr. Marty Peretz, Editor-in-Chief of *The New Republic*; Susan Petty, Chief Technology Officer, AltaRock Energy, Inc.; Dr. William Pizer, Senior Fellow at Resources for the Future; John Podesta, President and CEO of the Center for American Progress; George Polk, Founder and CEO of the Catalyst Project and Executive Committee Chairman of the European Climate Foundation; Dr. Henry Pollack; Lila Preston, Generation Investment Management; Bill Prindle, Vice President, ICF International; Bruce Radford, President of Public Utilities Reports Inc; Dr. Mario Ragwitz, Senior Scientist in the Department of Energy Policy and Energy Systems at the Fraunhofer Institute for Systems and Innovation Research (ISI); Dr. V. Ramanathan; Dr. Antonio Rangel, Associate Professor of Economics at the California Institute of Technology; Dan Rastler, Program Director at the Electric Power Research Institute; Bill Reinert, National Manager of Advanced Technology for Toyota Motor Sales; Michael Renner, Senior Researcher at the Worldwatch Institute; Dr. Heather Cox Richardson, Professor of History at the University of Massachusetts, Amherst; Dr. Rick Riman, Professor of Materials Science and Engineering at Rutgers University; Jim Robo, President and Chief Operating Officer of FPL Group, Inc; Dr. Joe Romm; Ted Roosevelt IV, Managing Director, Barclays Capital; Dr. Arthur Rosenfeld, Commissioner of the California Energy Commission; Niki Rosinski, Director, Generation Investment Management; Dr. Hans Rosling; Joe Rospars, Founding Partner of Blue State Digital; Dr. Jonathan Sackett, Former Associate Laboratory Director at the Argonne National Laboratory; Dr. William Schlesinger, President of the Cary Institute of Ecosystem Studies; Dr. Gavin Schmidt; Dr. Juliet Schor, Professor of Sociology at Boston College; Dr. Daniel Schrag; Allan Schurr, Vice President, Strategy and Development, IBM Global Energy and Utilities Industry; Larry Schweiger, President of the National Wildlife Federation; Brent Scowcroft, President of the Scowcroft Group; Tim Searchinger, Visiting Scholar and Lecturer in Public and International Affairs at Princeton's Woodrow Wilson School; Dr. Eldar Shafir, Professor of Psychology and Public Affairs at the department of psychology and the Woodrow Wilson School of Public and International Affairs at Princeton; Jigar Shah, Founder and Chief Strategy Officer at SunEdison; Dr. David Shearer, Cofounder and Chief Scientist of Full Circle Solutions; Chris Shelton, Head of Grid Stability and Efficiency at the AES Corporation and Vice Chairman of the Electricity Storage Association; Dr. Drew Shindell; Alison Silverstein; Cameron Sinclair, Cofounder and Eternal Optimist at Architecture for Humanity; Dr. Kirk Smith, Professor of Global Environmental Health at University of California, Berkeley's School of Public Health; Malcolm Smith, Director, Arup Urban Design; Dr. Anthony Socci, Senior Science and Communications Fellow with the American Meteorological Society; George Soros, Founder and Chairman of the Open Society Institute; Dr. KR Sridhar, Principal Founder and CEO of Bloom Energy; Paul Stamets, Mycologist, Author, and Founder of Fungi Perfecti; Dr. Robert Stavins, Albert Pratt Professor of Business and Government at Harvard University, Director of the Harvard Environmental Economics Program; Dr. Peter Stearns, Professor of History and Provost at George Mason University; Dr. Joe Stiglitz; James Stoppert, President and Chief Executive Officer, Segetis; Dr. Charlotte Streck, Director of Climate Focus; Bob Sussman, senior policy adviser, Environmental Protection Agency; Dr. Richard Swanson, Founder and President of SunPower Corporation; Trey Taylor, Cofounder and President of Verdant Power Inc; Dr. Jeff Tester; Christine Tezak, Senior Analyst in Energy and Environmental Policy Research at Robert W. Baird & Co; Lee Thomas, Retired President and COO of Georgia-Pacific Corp.; Dr. Tore Torp, Project Manager at StatoilHydro; Tom Tuchman, President, U.S. Forest Capital LLC; Dr. Laura Tyson; Philippine de T'Serclaes, Policy Analyst, Energy Efficiency and Climate Change Division, International Energy Agency; Dr. Francisco Valero, Distinguished Research Scientist at the Scripps Institution of Oceanography at the University of California, San Diego; Charles Vartanian, Project Manager for Distributed Energy Resources Development at Southern California Edison; Cronin Vining, ZT Services; Charles Visser, Technology Manager for the Geothermal Technology Program at the National Renewable Energy Laboratory; Dr. Kevin Volpp, Associate Professor of Medicine and Health Care Management at the University of Pennsylvania, Director of the Leonard Davis Institute of Health Economics Center for

Health Incentives; Dr. Diana Wais, Director and Faculty of the AEDP Institute; Kevin Wall, CEO of Control Room; Dr. Robert Watson, Former Head of the IPCC, Director of the International Assessment of Agricultural Science and Technology for Development, and Co-chair of the International Scientific Assessment of Stratospheric Ozone; Dr. Duncan Watts, Principal Research Scientist at Yahoo! Research, Adjunct Senior Research Fellow at Columbia University; Laurie Wayburn, Cofounder and President of the Pacific Forest Trust; Dr. Eicke Weber, Director of the Fraunhofer Institute for Solar Energy Systems ISE; Dr. David Weinberger, Fellow at the Harvard Berkman Center for Internet & Society; David Wells, Partner, Kleiner Perkins Caufield & Byers; Dr. Todd Werpy, Vice President of Chemicals & Biofuels at Archer Daniels Midland Company; Dr. Drew Westen, Professor of Psychology at Emory University; Dr. David Wheeler, Senior Fellow at the Center for Global Development; Mason Willrich, Chair, California ISO Board of Governors; Dr. Ryan Wiser, Staff Scientist, Electricity Markets and Policy Group, Lawrence Berkeley National Laboratory; Iain Wright, Project Manager at BP Alternative Energy; Dr. Charles Wyman, Ford Motor Company Chair in Environmental Engineering at the University of California, Riverside; David Yeh, Associate, Generation Investment Management; Linda Zall, The Central Intelligence Agency's MEDEA Program; Dr. Shi Zhengrong, Founder, Chairman, and CEO of Suntech; Dr. Qianlin Zhuang, Sales and Commercial Operations Leader (Asia) for GE Energy; and Cathy Zoi, Assistant Secretary for Energy Efficiency and Renewable Energy, Department of Energy.

I want to thank the Alliance for Climate Protection, headed by Maggie Fox (and headed by Cathy Zoi when this project began) for all of their help in sponsoring the Solutions Summits and in providing additional research help. I am proud to donate all of my earnings from this book to the Alliance, as I did with *An Inconvenient Truth*.

I also want to thank my partners at Generation Investment Management for their generous help in reviewing the outline for the book at the beginning, reading selected chapters for factual accuracy, and reading the entire manuscript toward the end.

Similarly, my partners at Kleiner Perkins Caufield & Byers (KPCB) have been of invaluable assistance in reviewing the outline and helping with the answers to detailed questions. Both Generation and KPCB have attended all of the Solutions Summits and have recruited CEOs and business leaders to join the scientists, engineers, and technology and policy experts in ensuring that the in-depth discussions benefited from their commercial and market insights and experience.

For purposes of full disclosure, some of the information that I gathered for this book outside of the Summits also resulted from my affiliations with the Alliance for Climate Protection, Generation Investment Management, and Kleiner Perkins Caufield & Byers. As an advocate and a businessman, I also invest in alternative energy companies.

I want to thank friend Steve Murphy, former CEO of Rodale, and Maria Rodale, CEO and Chairman of Rodale, for their belief in and commitment to this book, and for all the wonderful help Rodale has provided all along the way. I am especially grateful to my outstanding editor, Karen Rinaldi, who is also Senior Vice President, General Manager, and Publisher at Rodale, for her skill, dedication, and stamina—and for crucially important suggestions about the best way to present this material. Also: Colin Dickerman, Vice President and Publishing Director; Julie Geiringer, Sales Director; Yelena Gitlin, Publicity Director; Beth Lamb, Associate Vice President and Associate Publisher; Francesca Minerva, Director, Special Sales; Bob Niegowski, Rights Director; Ellie Prezant, Manager of Corporate Communications; Robin Shallow, Senior Vice President and Director of Corporate Communications; Malcolm J. Gross, Esq.; and Gena Smith, Assistant Editor.

I have once again enjoyed the chance to work with my friend Charles Melcher, who is the very best at book architecture, design, and production. It is difficult to adequately thank the men and women at Melcher Media who put in so many long hours—day and night—on the pictures, graphics, and related work, including: Kurt Andrews; Erin Barnes; Christopher Beha; Peter Bil'ak, Typotheque; Duncan Bock; Michael Brenner, mgmt. design; Adam Bright; David E. Brown; Dennis Bunnell; Alicia Cheng, mgmt. design; Amélie Cherlin; Daniel del Valle; Max Dickstein; Danielle Dowling; Bonnie Eldon; Alissa Faden; Don Foley, illustrator; Marilyn Fu; Sarah Gephart, mgmt. design; Sallie Gmeiner; Barbara Gogan; Filomena Guzzardi; Stephanie Heimann; John House; Coco Joly; Tami Kaufman; Terry Klockow; Eleanor Kung; Peter Lucas; Phil MacDonald; Lisa Maione; Parlan McGaw; Marie Mulcahy; Lauren Nathan; Brian Payne, Sr.; Richard Petrucci; Lia Ronnen; Holly Rothman; Jessi Rymill; Genevieve Smith; Erin Slonaker; Erin Sommerfeld; Lindsey Stanberry; Shoshana Thaler; Scott Travers; Rebecca Wiener; Lee Wilcox; Nancy Wolff; and Megan Worman.

Don Foley, in my opinion the best graphic artist in the world for this kind of material, has been skillful in creating the illustrations—and patient in modifying them through multiple iterations to align the details and conceptual proportions to the best available science and research. Thank you for a terrific job, Don! Charles Melcher and his team served up a rich menu of pictures from which to choose, and others offered suggestions for particular images. I am grateful to all of the photographers whose work appears in this book. I particularly want to thank my friend Yann Arthus-Bertrand, for generously providing several images from his spectacular body of work. In addition, I am once again grateful to National Geographic for donating the use of several of their wonderful photographs for this book. And thanks to Tom Mangelsen for his photo of penguins that opens the Introduction.

I want to thank my friend Natilee Duning for once again volunteering her time to line edit many of my rough chapter drafts; my friend and partner Joel Hyatt and my friend Mike Feldman for important advice; and my agent and friend, Andrew Wylie, for his advice and for once again skillfully working out the various arrangements crucial to this book's publication.

CREDITS

The author wishes to recognize the following individuals and companies for generously supporting the Alliance for Climate Protection by contributing photographs to this project:

The Associated Press; Argos Collectif; Yann Arthus-Bertrand; Aurora Photos; Edward Burtynsky; Robert Clark, Livia Corona; Hélène David; Envision Stock Photography; Mitch Epstein; Getty Images; Robert W. Ginn; Chris Jordan; Vince LaForet; Tony Law; Len Jenshel and Diane Cook; Alex S. MacLean; Magnum Photos, Tom Mangelsen, Sean Nolan; the National Geographic Society and its photographers—Jonathan Blair, Michael Melford, George F. Mobley, James C. Richardson, Tyrone Turner, Willis D. Vaughn; the *Los Angeles Times*; the *Syracuse Post-Dispatch*; OnAsia Images; Panos Pictures; Peter Arnold Inc.; Redux Images; Sipa Press; George Steinmetz; UNICEF; and Zuma Press.

Images are referenced by page number. All photographs and illustrations are copyright © by their respective sources.

Page 4: NASA; 10–11: Livia Corona; 13: Tom Mangelsen; 14: Noah Seelam/AFP/Getty; 16–17: Hélène David/Argos Collectif; 19: Mitch Epstein; 20: George Steinmetz; 22–23: Robert W. Ginn/Envision; 25: Tomas Bravo/Reuters; 26: Riccardo Venturi/Contrasto/Redux; 28–29: Sean Nolan/seannolanphotography.com; 30–31: Ralph Orlowski/Getty; 33: Ian Berry/Magnum Photos; 36–37: John Ulan; 38–39: Yann Arthus-Bertrand/Altitude; 40: David Allan Harvey/Magnum Photos; 41: Topham/The Image Works; 43: NASA; 48–49: Alexandre Meneghini/Associated Press; 50–51: Alex S. MacLean/alexmaclean.com; 53: Matthew Staver/*The New York Times*/Redux; 56–57: Michael Melford/National Geographic Stock; 62–63: Martin Bond/Still Pictures/Peter Arnold; 65: Nobesol; 66–67: Naturimages; 72–73: Andreas Solaro/AFP/Getty; 75: Alan K. Barley/Barley & Pfeiffer Architects; 76–77: Leah Nash/*The New York Times*/Redux; 79: Frank Huster/Aurora Photos; 82–83: David Hurn/Magnum Photos; 85: Paul Langrock/Zenit/Laif/Redux; 87: Tom Rielly/ted.com; 88–89: David McNew/Getty; 92–93: Palmi Gudmundson/Nordic Photos/Aurora Photos; 96–97: Jonathan Blair/National Geographic Stock; 99: Newscom; 102: Vilhelm Gunnarsson/WPN; 104: Geodynamics Ltd.; 106–7: Tim Graham/Getty; 108: George F. Mobley/National Geographic Stock; 112–13: Nelson Almeida/Getty; 115: Dr. Rob Stepney/SPL/Photo Researchers; 116–17: all Julia Knop/Laif/Redux; 118–19: Paulo Fridman/Polaris; 121: Chris Knapton/SPL/Photo Researchers;

126: Mike Derer/Associated Press; 128: Jessica McGowan/*The New York Times*/Redux; 131: Robert Clark; 132–33: Paul Langrock/Zenit/Laif/Redux; 134–35: Øyvind Hagen/StatoilHydro; 139: Paul Corbit Brown; 142: Peter Arnold Inc.; 144: Marguerite Holloway/*The New York Times*/Redux; 147: Ullstein/The Image Works; 148–49: Christoph Busse/Peter Arnold Inc.; 150–51: Peter Essick/Aurora Photos; 153: Peter Essick/Aurora Photos; 155: Sergei Supinsky/AFP/Newscom; 157: Yann Arthus-Bertrand/Altitude; 162–63: Baptiste Fenouil/REA/Redux; 166: Lawrence Livermore National Laboratory; 168–69: Peter Essick/Aurora Photos; 170–71: Yann Arthus-Bertrand/Altitude; 173: Tian Lee/Zuma Press; 176 (both): John Novis/Greenpeace/OnAsia.com; 177 (top): Vinai Dithajohn/OnAsia.com; 177 (bottom): John Novis/Greenpeace/OnAsia.com; 178–79: John Novis/Greenpeace/OnAsia.com; 180: Jay Ullal/Orangutan Outreach, redapes.org; 182–83: Kemal Jufri/Polaris; 184: Guenter Fischer/The World of Stock; 189: Vince LaForet; 190: Scripps Institution of Oceanography, UC-San Diego; 192–93: Stephen Shaver/Polaris; 195: James C. Richardson/National Geographic Stock; 196–97: James C. Richardson/National Geographic Stock; 200–201: James C. Richardson/National Geographic Stock; 202–3: Juan Silva/Getty; 204–5: Willis D. Vaughn/National Geographic Stock; 207: James C. Richardson/National Geographic Stock; 208: Brian Vander Brug/*Los Angeles Times*; 211: Steve Satushek/Getty; 213: David McLain/Aurora Photos; 214: Yann Arthus-Bertrand/Altitude; 216–17: James C. Richardson/National Geographic Stock; 218: Jeff Hutchens/Getty; 223: James C. Richardson/National Geographic Stock; 224–25: George Osod/Panos Pictures; 228–29 (left to right): Asad Zaidi/NYHQ2004-0209/UNICEF; Lynsey Addario/VII; Giacomo Pirozzi/NYHQ2004-0676/UNICEF; Richard Lord/The Image Works; 230: Andre Vieira; 232–33: Stuart Freedman/Panos Pictures; 237: Tony Law/Redux; 238–39: ArabianEye/Getty; 241: Abbas/Magnum Photos; 242–43: Siemens AG, Energy Sector; 245: Mark Heffron/The Kubala Washatko Architects; 248: Unkel/Ullstein Bild/Peter Arnold Inc.; 250–51: Chris Jordan; 256 (all): Harry Gruyaert/Magnum Photos; 259: Jeff Jacobson/Redux; 263: Tyrone Turner/National Geographic Stock; 264 (clockwise from top right): John Berry/*The Post Standard*; Jim West/Zuma Press; Toma Babovic/Laif/Redux; Nic Bothma/EPA/Zuma Press; Philip Hall/Sipa Press; Crista Jeremiason/*Santa Rosa Press Democrat*/Zuma Press; 266–67: Rony Zakaria/OnAsia.com; 268: Tyrone Turner/National

Geographic Stock; 271: Al Golub/Associated Press; 272–73: Mark Ralston/AFP/Getty; 276: Jeff Jacobson/Redux; 282–83: Paul Langrock/Zenit/Laif/Redux; 284: Jeffrey Sauger/General Motors; 287: Tesla; 293: Cook+Fox Architects; 294: James C. Richardson/National Geographic Stock; 297: Kevin Moloney/*The New York Times*/Redux; 298–99: Andrew Coleman/iStockphoto; 302–3: Ashley Cooper/GHG Photos/Aurora Photos; 306–7: TIPS Images; 310: Antonio Rangel; 312–13: Lyza Danger Gardner; 316–17: Adrian Bradshaw/Liaison/Getty; 318–19: Mitch Epstein; 321: Andrew Kornylak/Aurora Photos; 322–23: M. Flores/UNEP/Still Pictures/Peter Arnold Inc.; 326: Daniel Dancer/Still Pictures/Peter Arnold Inc.; 328–29: Edward Burtynsky, courtesy Hasted Hunt Kraeutler, New York/Nicholas Metivier Gallery, Toronto; 332–33: Wade Payne/Associated Press; 334: Eric Tourneret/Visuals Unlimited; 336–37 (left to right): Associated Press (x2), AFP/Getty, Associated Press, Time & Life Pictures/Getty, Associated Press, Steve McCurry/Magnum Photos, Associated Press, Robert Reed/U.S. Coast Guard/Associated Press; 338: Abbas/Magnum Photos; 340–41: Alex S. MacLean; 344: EPA Acid Rain Program; 346–47: Raf Madka/View/Artedia; 348–349: Mandel Ngan/AFP/Getty; 351: Frank Augstein/Associated Press; 353: Chip Somodevilla/Getty; 356: Stephen Crowley/*The New York Times*/Redux; 357: Alex Wong/Getty; 364: Ericka Ekstrom; 367: Alliance for Climate Protection; 368–69: L.M. Otero/Associated Press; 370–71: Bell Labs/Lumeta Corp.; 373: NASA; 375: Atmospheric Research Laboratory, Scripps Institution of Oceanography, UC-San Diego; 378: The Vulcan Project/Dr. Kevin Guerney and Purdue University; 380 (both): U.S. Geological Survey, National Civilian Applications Program; 381 (left): Associated Press; 382–83: Jetta Productions/Getty; 384: Mike Derer/Associated Press; 387: Eric Lee/Paramount Classics; 389 (left to right): Eugene Garcia/*The Orange County Register*/Zuma Press; courtesy of the Energy Retail Association; 390 (both): courtesy Redlands Police Department; 392–93: Jon Holloway/Stock Connection/Aurora Photos; 395: Isaac Brekken/Associated Press; 396–97: Diane Cook and Len Jenshel; 400: Paul Langrock/Zenit/Laif/Redux; 403 (top to bottom): Amy Scaife/Press Association Images; Oli Scarff/Getty; 404–5: Andrea Gjestvang/Moment.

Scripture on page 309 taken from the New King James Version. Copyright © 1982 by Thomas Nelson, Inc. Used by permission. All rights reserved.

ABOUT THIS BOOK

I have worked to ensure that this book contributes as little as possible to our greenhouse gases and, indeed, represents some of the actions we can take to lessen our impact on the atmosphere.

The North American editions of *Our Choice* are printed on a 100 percent recycled paper that was custom-made for this book by Newton Falls Fine Paper in Newton Falls, New York. It contains 10 percent post-consumer waste, with the balance coming from industrial waste streams. Both sources of pulp recycle paper that might otherwise go into landfills. Compared with paper made of virgin fiber, the stock used in *Our Choice* conserved 1,153 trees and reduced greenhouse gas emissions by about 110,000 pounds of CO_2. This also saved 366 million BTUs of energy and reduced wastewater by 528,000 gallons. By using a lighter 68-pound paper, rather than the standard 80-pound thickness, an additional 15 percent of paper fiber, CO_2 emissions, energy, and water was saved.

This book was printed and bound by World Color in Taunton, Massachusetts. The printer was chosen, in part, because of its proximity to the paper mill and the publisher's warehouse, thus minimizing the amount of fuel used in shipping and the related CO_2 emissions.

Our Choice is a CarbonNeutral publication. The carbon emissions that resulted from the manufacturing of this book were calculated by the CarbonNeutral Company. For every ton of CO_2 produced, the CarbonNeutral Company has arranged for a ton of CO_2 to be saved by climate-friendly projects.

This book was published by Rodale. For more than six decades, Rodale has been at the forefront of environmental responsibility. The company's founder, J.I. Rodale, was the founder of the organic movement in America and pioneered the idea that food grown without harmful chemicals was better for the soil, better for the environment, and better for people. Ever since, Rodale has been promoting "healthy living on a healthy planet"—both through the content in its publications and the way it does business.

In order to avoid the destruction of our forests, it is important that we use paper pulp sources that are responsibly managed and sustainable. For this reason, I have made sure that all of the suppliers I worked with on this project, the paper company, the printer, even the book producer, Melcher Media, have demonstrated their commitment to sustainable forestry practices by gaining certification from the Forestry Stewardship Council (FSC). This book is FSC certified.

The next step in reducing this book's effects on the environment is yours. When you're finished reading *Our Choice,* please don't throw it away; rather, pass it along to a friend, to a library, or to someone that you feel needs to read it.

CarbonNeutral.com

CO₂ emissions reduced to net zero in accordance with The CarbonNeutral Protocol

Mixed Sources
Product group from well-managed forests, controlled sources and recycled wood or fiber
www.fsc.org Cert no. BV-COC-031730
© 1996 Forest Stewardship Council